The Concise Gray's Anatomy

The Concise Gray's Anatomy

—

Professor C. H. Leonard AMA, MD

with a foreword by

THE LORD REA MD, MRCGP

Wordsworth Reference

This edition published 1997 by Wordsworth Editions Ltd,
Cumberland House, Crib Street, Ware, Hertfordshire SG12 9ET

From the 16th edition of *The Pocket Anatomist*,
published 1889, founded on *Gray's Anatomy*

ISBN 1 85326 394 X

Printed and bound in Great Britain by
Mackays of Chatham plc, Chatham, Kent

FOREWORD

by The Lord Rea MD, MRCGP

Human Anatomy as a basic life science now concentrates less on minute detail and more on principles. Most modern anatomy textbooks for undergraduates are illustrated by diagrams which simplify the structures described in order to increase clarity and understanding. But while much is gained by this approach, much is lost. The anatomists of the nineteenth century, of whom Gray though the best known was only one, formed part of the great descriptive and classifying period of natural science. This was characterised by meticulous drawing and attention to exact detail so that the scientific illustrations became works of art in themselves.

This book is a worthy product of that era though it is greatly compressed as befits a popular pocket volume. The 193 illustrations are all engravings on which many painstaking hours were obviously spent. They represent actual views of all the important structures of the human body as revealed by careful dissection. The book was in fact used by medical students as a dissecting room companion or guide as well as an abbreviated text useful before exams. By reissuing this small Victorian gem as a commemorative edition the publishers have done a great service to all students of human anatomy, whether these be budding doctors or artists, or those from all walks of life who find the topic of human form fascinating for its own sake. In time it may well be regarded as a collector's piece in its own right.

Professor Leonard in his own preface to the fourteenth edition acknowledges his debt to Gray and explains how his book is in many ways a compressed version of its larger brother, but he has added one or two ingenious sections of his own.

London June 1983

PREFACE TO THE FOURTEENTH EDITION, 1888

This little book, virtually a Dissecting-room Companion, has been, from its rapid and large sales, a source of surprise and pardonable pride to the author. It is now in its fourteenth edition, five of these editions having been sold in London alone. Then, too, he has been complimented by having a copy of the London edition put into plates by a large American publishing house, before it was discovered that the work was from an American author, and could not be re-published here.

This fourteenth edition has been increased in size by the addition of over 100 pages of text and 100 engravings; the page of the book has also been somewhat enlarged, to better accommodate the engravings. The Brain and its Membranes, the Eye, Ear and Throat, in fact the entire Viscera and the Generative Organs of both Sexes, form the new subject matter in this edition.

While the larger portion of the work is simply "Gray" condensed and transposed, the author has also used, quite liberally, in the preparation of the new section on "Triangles and Spaces," Brown's *Aid to Anatomy* and Bryant's *System of Surgery*. In the Gynæcological section, the works of Gray, Savage, Hirschfeld, Barnes and Playfair have all been consulted. The section "Points Worth Remembering" the author hopes will prove of use, as well as interest, to the medical student, for whom the little book is mainly intended.

C.H.L.

DISSECTION HINTS

THE HEAD AND NECK.

Cranial Region (Fig.1). The head being shaved, and a block placed beneath the back of the neck, make a vertical incision through the skin from before backwards,

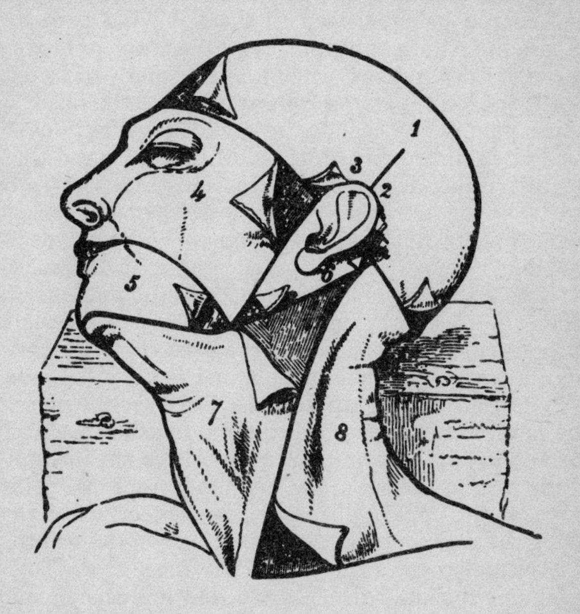

1. Dissection of scalp. 2, 3, of auricular region.
4, 5, 6, of face. 7, 8, of neck.

Fig.1 Dissection of the head, face, and neck.

commencing at the root of the nose in front, and terminating behind at the occipital protuberance; make a second incision in a horizontal direction along the forehead and round the side of the head, from the anterior to the posterior extremity of the preceding. Raise the skin, in front, from the subjacent muscle from below upwards; this must be done with extreme care, on account of their intimate union. The tendon of the muscle is best avoided by removing the integument from the outer surface of the vessels and nerves which lie between the two.

The *superficial fascia* in the epicranial region is a firm, dense layer, intimately adherent to the integument, and to the Occipito-frontalis and its tendinous aponeurosis; it is continuous, behind, with the superficial fascia at the back part of the neck; and, laterally, is continued over the temporal aponeurosis; it contains between its layers the small muscles of the auricles and the superficial temporal vessels and superficial nerves.

Auricular Region. This requires considerable care, and should be performed in the following manner: To expose the Attollens aurem, draw the pinna or broad part of the ear downwards, when a tense band will be felt beneath the skin, passing from the side of the head to the upper part of the concha; by dividing the skin over the tendon, in a direction from below upwards, and then reflecting it on each side, the muscle is exposed. To bring into view the Attrahens aurem, draw the helix backwards by means of a hook, when the muscle will be made tense, and may be exposed in a similar manner to the preceding. To expose the Retrahens aurem, draw the pinna forwards when the muscle being made tense may be felt beneath the skin, at its insertion into the back of the concha, and may be exposed in the same manner as the other muscles.

Palpebral Region. In order to expose the muscles of the face, continue the longitudinal incision made in the dissection of the Occipito-frontalis, down the median line of the face to the tip of the nose, and from this point onwards to the upper lip; another incision should be carried along the margin of the lip to the angle of the mouth, and transversely across the face to the angle of the jaw. The integument should also be divided by an incision made in front of the external ear, from the angle of the jaw, upwards, to the transverse incision made in exposing the Occipito-frontalis. These incisions include a square-shaped flap which should be carefully removed in the direction marked in the figure, as the muscles at some points are intimately adherent to the integument.

Orbital Region. To open the cavity of the orbit, the skull-cap and the brain should be first removed; then saw through the frontal bone at the inner extremity of the supra-orbital ridge, and externally at its junction with the malar. The thin roof of the orbit should then be comminuted by a few slight blows with the hammer, and the superciliary portion of the frontal bone driven forwards by a smart stroke; but it must not be removed. The several fragments may then be detached, when the periosteum of the orbit will be exposed: this being removed, together with the fat which fills the cavity of the orbit, the several muscles of this region can be examined. To facilitate their dissection, the globe of the eye should be distended; this may be effected by puncturing the optic nerve near the eyeball, with a curved needle, and pushing it onwards into the globe. Through this aperture the point of a blow-pipe should be inserted, and a little air forced into the cavity of the eyeball; then apply a ligature round the nerve, so as to prevent the air escaping. The globe should now be drawn forwards, when the muscles will be put upon the stretch.

Inf. Maxillary Region. The Muscles in this region may be dissected by making a vertical incision through the integument from the margin of the lower lip to the chin: a second incision should then be carried along the margin of the lower jaw as far as the angle, and the integument carefully removed in the direction shown in figure 1.

Inter-Maxillary Region. The dissection of these muscles may be considerably facilitated by filling the cavity of the mouth with tow, so as to distend the cheeks and lips; the mouth should then be closed by a few stitches, and the integument carefully removed from the surface.

Tempero-Maxillary Region. In order to expose the Temporal muscle, this fascia should be removed; this may be effected by separating it at its attachment along the upper border of the zygoma, and dissecting it upwards from the surface of the muscle. The zygomatic arch should then be divided, in front, at its junction with the malar bone, and, behind, near the external auditory meatus, and drawn downwards with the Masseter, which should be detached from its insertion into the ramus and angle of the jaw. The whole extent of the Temporal muscle is then exposed.

Pterygo-Maxillary Region. The Temporal muscle having been examined, the muscles in the pterygo-maxillary region may be exposed by sawing through the base of the coronoid process, and drawing it upwards, together with the Temporal muscle, which should be detached from the surface of the temporal fossa. Divide the ramus of jaw just below the condyle, and, also, by a transverse incision extending across the commencement of its lower third, just above the dental foramen; remove the fragment, and the Pterygoid muscles will be exposed.

Superficial Cervical Region. A box having been placed at the back of the neck, and the face turned to the side opposite to that to be dissected, so as to place the parts upon the stretch, two transverse incisions are to be made: one from the chin, along the margin of the lower jaw, to the mastoid process, and the other along the upper border of the clavicle. These are to be connected by an oblique incision made in the course of the Sterno-mastoid muscle, from the mastoid process to the sternum; the two flaps of integument having been removed in the direction shown in figure 1, the superficial fascia will be exposed.

Infra-hyoid Region. The muscles in this region may be exposed by removing the deep fascia from the front of the neck. In order to see the entire extent of the Omohyoid, it is necessary to divide the Sterno-mastoid at its centre, and turn its ends aside, and to detach the Trapezius from the clavicle and scapula, if this muscle has been previously dissected; but not otherwise.

Supra-hyoid Region. To dissect these muscles, a block should be placed beneath the back of the neck, and the head drawn backwards, and retained in that position. On the removal of the deep fascia, the muscles are at once exposed.

The Mylo-hyoid should now be removed, in order to expose the muscles which lie beneath; this is effected by detaching it from its attachments to the hyoid bone and jaw, and separating it by a vertical incision from its fellow of the opposite side.

Lingual Region. After completing the dissection of the preceding muscles, saw through the lower jaw just external to the symphysis. The tongue should then be drawn forwards with a hook, and its muscles, which are thus put on the stretch, may be examined.

Pharyngeal Region. In order to examine the muscles of the pharynx, cut through the trachea and œsophagus just above the sternum, and draw them upwards by dividing the loose areolar tissue connecting the pharynx with the front of the vertebral column. The parts being drawn well forwards, apply the edge of the saw immediately behind the styloid processes, and saw the base of the skull through from below upwards. The pharynx and mouth should then be stuffed with tow, in order to distend its cavity and render the muscles tense and easier of dissection.

Palatal Region. Lay open the pharynx from behind, by a vertical incision extending from its upper to its lower part, and partially divide the occipital attachment by a transverse incision on each side of the vertical one; the posterior surface of the soft palate is then exposed. Having fixed the uvula so as to make it tense, the mucous membrane and glands should be carefully removed from the posterior surface of the soft palate, and the muscles of this part are at once exposed.

The Palato-glossus and Palato-pharyngæous are exposed by removing the mucous membrane which covers the pillars of the soft palate throughout nearly their whole extent.

THE BACK

The body should be placed in the prone position, with the arms extended over the sides of the table, and the chest and abdomen supported by several blocks, so as to render the muscles tense. An incision should then be made along the middle line of the back, from the occipital protuberance to the coccyx. From the upper end of this, a transverse incision should extend to the mastoid process; and from the lower end, a third incision should be made along the crest of the ilium to about its middle. This large intervening space, for convenience of dissection, should be subdivided by a fourth incision, extending obliquely from the spinous process of the last dorsal ver- tebra, upwards and outwards, to the acromion process. This incision corresponds with the lower border of the Trapezius muscle. The flaps of integument should then be removed in the direction shown in figure 2.

Second Layer. The Trapezius must be removed in order to expose the next layer; to effect this, detach the muscle from its attachment to the clavicle and spine of the scapula, and turn it back towards the spine.

Third Layer. The third layer of muscles is brought into view by the entire removal of the preceding, together with the Latissimus dorsi. To effect this, the Levator anguli scapulæ and Rhomboid muscles should be detached near their insertion, and reflected upwards, thus exposing the Serratus posticus superior; the Latissimus dorsi should then be divided in the middle by a vertical incision carried from its upper to its lower part, and the two halves of the muscle reflected.

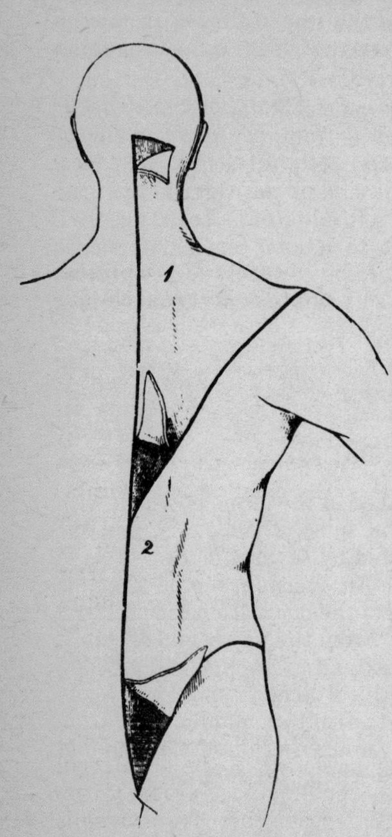

Fig.2 Dissection of the muscles of the back.

The Serratus posticus superior should now be detached from its origin and turned outwards, when the Splenius muscle will be brought into view.

Fourth Layer. To expose the muscles of the fourth layer, remove entirely the Serrati and vertebral aponeurosis. Then detach the Splenius by separating its attachment to the spinous process, and reflecting it outwards.

Fifth Layer. The muscles of the preceding layer must be removed by dividing and turning aside the Complexus; then detach the Spinalis and Longissimus dorsi from their attachments, and divide the Erector spinæ at its connection below to the sacral and lumbar spines, and turn it outward. The muscles filling up the interval between the spinous and transverse processes are then exposed.

THE ABDOMEN

To dissect the abdominal muscles, see figure 3, make a vertical incision from the ensiform cartilage to the pubes; a second incision from the umbilicus obliquely upwards and outwards to the surface of the chest, as high as the lower border of the fifth or sixth rib; and a third, commencing midway between the umbilicus and pubes, transversely outwards to the anterior superior iliac spine, and along the crest of the ilium as far as its posterior third. Then reflect the three flaps included between these incisions from within outwards, in the line of direction of the muscular fibres. If necessary, *the abdominal muscles may be made tense by inflating the peritoneal cavity through the umbilicus.*

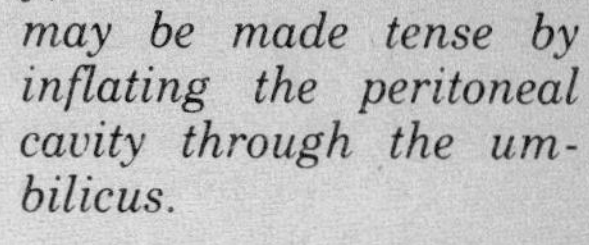

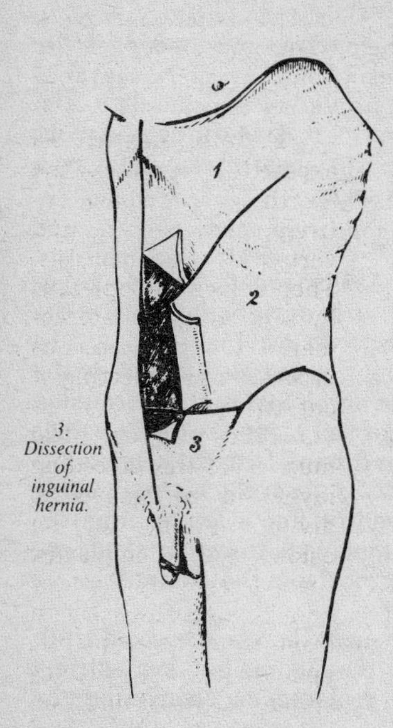

Fig.3 Dissection of abdomen.

Internal Oblique, etc. The External oblique should now be detached by dividing across, just in front of its attachment to the ribs, as far as its posterior border, by separating it below from the crest of the ilium as far as the spine; the muscle should then be carefully separated from the Internal oblique, which lies beneath, and turned towards the opposite side.

Transversalis, etc. Detach the Internal oblique in order to expose the Transversalis beneath. This may be effected by dividing the muscle, above, at its connection with Poupart's ligament and the crest of the ilium; and behind, by a vertical incision extending from the last rib to the crest of the ilium. The muscle should previously be made tense by drawing upon it with the fingers of the left hand, and if its division is carefully effected, the cellular interval between it and the Transversalis, as well as the direction of the latter muscle, will afford a clear guide to their separation; along the crest of the ilium the circumflex iliac vessels are interposed between them, and form an important guide in separating them. The muscle should then be thrown forwards towards the linea alba.

Rectus, etc. To expose the Rectus muscle, its sheath should be opened by a vertical incision from the margin of the thorax to the pubes, the two portions should then be reflected from the surface of the muscle, which is easily effected, excepting at the lineæ transversæ, where so close an adhesion exists, that the greatest care is requisite in separating them. The outer edge of the muscle should now be raised, when the posterior layer of the sheath will be seen. By dividing the muscle in the centre, and turning its lower part downwards, the point where the posterior wall of the sheath terminates in a thin curved margin will be seen.

UPPER EXTREMITY

Pectoral Region and Axilla (Fig. 4). The arm being drawn away from the side nearly at right angles with the trunk, and rotated outwards, a vertical incision should be made through the integument in the median line of the chest, from the upper to the lower part of the sternum; a *second* incision should be carried along the lower border of the Pectoral muscle, from the ensiform cartilage to the outer side of the axilla; a *third*, from the sternum along the clavicle, as far as its centre; and a *fourth*, from the middle of the clavicle obliquely downwards, along the inter-space between the Pectoral and Deltoid muscles, as low as the fold of the arm-pit.

The flap of integument may then be dissected off in the direction indicated in the figure, but not entirely removed, as it should be replaced on completing the dissection.

If a transverse incision is now made from the lower end of the sternum to the side of the chest, as far as the posterior fold of the arm-pit, and the integument reflected outwards, the axillary space will be more completely exposed.

Deep Pectoral Layers. Detach the Pectoralis major by dividing the muscle along its attachment to the clavicle, and by making a vertical incision through its substance a little external to its line of attachment to the sternum and costal cartilages. The muscle should then be reflected outwards, and its tendon carefully examined.

The Pectoralis minor is now exposed, and immediately above it, in the interval between its upper border and the clavicle, a strong fascia, the costo coracoid membrane.

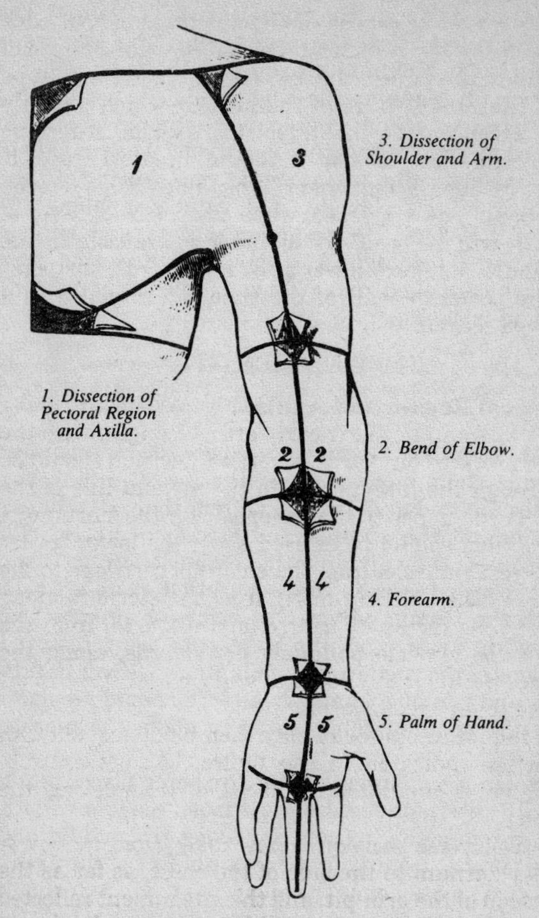

Fig.4. Dissection of upper extremity.

The costo-coracoid membrane should now be removed, when the Subclavius muscle will be seen.

If the costal attachment of the Pectoralis minor is divided across, and the muscle reflected outwards, the axillary vessels and nerves are brought fully into view, and should be examined.

Lateral Thoracic Region. After completing the dissection of the Axilla, if the muscles of the back have been dissected, the upper extremity should be separated from the trunk. Saw through the clavicle at its centre, and then cut through the muscles which connect the scapula and arm with the trunk, viz., the Pectoralis minor, in front, Serratus magnus, at the side, and the Levator anguli scapulæ, the Rhomboids, Trapezius, and Latissimus dorsi behind. These muscles should be cleaned and traced to their respective insertions. Then make an incision through the integument, commencing at the outer third of the clavicle, and extending along the margin of that bone, the acromion process, and spine of the scapulæ; the integument should be dissected from above downwards and outwards, when the fascia covering the Deltoid is exposed.

Divide the Deltoid across, near its upper part, by an incision carried along the margin of the clavicle, the acromion process, and spine of the scapula, and reflect it downwards; the bursa will be seen on its under surface, as well as the circumflex vessels and nerves. The insertion of the muscle should be carefully examined.

Post. Scapular Region. To expose the muscles, and to examine their mode of insertion into the humerus, detach the Deltoid and Trapezius from their attachment to the spine of the scapula and acromion process. Remove the clavicle by dividing the ligaments connecting it with the coracoid process, and separate it at its articulation with the scapula; divide the acromion process near its root with a saw, and, the fragments being removed, the tendons of the posterior Scapular muscles will be fully exposed, and can be examined. A block should be placed beneath the shoulder-joint, so as to make the muscles tense.

Ant. Humeral Region. The arm being placed on the table, with the front surface uppermost, make a vertical incision through the integument along the middle line, from the outer extremity of the anterior fold of the axilla, to about two inches below the elbow joint, where it should be joined by a transverse incision, extending from the inner to the outer side of the forearm; the two flaps being reflected on either side, the fascia should be examined.

Forearm. To dissect the forearm, place the limb in the position in figure 4; make a vertical incision along the middle line from the elbow to the wrist, and a transverse incision at each extremity of this; the flaps of integument being removed, the fascia of the forearm is exposed.

Ant.Brachial, Deep Layer. Divide each of the superficial muscles at its centre, and turn either end aside; the deep layer of muscles, together with the median nerve and ulnar vessels, will then be exposed.

Radial Region. Divide the integument in the same manner as in the dissection of the anterior brachial region; and after having examined the cutaneous vessels and nerves and deep fascia, remove all those structures. The muscles will then be exposed. The removal of the fascia will be considerably facilitated by detaching it from below upwards. Great care should be taken to avoid cutting across the tendons of the muscles of the thumb, which cross obliquely the larger tendons running down the back of the radius.

The Hand. Make a transverse incision across the front of the wrist, and a second across the heads of the metacarpal bones, connect the two by a vertical incision in the middle line, and continue it through the centre of the middle finger. The anterior and posterior annular ligaments, and the palmar fascia, should first be dissected.

LOWER EXTREMITY

(Anterior Portion.)

Dissection. To expose the muscles and fasciæ in this region, make an incision along Poupart's ligament (Fig. 5) from the spine of the ilium to the pubes; a vertical incision from the centre of this, along the middle of the thigh to below the knee joint, and a transverse incision from the inner to the outer side of the leg, at the lower end of the vertical incision.

The flaps of integument having been removed, the superficial and deep fasciæ should be examined.

The more advanced student should commence the study of this region by an examination of the anatomy of femoral hernia, and Scarpa's triangle, the incisions for the dissection of which are marked out in figure 5.

Iliac Region. No detailed description is required for the dissection of these muscles. They are exposed after the removal of the viscera from the abdomen, covered by the peritoneum and a thin layer of fascia, the fascia iliaca.

Internal Femoral Region. These muscles are at once exposed by removing the fascia from the fore part and inner side of the thigh. The limb should be abducted, so as to render the muscles tense, and easier of dissection.

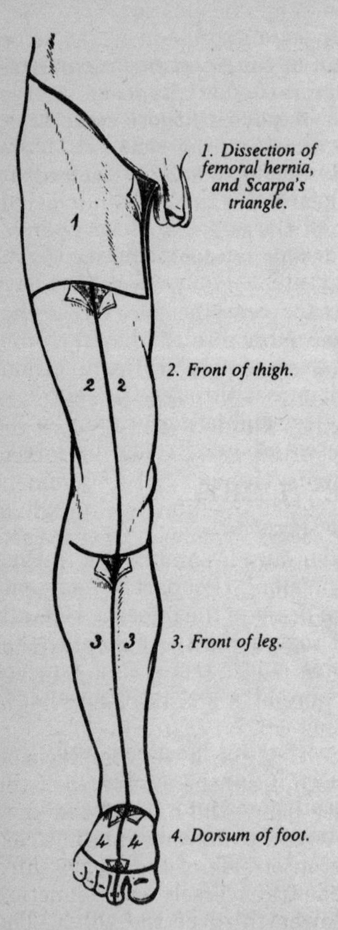

Fig.5 Dissection of lower extremity. Front view.

The Pectineus and *Adductor longus* should now be divided near their origin, and turned downwards, when the Adductor brevis and Obdurator externus will be exposed.

The Adductor brevis should now be cut away near its origin, and turned outwards, when the entire extent of the Adductor magnus will be exposed.

Gluteal Region. Divide the Gluteus maximus near its origin, by a vertical incision carried from its upper to its lower border; a cellular interval will be exposed, separating it from the Gluteus medius and External rotator muscles beneath. The upper portion of the muscle is to be altogether detached, and the lower portion turned outwards; the loose areolar tissue filling up the interspace between the trochanter major and tuberosity of the ischium being removed, the parts already enumerated as exposed by the removal of this muscle will be seen.

This muscle should now be divided near its insertion, and turned upwards, when the *Gluteus minimus* will be exposed.

Obturator Internus. This muscle, as well as the origin of the Pyriformis, can only be seen when the pelvis is divided, and the visceria contained in this cavity removed.

In order to display the peculiar appearances presented by the tendon of this muscle, it should be divided near its insertion and reflected outwards.

Obturator Externus. In order to expose this muscle, it is necessary to remove the Psoas, Iliacus, Pectineus, and Adductor brevis and Adductor longus muscles, from the front and inner side of the thigh; and the Gluteus maximus and Quadratus femoris, from the back part. Its dissection should consequently be postponed until the muscles of the anterior and internal femoral regions have been examined.

(Posterior Portion.)

Dissection (Fig. 6). The subject should be turned on its face, a block placed beneath the pelvis to make the buttocks tense, and the limbs allowed to hang over the end of the table, the foot inverted, and the limb abducted. An incision should be made through the integument along the back part of the crest of the ilium and margin of the sacrum to the tip of the coccyx, from which point a second incision should be carried obliquely downwards and outwards to the outer side of the thigh, four inches below the great trochanter. The portion of integument included between these incisions, together with the superficial fascia, should be removed in the direction shown in the figure, when the Gluteus maximus and the dense fascia covering the Gluteus medius will be exposed.

Post. Femoral Region. Make a vertical incision along the middle of the thigh, from the lower fold of the nates to about three inches below the back of the knee joint, and there connect it with a transverse incision, carried from the inner to the outer side of the leg. A third incision should then be made transversely at the junction of the middle with the lower third of the thigh. The integument having been removed from the back of the knee, and the boundaries of the popliteal space examined, the removal of the integument from the remaining part of the thigh should be continued, when the fascia of this region will be exposed.

Muscles, etc., of Leg. The knee should be bent, a block placed beneath it, and the foot kept in an extended position; an incision should then be made through the integument in the middle line of the leg to the ankle, and continued along the dorsum of the foot to the toes. A second incision should be made transversely across the ankle, and a third in the same direction across the bases of the toes; the flaps of integument included between these incisions should be removed, and the deep fascia of the leg examined.

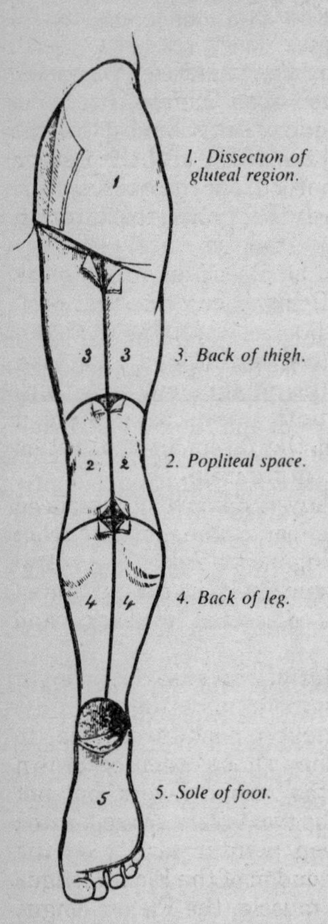

Fig.6 Dissection of lower extremity. Posterior view.

The fascia should now be removed by dividing it in the same direction as the integument, excepting opposite the ankle, where it should be left entire. The removal of the fascia *should be commenced from below*, opposite the tendons, and detached in the line of direction of the muscular fibres.

Post. Tibio-Fibular Region. Make a vertical incision along the middle line of the back of the leg, from the lower part of the popliteal space to the heel, connecting it by a transverse incision extending between the two malleoli; the flaps of integument being removed, the fascia and muscles should be examined.

The Gastrocnemius should be divided across, just below its origin, and turned downwards, in order to expose the *Soleus, Plantaris,* etc.

Deep Layer. Detach the Soleus from

its attachment to the fibula and tibia, and turn it downwards, when the deep layer of muscles is exposed, covered by the deep fascia of the leg.

This fascia should now be removed, commencing from below, opposite the tendons, and detaching it from the muscles in the direction of their fibres.

Tibular Region. These muscles are readily exposed, by removing the fascia covering their surface, *from below upwards*, in the line of direction of their fibres.

The Foot. The fibrous bands which bind down the tendons in front of and behind the ankle in their passage to the foot, should now be examined; they are termed the *annular ligaments*, and are three in number, the anterior, internal, and external.

The Sole. The foot should be placed on a high block with the sole uppermost, and firmly secured in that position. Carry an incision round the heel and along the inner and outer borders of the foot to the great and little toes. This incision should divide the integument and thick layer of granular fat beneath, until the fascia is visible; it should then be removed from the fascia in a direction from behind forwards, as seen in figure 6.

Plantar Muscles—*First Layer.* Remove the fascia on the inner and outer sides of the foot, commencing in front over the tendons, and proceeding backwards. The central portion should be divided transversely in the middle of the foot, and the two flaps dissected forwards and backwards.

Second Layer. The muscles of the superficial layer should be divided at their origin by inserting the knife beneath each, and cutting obliquely backwards, so as to detach them from the bone; they should then be drawn forwards, in order to expose the second layers, but not separated at their insertion. The two layers are separated by a thin membrane, the deep plantar fascia, on the removal of which are seen the tendon of the Flexor longus digitorum, with its Accessory muscle, the Flexor longus pollicis, and the Lumbricales. The long flexor tendons cross each other at an acute angle, the Flexor longus pollicis running along the inner side of the foot, on a plane superior to that of the Flexor longus digitorum, the direction of which is obliquely outwards.

Third Layer. The Flexor tendons should be divided at back of the foot, and the Flexor accessorius at its origin, and drawn forwards, in order to expose the third layer.

Fourth Layer: the Dorsal and the Plantar Interossei.

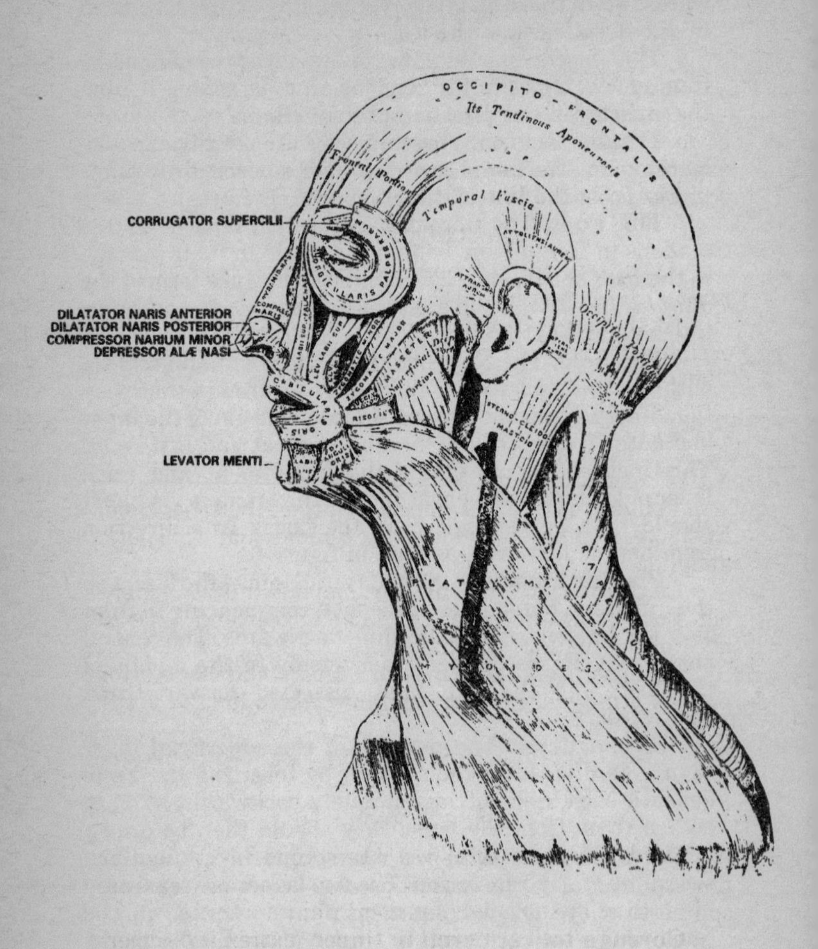

Muscles of the head, face, and neck.

HEAD AND NECK.

MUSCLES OF THE HEAD—(*11 Regions, 39 Muscles*).

The nervous supply is indicated by [ ] brackets.
The – divides the origin from the insertion.

(REGION 1) EPICRANIAL REGION. 1 MUSCLE.

Occip´ito-fronta´lis: outer ⅔ superior curved line of occiput, and mastoid process—frontal quadrilateral expansion to the facial muscles. [Supra-orbital, facial, occipital, posterior auricular.]

(2) AURICULAR REGION. 3.

Attol´lens au´rem: occipital fascia—upper part of pinna. [Small occipital.]

At´trahens au´rem: lateral edge aponeuro´sis of occipito-frontalis—front of helix. [Facial, inferior maxillary.]

Ret´rahens au´rem: mastoid process—lower cranial surface of the concha. [Facial.]

(3) INTRA-AURICULAR REGION. 4.

Ten´sor tym´pani: inferior surface petrous bone, Eustachian tube—backwards to handle malleus. [Otic ganglion.]

Laxa´tor tym´pani ma´jor: spinous process sphenoid, Eustachian tube—back through Glaserian fissure to neck of the malleus. [Facial.]

Laxa´tor tym´pani mi´nor: superior and posterior part external meatus—for- and inwards to handle of the malleus. [Facial.]

Stape´dius: interior of pyramid—forward to neck of stapes. [Facial.]

(4) PALPEBRAL REGION (4). 3.

Orbicula´ris palpebra´rum: internal angular process frontal bone, nasal process superior maxilla. sphineter of eye. [Facial, supra-orbital.]

Corruga´tor supercil´ii: inner extremity superciliary ridge—under surface orbicularis, opposite the middle of the orbital arch. [Facial and supra-orbital.]

Ten´sor tar´si: crest of os lachrymalis—tarsal cartilage near puncta; covers in lachrymal canals. [Facial.]

(5) ORBITAL REGION. 7.

Lava´tor pale´bræ superio´ris: inferior surface lesser wing of sphenoid, anteriorly to foramen opticum—upper border superior tarsal cartilage. [IIId.]

Rec´tus supe´rior: margin optic foramen—sclerotica. [IIId.]

Rec´tus infe´rior: optic foramen—sclerotica. [IIId.]

Rec´tus inter´nus: optic foramen—sclerotica. [IIId.]

Rec´tus exter´nus: 2 heads between which pass IIId., nasal branch of Vth, and VIth nerves and ophthalmic vein; *upper* from outer margin optic foramen, *lower* from ligament of Zinn and process of bone at sphenoidal fissure—sclerotica. [VIth.]

Obliq´uus supe´rior: near optic foramen—"pulley" thence at right angles to sclerotica. [IVth.]

Obliq´uus infe´rior: depression in orbital plate in superior maxilla—sclerotica, outer surface. [IIId.]

(6) NASAL REGION. 7.

Pyramida´lis na´si: occipito-frontalis—compressor naris. [Facial.]

Leva´tor la´bii superio´ris alæ´que na´si: nasal process superior maxilla—cartilage of the ala and lip. [Facial.]

Dila´tor na´ris ante´rior: Cartilage ala—inner border integument ala. [Facial.]

Dila´tor na´ris poste´rior: nasal notch superior maxilla—skin at inner margin nostril. [Facial.]

Compres´sor na´ris: above incisive fossa superior maxilla—pyramidalis nasi, nasal fibro-cartilage, its fellow opposite side. [Facial.]

Compres´sor na´rium mi´nor: alar cartilage—skin at the end of the nose. [Facial.]

Depres´sor a´læ na´si: incisive fossa superior maxilla—septum and ala nasi. [Facial.]

(7) SUPERIOR MAXILLARY REGION. 4.

Leva´tor la´bii superior´ris: lower margin orbit—lip. [Facial.]

Leva´tor an´guli o´ris: canine fossa superior maxilla—angle mouth. [Facial.]

Zygomat´icus ma´jor: in front zygoma—angle oris. [Facial.]

Zygomat´icus mi´nor: malar bone near maxillary suture—angle oris. [Facial.]

(8) INFERIOR MAXILLARY REGION. 3.

Leva´tor la´bii Inferio´ris, or **Leva´tor men´ti:** incisive fossa inferior maxilla—skin of chin.

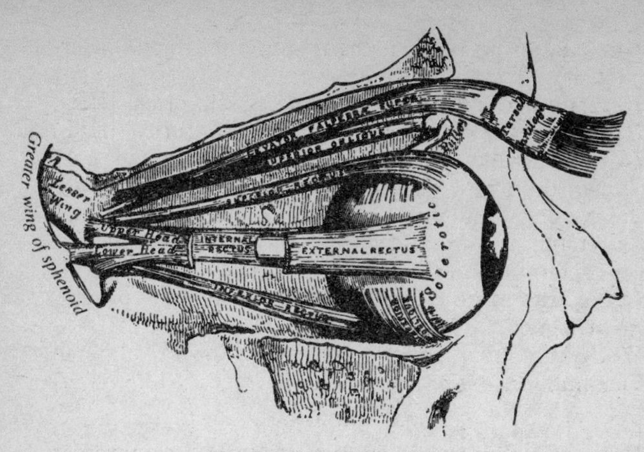

Muscles of the right orbit.

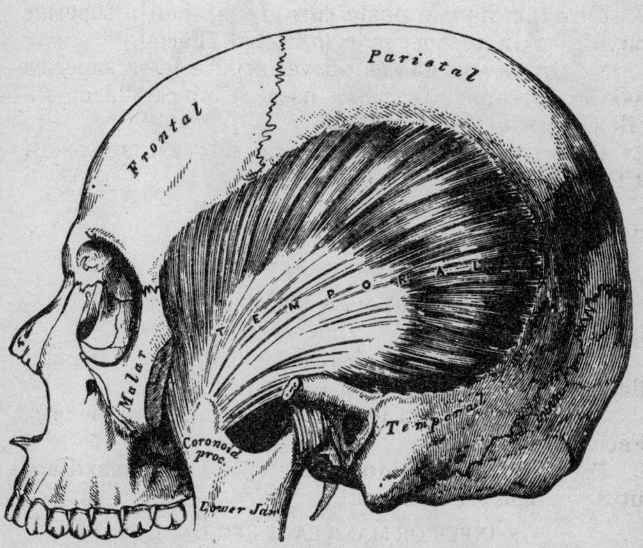

The Temporal muscle, the zygoma and Masseter having been removed.

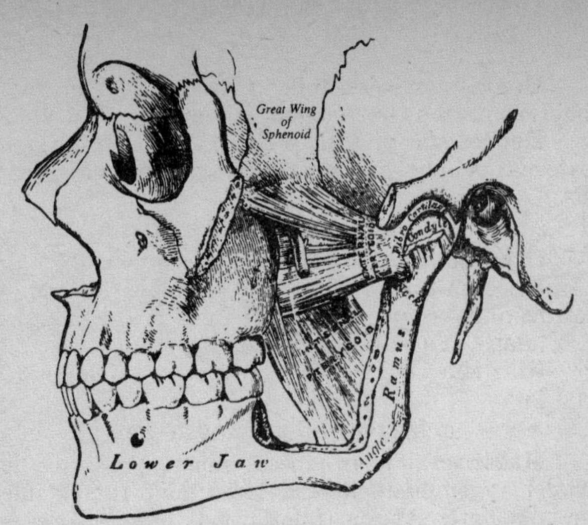

The Pterygoid muscles, the zygomatic arch and a portion of the ramus of the jaw having been removed.

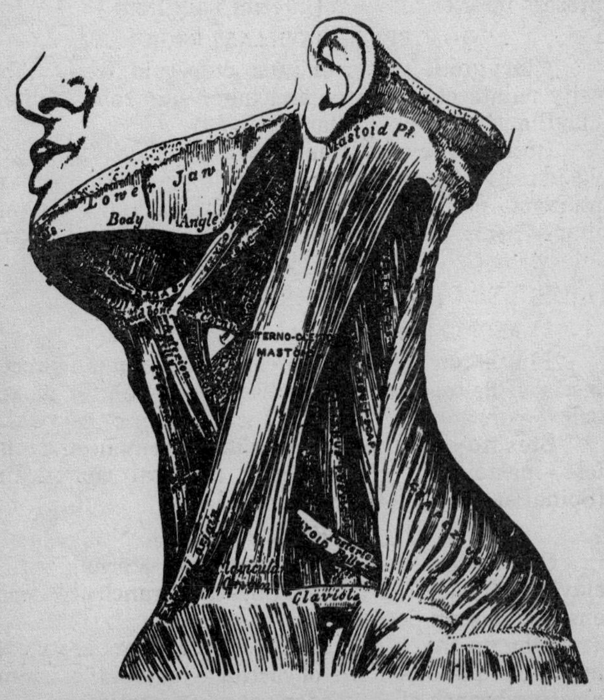

Muscles of the neck and boundaries of the triangles.

Depres´sor la´bii inferio´ris: external oblique line inferior maxilla—integument of lower lip. [Facial.]

Depres´sor an´guli o´ris, or **Triangular men´ti:** external oblique line inferior maxilla—angle oris. [Facial.]

(9) INTER-MAXILLARY REGION, 3.

Orbicula´ris o´ris: sphineter oris. [Facial.]

Buccina´tor: alveolar processes superior and inferior maxillæ—converges, to the angle of the mouth, and orbicularis. [Facial, inferior maxillary.]

Riso´rius: fascia above masseter—angle oris. [Facial.]

(10) TEMPORO-MAXILLARY REGION, 2.

Masse´ter: malar process superior maxilla, lower border zygoma—angle and lower half ramus inferior maxilla, outer surface. [Inferior maxillary.]

Tempora´lis: temporal fossa, curved line of frontal and parietal bones, pterygoid ridge of sphenoid—coronoid process inferior maxilla. [Inferior maxillary.]

(11) PTERYGO-MAXILLARY REGION, 2.

Pterygoide´us inter´nus: pterygoid fossa, tuberosity palate bone—lower and inner side ramus inferior maxilla. [Inferior maxillary.]

Pterygoide´us exter´nus: *upper* head from pterygoid ridge of great wing of sphenoid; *lower* from external pterygoid plate, tuberosity of palate, and superior maxillary bones—pterygoid depression in front condyle inferior maxilla. [Inferior maxillary.]

MUSCLES OF THE NECK—(*11 Regions, 42 Muscles*).

(REGION 1) SUPERFICIAL CERVICAL REGION, 2.

Platys´ma myoide´us: clavicle, acromian process, fascia of deltoid and pectoralis major—inferior maxilla below external oblique line. [Facial, superior cervical.]

Ster´no-clei´do mastoide´us: sternum and clavicle—mastoid process, superior occipital curved line. [Spinal accessory, cervical plexus.]

(2) INFRA-HYOID REGION, 4.

Sterno-hyoide´us: sternum and sternal end of clavicle—hyoid bone. [Communicating branch of descendens and communicans noni.]

Ster´no-thyroide´us: upper posterior edge sternum—oblique line of ala of cartilage (thyroid). [Communicating branch of descendens and communicans noni.]

Thy´ro-hyoide´us: oblique line of thyroid cartilage—body and greater cornu hyoid bone. [Hypoglossal.]

O´mo-hyoide´us: upper border scapula (bound down to clavicle by cervical fascia)—hyoid bone. [Communicating branch of descendens and communicans noni.]

(3) SUPRA-HYOID REGION, 4.

Digas´tricus: mastoid process of temporal (ligament binding hyoid bone)—symphysis inferior maxilla. [Facial, inferior dental.]

Sty´lo-hyoide´us: outer surface, middle styloid process—body hyoid bone, perforated by digastricus. [Facial]

My´lo-hyoide´us: (forms floor of mouth) mylo-hyoid ridge of inferior maxilla—body of os hyoides. [Inferior dental.]

Ge´nio-hyoide´us: inferior genial tubercle of inferior maxilla—body os hyoides. [Hypoglossal.]

(4) LINGUAL REGION (5). 4.

Ge´nio-hyo-glos´sus: superior genial tubercle of inferior maxilla—os hyoides and whole length inferior surface tongue. [Hypoglossal.]

Hyo-glos´sus: side of body and greater and lesser cornua hyoid—back and side of tongue. [Hypoglossal.]

Lingua´lis: under surface glossa from base to tip, between hyo-glossus and genio-hyo-glossus. [Chorda tympani.]

Sty´lo-glos´sus: outer and anterior center stylid process—side of tongue. [Hypoglossal.]

(5) PHARYNGEAL REGION (5), 4.

Constric´tor infe´rior: sides of cricoid and thyroid cartilages—fibrous raphé of posterior median line of pharynx. [Pharyngeal plexus, glosso-pharyngeal, external laryngeal.]

Constric´tor me´dius: greater and lesser cornua hyoid—posterior median pharyngeal raphé. [Glosso-pharyngeal, pharyngeal plexus.]

Constric´tor supe´rior: lower 3d of the margin of internal pterygoid plate, palate and contiguous palatal muscles—posterior median pharyngeal raphé and occipital pharyngeal spine. [Glosso-pharyngeal, pharyngeal plexus.]

Sty´lo-pharynge´us: inner side base of styloid process—constrictor muscles and upper border thyroid cartilage. [Glosso-pharyngeal and pharyngeal plexus.]

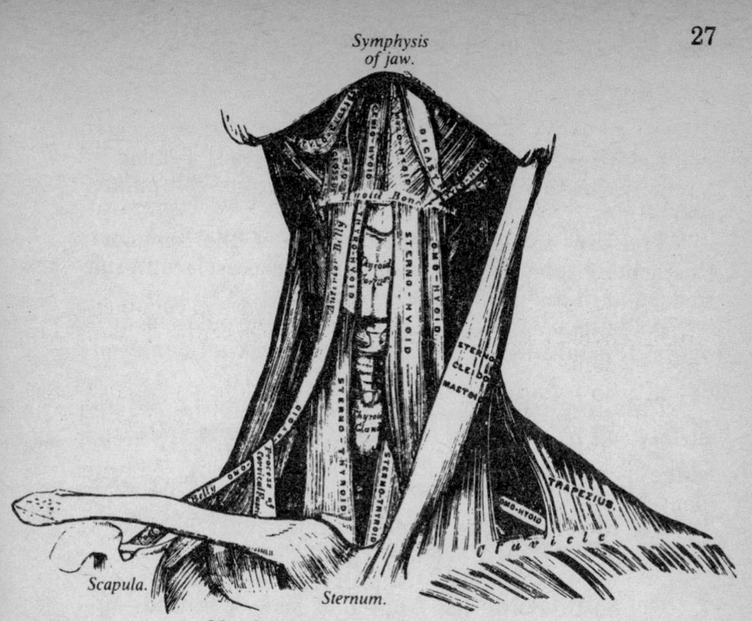

Muscles of the neck. Anterior view.

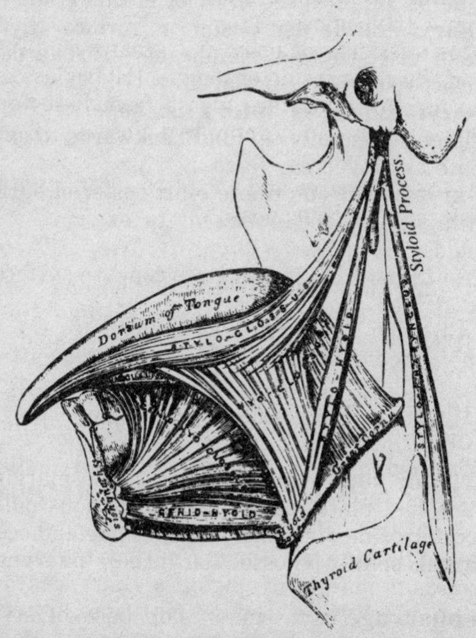

Muscles of the tongue. Left side.

(6) PALATAL REGION, 5.

Leva´tor pala´ti: under surface petrous portion of temporal Eustachian tube—posterior surface soft palate. [Facial.]

Ten´sor pala´ti: scaphoid fossa of the sphenoid, Eustachian tube (bound to hamular process)—anterior surface hard and soft palate. [Otic ganglion.]

Az´ygos uv´ulæ: posterior nasal spine palate bone—uvula. [Facial.] (Is not a *single* muscle as its name implies.)

Pala´to-glos´sus: (*anterior pillar*) anterior lateral surface soft palate—side and dorsum of tongue. (Meckel's ganglion.]

Pala´to-pharynge´us: (*posterior pillar*) soft palate—joins stylo-pharyngeus to be inserted into posterior border thyroid cartilage. [Meckel's ganglion.]

(7) INTRA-LARYNGEAL REGION, 5.

Cri´co-thyroide´us: front and side of cricoid—up- and outwards to lower border thyroid cartilage. [Laryngeal nerve supplies the muscles of this group.]

Thy´ro-arytænoide´us: posterior surface thyroid cartilages and crico-thyroid membrane—backwards to anterior surface arytenoid cartilage.

Cri´co-arytænoide´us latera´lis: superior border cricoid cartilage—obliquely up- and backwards to external angle base arytenoid cartilage.

Cri´co-arytænoide´us poste´rior: posterior surface cricoid cartilage—up- and outwards to external angle base arytenoid.

Arytenoide´us: fills up posterior concave surface of Arytenoid cartilage.

Vocal Chords: the INFERIOR or TRUE are the *inf. thyro-arytenoid ligaments* of yellow elastic tissue attached, in front, to depression between the two alæ of the thyroid cartilage—ant. angle of base of arytenoid cartilage. The SUPERIOR or FALSE are the *Sup. thyro-arytenoid ligaments* attached, in front, to angle of thyroid cartilage, below epiglottis—anterior surface of arytenoid cartilage.

(8) EPIGLOTTIDIAN REGION, 3.

Thy´ro-epiglottide´us: internal surface thyroid cartilage—upwards to margin of epiglottis. [Laryngeal.]

Arytæ´no-epiglottide´us supe´rior: apex arytenoid cartilage—to fold mucous membrane between cartilage and side of epiglottis. [Laryngeal.]

Arytæ´no-epiglottide´us infe´rior: arytenoid cartilage just above superior vocal chord—forwards and upwards to the margin of the epiglottis. [Laryngeal.]

(9) ANTERIOR VERTEBRAL REGION, 4.

Rec´tus cap´itis anti´cus ma´jor: (continuation scalenus anticus) 4 slips from anterior tubercles transverse processes 3d, 4th, 5th and 6th cervical vertebræ—basilar process occipital bone. [Suboccipital, and cervical plexus.]

Rec´tus cap´itis anti´cus mi´nor: anterior surface lateral mass of atlas and its transverse process—basilar process occipital. [Suboccipital, cervical plexus.]

Rec´tus latera´lis: upper surface transverse process atlas—jugular process occipital. [Suboccipital.]

Lon´gus col´li: *lst portion* from anterior tubercles transverse processes of 3d, 4th and 5th cervical vertebræ— tubercle of anterior arch of atlas; *2d portion* from 1st, 2d (and 3d) dorsal—transverse processes 5th and 6th cervical vertebræ; *3d portion* from 1st, 2d, 3d dorsal and 7th, 6th, 5th cervical—bodies 2d, 3d and 4th cervical vertebræ. [Lower cervical branches.]

(10) LATERAL VERTEBRAL REGION, 3.

Scale´nus anti´cus: inner border and superior surface lst rib—anterior tubercles transverse processes 3d, 4th, 5th and 6th cervical vertebræ. [Branches lower cervical.]

Scale´nus me´dius: behind groove for subclavian artery on lst rib—posterior tubercles transverse processes lower 6 cervical vertebræ. [Branches lower cervical.]

Scale´nus posti´cus: 2d rib, outer surface—transverse processes lower 3 cervical vertebræ. [Branches lower cervical.]

(11) POSTERIOR VERTEBRAL REGION, 4.

Rec´tus cap´itis posti´cus ma´jor: spinous process axis—inferior occipital curved line. [Occipital.]

Rec´tus cap´itis posti´cus mi´nor: tubercle posterior arch atlas—beneath insertion of above. [Occipital.]

Obliq´uus infe´rior: spinous process axis—horizontally to transverse process atlas. [Occipital.]

Obliq´uus supe´rior: transverse process atlas—occipital bone, between curved lines. [Occipital.]

ARTERIES OF THE HEAD AND NECK.

CARO´TIS COMMU´NIS: arises on right side, from innominate, behind sterno-clavicular articulation; on left side, from arch of aorta, highest part, and is more deeply

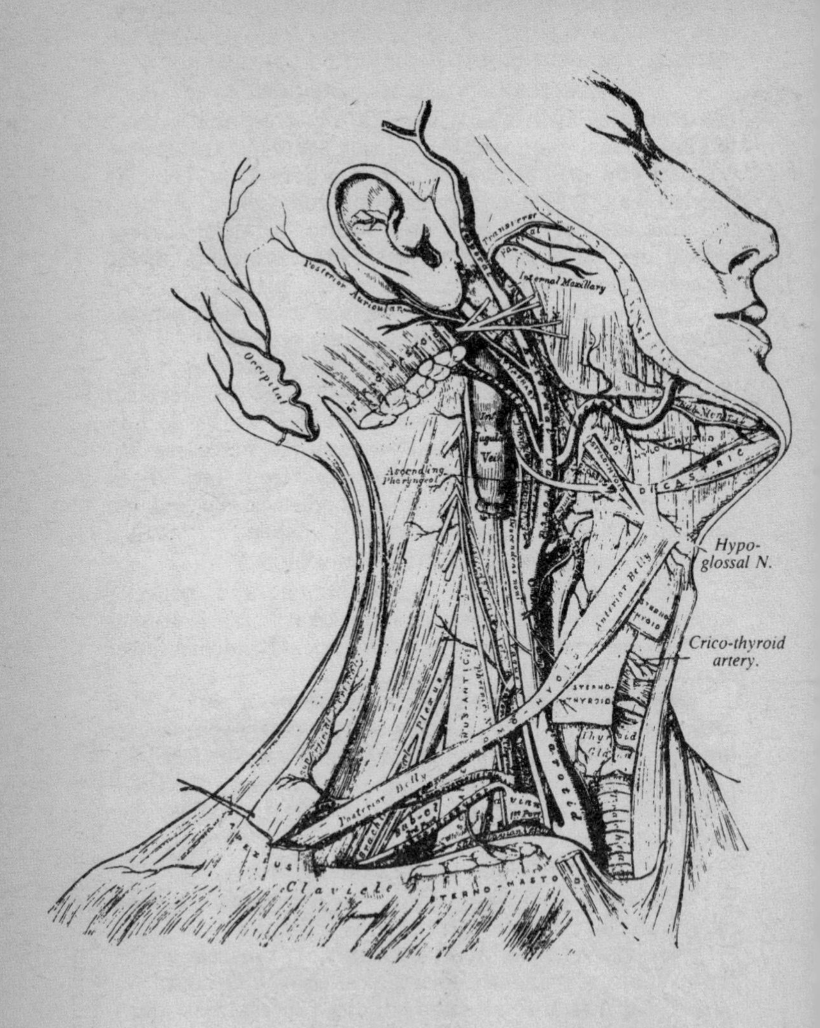

Surgical anatomy of the arteries of the neck, showing the carotid and subclavian arteries.

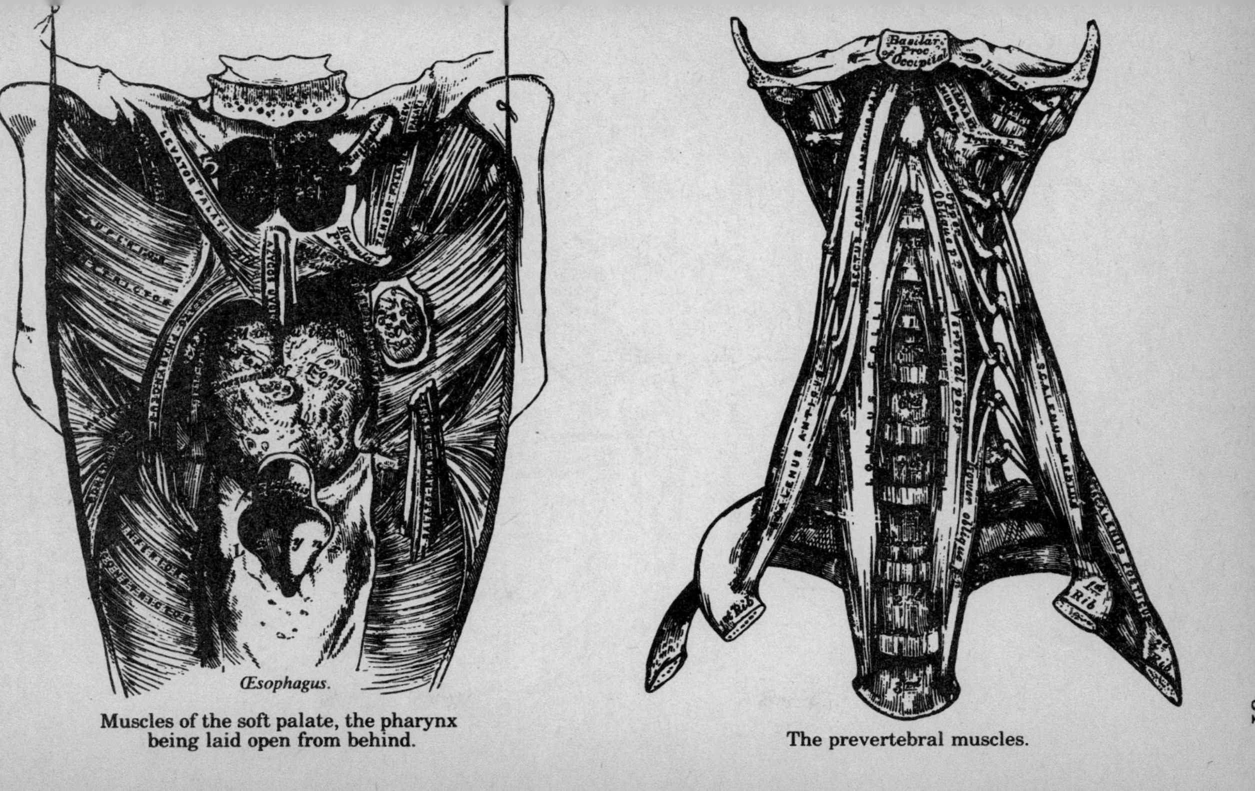

Muscles of the soft palate, the pharynx being laid open from behind.

The prevertebral muscles.

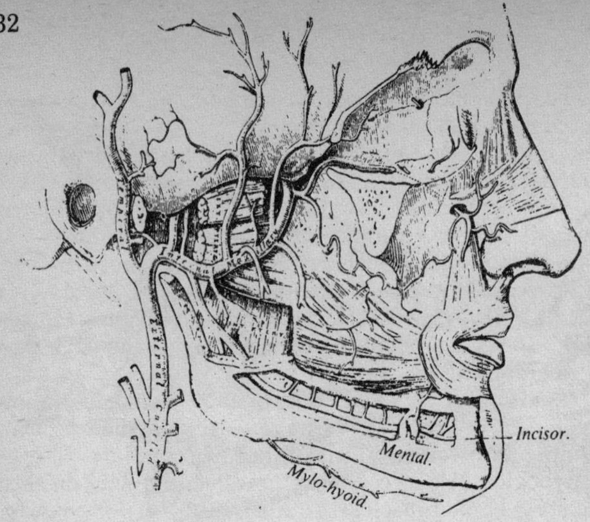

The internal maxillary artery and its branches.

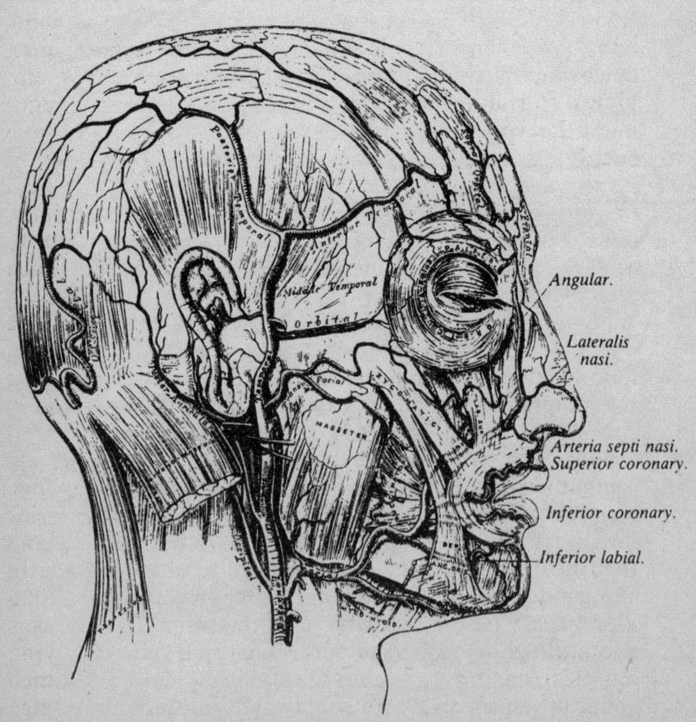

The arteries of the face and scalp.

placed than the right and passes obliquely outwards to root of neck behind sterno-hyoid and sterno-thyroid muscles, innominate vein and thymus gland. Starting now from each side of neck, each passes up- and outwards to superior border of thyroid cartilage, there dividing into external and internal carotid. Course indicated by line from sternal end clavicle to point midway between mastoid process and angle of interior maxilla. Vein lies to outside, pneumogastric nerve on posterior plane between them, the three being enveloped by same sheath of cervical fascia. No branches but terminal.

CARO´TIS EXTER´NA: (8 brs., see above) up between neck of inferior maxilla and external meatus, there dividing into temporal and internal maxillary. Crossed by hypoglossal nerve, lingual and facial veins, digastric and stylo-hyoid muscles. Is quite superficial. **Thyroide´a supe´rior:** greater cornu hyoid, curving down to thyroid gland, *anas*, with its fellow of opposite side and inferior thyroid. *Hyoide´a*, runs along inferior border of bone, *anas*, with opposite fellow. *Descen´dens superficialis*, down- and outwards across sheath common carotid supplying sterno-mastoid and adjacent muscles and integument. *Larynge´a supe´rior* pierces thyro-hyoid membrane supplying muscles, mucous membrane, glands, etc., of larynx and epiglottis, *anas*, with opposite fellow. *Cri´co-thyroide´a,* transversely across crico-thyroid membrane, *anas*, with opposite fellow. **Lingua´lis:** up- and inwards to under surface of tongue (ranine); runs parallel with hypoglossal nerve. *Hyoide´a,* along superior border bones, supplying muscles, *anas*, with opposite fellow. *Dorsa´lis linguæ,* ascends to dorsum tongue, *anas*, with opposite fellow, supplying mucous membrane, tonsil, epiglottis, soft palate, etc. *Sublingua´lis* runs for- and outwards in sublingual gland, supplies it, adjacent muscles, membranes, etc. *Rani´na*, on lingualis to tip of tongue, accompanied by gustatory nerve, *anas*, with opposite fellow, supplying adjacent parts. **Facia´lis:** near angle inferior maxillary obliquely for- and upwards to maxillary gland then up over jaw, up- and forwards to angle of mouth, along side of nose to inner canthus of eye (angular.) CERVICAL BRS.: *Palati´na ascen´dens*, between stylo-glossus and stylo-pharyngeus, to outer side pharynx, supplying muscles, tonsil, Eustachian tube, etc.; divides, one branch going up tensor palati to supply soft palate, glands, etc.; the other branch goes to tonsil, *anas*, with tonsillar. *Anas*, posterior palatine of internal maxillary. *Tonsillaris*, up to supply this gland and root of tongue. *Submaxilla´res* (3 or 4), supplying this gland and adjacent parts. *Submentalis*, off just as facial quits submaxillary gland, running forwards upon mylo-hyoid, supplying it and digastric (*anas.* with sublingual) to symphysis; the superficial branch turns round the chin, passing up to *anas.* with inferior labial, supplying muscles and integument; the deep branch runs up on bone to supply deep muscles and lip, *anas.* with inferior labial and mental. FACIAL BRS.: *Muscula´res*, to internal pterygoid, masseter, buccinator. *Labia´lis infe´rior*, beneath depressor anguli oris to lower lip, *anas.* inferior coronary, mental branch of dental, etc. *Corona´ria infe´rior* beneath depressor anguli oris along edge lower lip, supplying adjacent parts, and *anas.* with opposite fellow, inferior labial, and mental branch of inferior dental. *Corona´ria supe´rior* along edge of upper lip, *anas.* with opposite fellow, supplying adjacent parts, septum and ala of nose. *Latera´lis na´si* supplying side and dorsum of nose, septum, *anas.* opposite fellow, infra-orbital and nasal branch ophthalmic. *Angula´ris*, terminal branch, ascends up to inner canthus, *anas.* with nasal branch ophthalmic. **Occipita´lis:** from posterior part near inferior margin of digastricus, up between atlas and mastoid process, horizontally across occiput, then up to vertex, then dividing into numerous branches. *Muscula´res*, to digastricus stylo-hyoid, stylo-mastoid, splenius capitis, trachelo-mastoid. *Auricula´ris*, to posterior surface concha. *Meninge´a infe´rior* along side internal jugular vein through foramen lacerum to dura in posterior fossa. *Arte´ria prin´ceps cervi´cis*, descends back part neck, superficial branch supplying splenius and trapezius, *anas*, with superficial cervical; the deep branch *anas.* with vertebral and cervical branch superior intercostal; supplies adjacent parts. *Crania´les*, to muscles and integument of posterior surface cranium. **Auricula´ris poste´rior:** from above stylo-hyoid, ascends beneath parotid gland, to groove between mastoid process and ear cartilage, dividing into anterior and posterior branches, the former passes forwards to *anas.* with temporal, the other back to *anas.* with occipital. *Sty´lo-mastoide´a*, enters foramen supplying cells, tympanum and semi-circular canals. *Auricula´ris*, to back part of cartilage of ear, and penetrating to its anterior surface. **Pharynge´a ascen´-dens:** (smallest branch) deep seated, arising near commencement external carotid, up between internal carotid and pharynx, to base of skull. *External branches*, to recti antici muscles, glands of neck, sympathetic, pneumogastric and hypoglossal nerves; *anas.* with ascending cervical. *Pharynge´æ* (3 or 4) to parts of pharynx and adjacent muscles, etc. *Meninge´æ* backwards through foramen lacerum posterius, another branch through foramen lacerum basis cranii, another through anterior condyloid foramen to dura mater. **Tempora´lis:** from parotid gland up to root zygoma, dividing into anterior and posterior. *Transver´sa facie´i*, in parotid gland, runs across face, supplying glands, integument and muscles, *anas.* with facial and infra-orbital. *Tempora´lis me´dia*, above zygoma to temporal muscle and orbicularis, *anas.* with lachrymal and palpebral branches of ophthalmic and deep temporal branches of internal maxillary. *Aricula´res anterio´res*, to anterior ear, *anas.* with posterior auricular. *Tempora´lis ante´rior* forwards over forehead, supplying integument, muscles, etc., *anas.* with frontal and supra-orbital. *Tempora´lis poste´rior*, up- and backwards over side of head, *anas.* with opposite fellow posterior auricular and occipital. **Maxilla´ris inter´na:** (see external carotid) inwards to inner side of condyle inferior maxilla into spheno-maxillary fossa, to supply deep structures of the face. MAXILLARY PORTION: *Cavi tympani* (tympanic) up thru fissura Glaseri, supplying membrani tympani, laxator tympanus; *anas.* with stylo-mastoid and Vidian. *Meninge´a me´dia*, from internal lateral ligament of jaw up through foramen spinosum, dividing into anterior and posterior branches, supplying anterior and posterior surface of dura and bones, facial nerves, and branches to other parts; *anas.* with opposite fellow, anterior and posterior meningeal. *Meninge´a par´va*, through foramen ovale to Casserian ganglion and dura; also to nasal fossa and soft palate. *Alveola´ris infe´rior*, (inf. dental) with dental nerve to foramen on ramus, then along dental canal supplying teeth, etc., till opposite bicuspid tooth, then divides into incisor and mental branches, the former to incisor teeth, *anas.* with opposite fellow; the latter passes out mental foramen, *anas.* with inferior labial, inferior coronary, submental and supplies adjacent parts. Mylo-hyoid branch given off just as artery enters inferior dental foramen; it runs in its groove to its muscle. PTERYGOID PORTION: *Tempora´lis profund´æ* (2) anterior and posterior branches up to temporal muscle. *Ptery-goide´æ*, to do muscles. *Masseter´ica*, to do muscles. *Bucca´lis*, to do muscles. SPHENO-MAXILLARY PORTION: *Alveola´ris*, common branch with following, supplying (superior dental) teeth, antrum and gums. *Infra-orbita´lis*, continuation of main artery, along infra-orbital canal, and out infra-orbital foramen, supplying inferior rectus and inferior oblique, antrum, front teeth, lachrymal sac, etc., *anas.* with facial, buccal, nasal branch ophthalmic, etc. *Palatina Descen´dens*, down posterior palatine canal to gums, mucous membrane, palate, etc. *Vidia´na*, through its canal, with nerve to pharynx, Eustachian tube and tympanum. *Pterygo-palati´na*, to upper part pharynx and Eustachian tube. *Spheno-palatina* (nasal) to mucous membrane of nose, septum, antrum, ethmoid and sphenoid cells.

CARO´TIS INTER´NA: (8 brs.) Superior border thyroid cartilage up through carotid foramen in temporal bone; in the skull it runs forwards in a course represented by *f* [italic f laid horizontally.] No branches from cervical part. Tonsil is internal to it. *Tympani´ca*: to tympanum. **Receptac´ulæ:** small branches to cavernous sinus, pituitary body, Casserian ganglion, etc. **Ophthal´mica:** at inside anterior clinoid process, forwards through optic foramen to inner canthus, dividing into frontal and nasal. *Lachryma´lis*, to lachrymal gland, conjunctiva; malar and meningeal branches; *anas.* freely with temporal, palpebral, etc. *Supra-orbita´lis*, out supra-orbital foramen to muscles and skin of forehead and pericranium; *anas.* with temporal, facial, etc. *Ethmoida´les*, (2) anterior and posterior to ethmoidal cells and meninges. *Palpebra´les*, (2) superior and inferior, encircle eyelids, down nasal duct, *anas.* with temporal, inferior orbital, etc. *Fronta´lis*, out inner angle orbit to forehead, supplying adjacent parts, *anas.* with supra-orbital. *Nasa´lis*, to lachrymal sac, then down the nose, supplying whole surface; *anas.* with facial, etc. *Cilia´res bre´ves*, (12-15) supply choroid and ciliary processes. *Cilia´res lon´gæ*, (2) ciliary ligament and iris. *Cilia´res anterio´res*, from muscular branches, to iritic arterial circle. *Centra´lis ret´inæ* pierces optic nerve and runs in it to retina. *Musculares*, (2) superior and inferior to muscles of eye. **Cere´bri arte´ria ante´rior:** at fissure of Sylvius forward in the great longitudinal fissure, *anas.* with its fellow by *ante´rior commu´nicans*; curves round anterior border corpus callosum, running back to its posterior part to *anas.* with posterior cerebral supplying olfactory and optic nerves, inferior surface anterior lobes, 3d venticle, anterior perforated space, corpus callo-

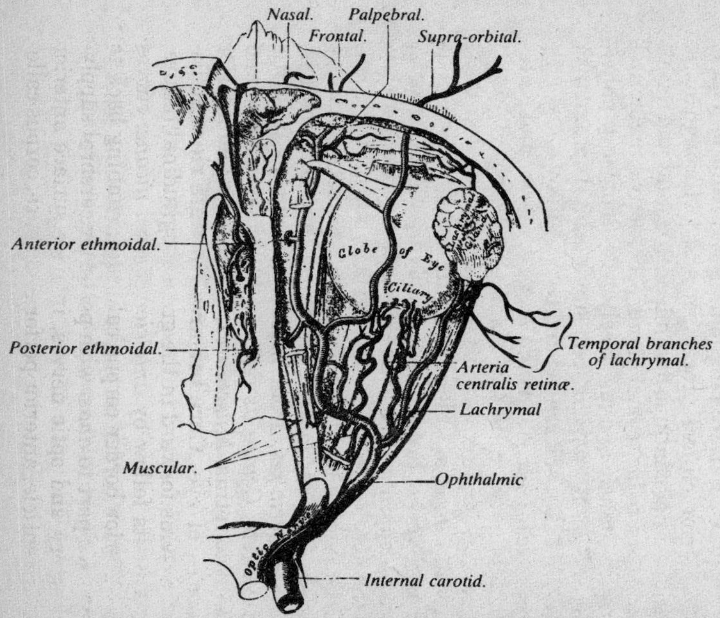

The ophthalmic artery and its branches, the roof of the orbit having been removed.

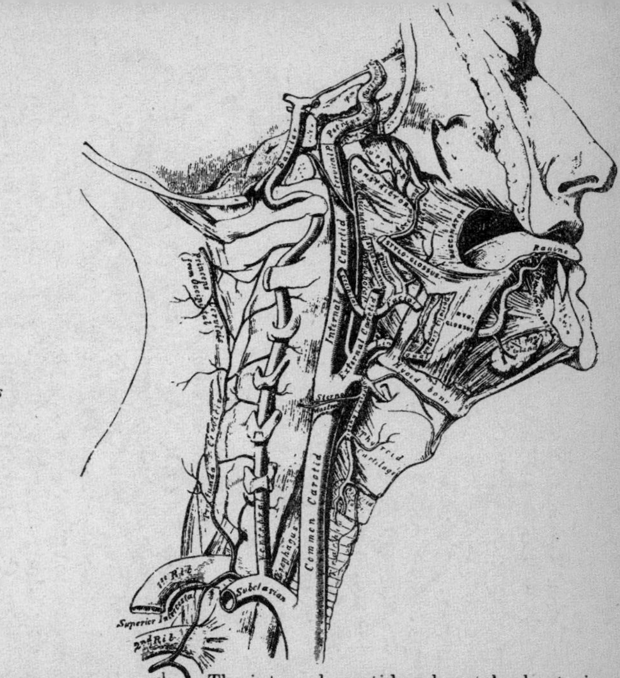

The internal carotid and vertebral arteries. Right side.

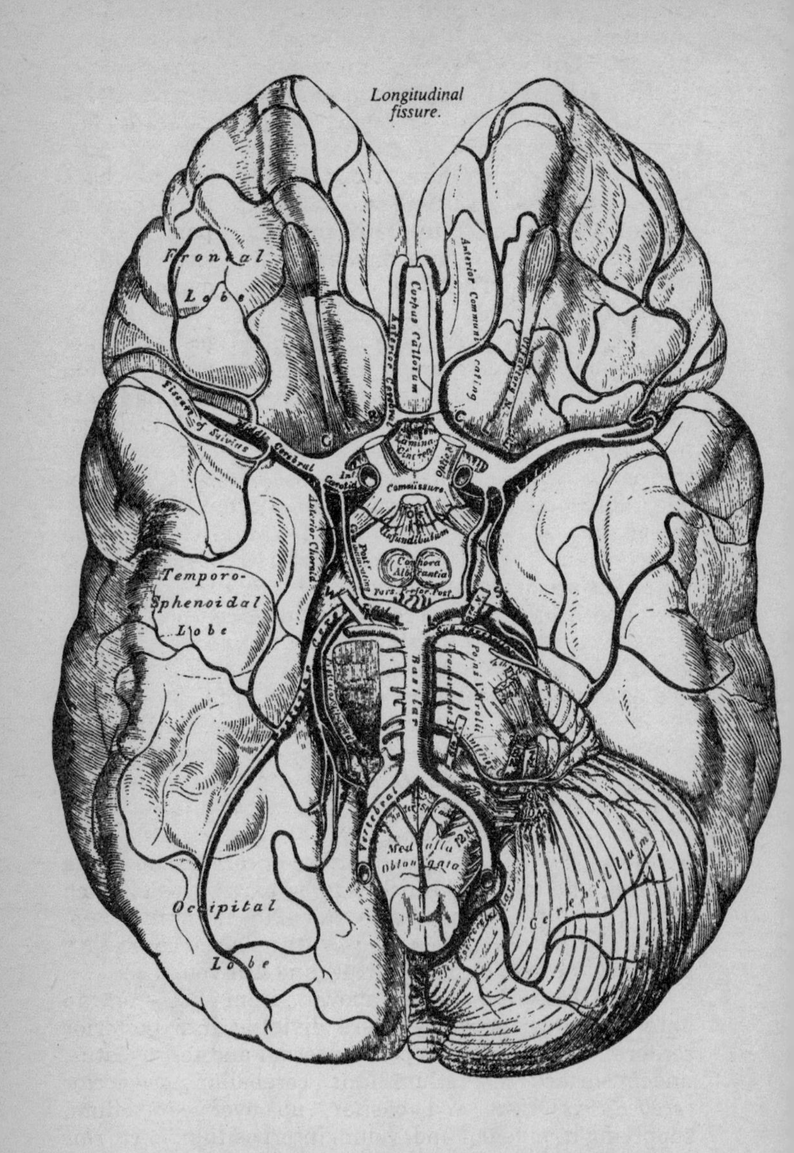

The arteries of the base of the brain. The right half of the cerebellum and pons have been removed.

sum and inner surface of hemispheres. **Cere´bri arte´ria me´dia:** (largest branch), obliquely outwards along fissure of Sylvius, dividing into *anterior* branch to pia of anterior lobe, *median* branch to small lobe at extremity of Sylvian fissure; *poste´rior* branch which supplies middle lobe; *small* branches to corpus striatum through substantia perforata. **Commu´nicans poste´rior:** from back part of artery backwards, *anas.* with posterior cerebral of basilar. **Choroide´a ante´rior:** from back part of artery back- and outwards, entering descending horn of lateral ventricle; is distributed to hippocampus major, corpus fimbriatum and choroid plexus.

VERTEBRA´LIS: (6 brs.) lst and largest branch of subclavian. Enters foramen in transverse process of 6th cervical vertebra and ascends in the vertebral foramina to the axis, then outwards, piercing occipito-ataloid ligament and dura, passing through foramen magnum along in front of medulla, unites with opposite fellow to form basilar. *Spina´les latera´les*, enter spinal canal through the intervertebral foramina and supply (anterior branches) the cord and membranes and (posterior branches) posterior surface of vertebral bodies. *Muscula´res*: deep muscles of neck, *anas.* with occipital and deep cervical. *Posterio´res meninge´æ*, (2) to falx cerebelli. *Spina´lis ante´rior*, given off near termination, unites with opposite fellow, and descends on cord, *anas.* with spinal branches through the intervertebral foramina down to sacrum. Supplies pia of cord (being placed beneath it) and cord. *Spina´lis poste´rior*, arises at side of medulla and passes down posterior surface of cord, being reinforced similarly to the anterior spinal, to sacrum. *Infe´rior cerebella´ris*, winds back over medulla, to under surface of cerebellum, there dividing, the inferior branches going backwards to notch between the two hemispheres, the external branch supplying the inferior surface, *anas.* with superior cerebellar; branches, also, to choroid plexus, and 4th ventricle.

BASILLA´RIS: (see above); from posterior to anterior border of pons, there dividing into posterior cerebral. *Transver´sæ*, to pons, internal auditory meatus, under surface cerebellum (ant. cerebellar.) *Supe´rior cerebella´ris*, near end basilar, up over cerebellum, supplying it, pineal gland, velum interpositum. *Poste´rior cerebra´lis*, winds round crus cerebri to inferior surface of posterior cerebral lobes, supplying them, and choroid plexus, *anas.* with anterior and middle cerebral.

Circle of Willis: (10 vessels); forward, from behind forwards, by basilar, 2 posterior cerebral, 2 posterior communicating, 2 internal carotids, 2 anterior cerebral, anterior communicating.

Infe´rior Thyroide´a: (see arteries of upper extremity); branch of thyroid axis, up behind sheath of common carotid and sympathetic nerve to under surface of thyroid gland, *anas.* with opposite fellow, and superior thyroid. *Laryngea´lis*, to back part larynx. *Trachea´lis*, to trachea, *anas.* with bronchial. *Œsophagea´les*, to the œsophagus. *Cervica´lis ascendens*, up neck, supplying muscles, vertebræ, cord and membranes.

Cervi´cis profun´da: (see arteries of upper extremity); branch of superior intercostal, ascends back part of neck, below complexus, to axis, supplying adjacent parts, and *anas.* with branches of vertebral and princeps cervicis of occipital.

VEINS OF THE HEAD AND NECK.

Ve´næ Dip´loes: walls only of epithelium, with many *cul-de-sacs*. *Fronta´lis*, opens into supra-orbital through supra orbital notch. *Tempora´lis ante´rior* opens into deep temporal. *Tempora´lis poste´rior* confined to parietal region, opens into lateral sinus. *Occipita´lis*, opens into occipital vein or sinus.

Cerebra´les: noted for their thin coats, muscular tissue and absence of valves. *Superio´res*, (7 or 8 on each side) for- and inwards to superior longitudinal sinus, there receiving interior cerebral which drain the same hemisphere. *Inferio´res anterio´res*, under surface of anterior lobes; terminate in cavernous sinus. *Inferio´res latera´les*, (3 to 5) terminate in lateral sinus. *Inferio´res Me´diæ*, from posterior lobe, etc., to straight sinus behind venæ Galeni. **Ve´næ Gale´ni** (2, one from right, one from left ventricle); formed by ve´na cor´poris stria´ti and ve´na choroide´a, pass back and out of transverse fissure to straight sinus. *Cerebella´res*, superior, inferior and lateral sets; the lst open into straight, the 2d into lateral, the 3d into superior petrosal sinus.

Sinus: (16 in No.) *Supe´rior longitudina´lis*, begins at crista Galli, runs back over cerebrum to torcular Herophili; receives superior cerebral and parietal veins. *Infe´rior longitudina´lis*, along posterior part of free margin of falx cerebri to straight sinus. *Tento´rii* (straight), junction

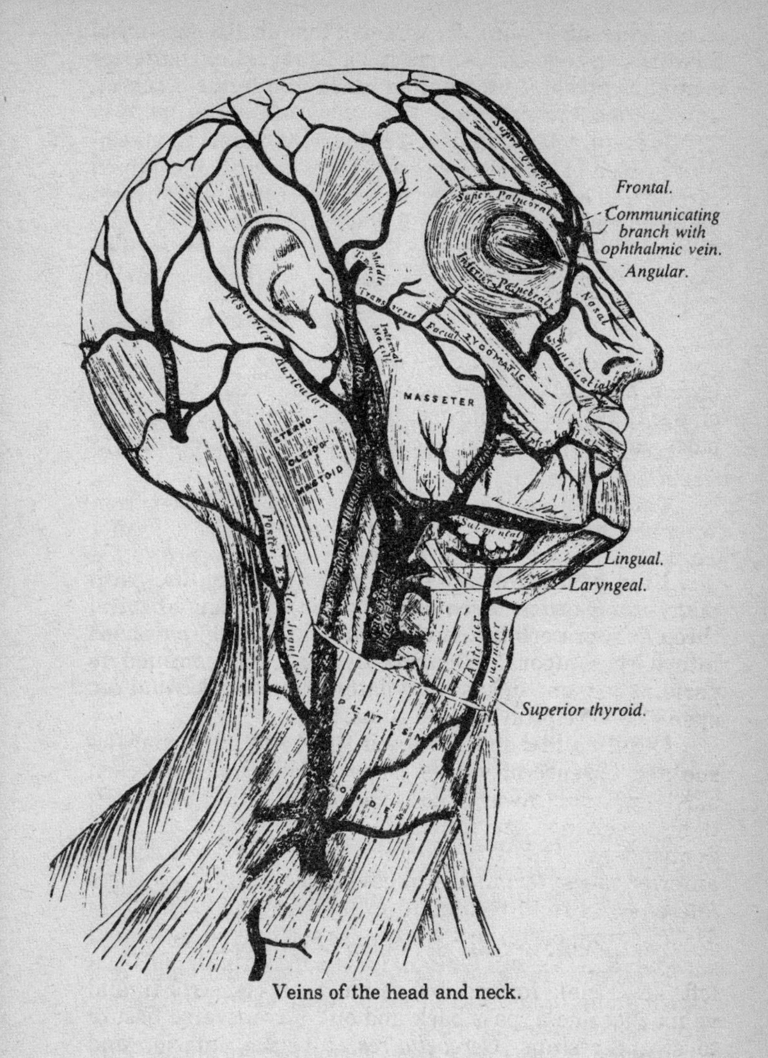

Veins of the head and neck.

of tentorium and falx cerebri to torcular Herophili; receives inferior longitudinal sinus, venæ Galeni, inferior median cerebral, and superior cerebellar veins. *Latera´les*, (2) from torcular to foramen lacerum posterius, into internal jugular vein; receives straight and occipital sinus, etc. *Occipitales*, (2) smallest; posterior margin of foramen magnum to torcular. *Caver´ni*, (2) sides of sella Turcica from sphenoid fissure to apex petrous part of temporal. Receives ophthalmic vein connecting the frontal with these sinus; also inferior anterior cerebral veins. *Circula´ris* surrounds pituitary body, communicates with each cavernous. *Inferio´res petrosa´les*, (2) termination of cavernous to internal jugular vein. *Transver´sus*, connects the inferior petrosales across basilar process of occipital. *Superio´res petrosa´les*, (2) on superior border petrous part of temporal, connecting lateral and cavernous; receives inferior lateral cerebral and anterior lateral cerebellar veins.

Ve´na Facia´lis: obliquely across side face from inner canthus, to unite, under inferior maxilla, to form a trunk for internal jugular. Receives *supra-orbita´lis, supra-palpebra´lis, nasa´lis, infe´rior palpebra´lis, fronta´lis, supra-orbita´lis, supra-labia´lis, infe´rior labia´lis, bucca´lis, masseter´ica, submenta´lis, infe´rior palati´na* (which arises from plexus about tonsil, etc.), *submaxilla´ris, rani´na* veins; also communicates with ophthalmic (see cavernous sinus).

Tempora´lis: from side and vertex of head, uniting with internal maxillary, to form temporo-maxillary. Receives *parotide´æ auricula´res anterio´res, transver´sa facie´i.*

Maxilla´ris Inter´na: *me´diæ meninge´æ, tempora´lis profun´da, pterygoide´a, masseter´ica, bucca´lis, palati´næ, infe´rior denta´lis*, forms, with above, temporo-maxillary.

Temporo-Maxilla´ris: union of temporal and internal maxillary; descends in parotid gland and divides, one branch going to join facial, the other to external jugular. Receives posterior auricular.

Auricula´ris Poste´rior: formed from plexus side of head; receives *stylo-mastoide´a* and branches from external ear; empties into temporo-maxillary.

Occipita´lis: from plexus, back part vertex of skull, descends deeply between muscles of neck, lying in course of artery, to internal jugular. Receives *mastoide´a*, which communicates with lateral sinus.

Jugula´ris Exter´na: from temporo-maxillary near

angle lower jaw, down into subclavian; accompanied by auricularis magnus nerve. Has 2 pairs of valves. Receives *occipita´lis, poste´rior jugula´ris exter´na* (draining superficial muscles of back of neck), *supra-scapula´ris, transver´sa cervi´cis* veins.

Jugula´ris Ante´rior: drains integument and superficial muscles of anterior and middle portion of neck, emptying into subclavian. No valves.

Jugula´ris Inter´na: from jugular foramen at junction of lateral and inferior petrosal sinus, vertically down the side of neck (outer side of main arteries), uniting with subclavian to form vena innominata; 1 pr. valves ¾ inch above termination. Receives *facialis, lingua´lis, pharynge´æ, supe´rior thyroide´a,* and *me´dia thyroide´a.*

Vertebra´lis: drains occipital region and deep muscles of back of neck; enters foramen in transverse process of atlas, down through similar foramina of the cervical vertebræ to 6th (or 7th) where it passes out to enter v. innominata. Receives *poste´rior condyloi´da, muscula´res, dorso-spina´les, menin´gio-rachidia´næ, ascen´dens and profun´da cervica´les;* 1 pr. valves guard its mouth.

NERVES OF THE HEAD AND NECK.

CRANIAL. lst or **Ner´vus olfac´tus.**—From corpus striatum, middle and anterior lobes of cerebrum. *Supplies* the Schneiderian membrane. Special function, *smelling.*

2d or **Op´ticus.**—From optic thalami and the corpora geniculata et quadrigemina, out through optic foramen to retina. Special function, *sight.*

3d or **Moto´rius Oc´uli.**–From crus cerebri and pons (?) out through foramen lacerum anterius to all the muscles of the orbit, save the superior oblique and external rectus; a few filaments pass to the iris. Is a motor nerve.

4th or **Pathet´icus.**—From valve of Vieussens, thru' foramen lacerum anterius to superior oblique. Is a motor.

5th or **Trigem´inus.**—The *sensory,* or posterior root, from the lateral tract of the medulla, the pons, and cerebellum (middle peduncle). The *motor* root from the pyramidal body. The *sensory* supplies are to the eye-ball (iris, ciliary body, etc.), lachrymal gland, conjunctiva, Schneiderian membrane, all the muscles and integument about the eye ball, orbit, os frontalis, nose, mouth, cheek, lips, temple, superior portion of pharynx, tongue, gums and teeth. *Motor* filaments are given to the external and internal pterygoid, temporal, buccinator, and masseter muscles. *Special* sensation (*taste*) to mucous membrane of

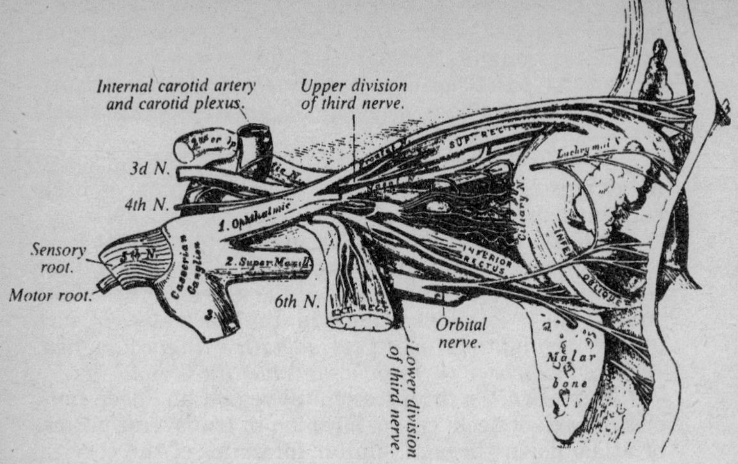

Nerves of the orbit and ophthalmic ganglion. Side view.

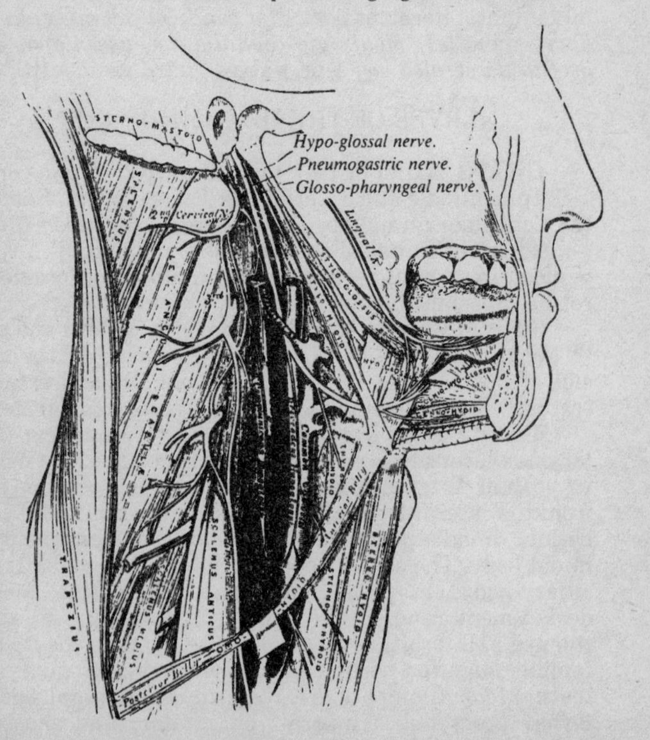

Hypoglossal nerve, cervical plexus, and their branches.

mouth, gums, tongue (anterior and middle portion), sublingual gland, conical and fungiform papillæ. BRS.—I. OPHTHAL´MICUS: *sensory*; forward through sphenoidal fissure from Casserian ganglion, joined by cavernous plexus of sympathetic. *Lachryma´lis. Fronta´lis*; (*a*) supratrochlea´ris, (*b*) supraorbita´lis. *Nasa´lis*, ganglionic, long ciliary (2 or 3), infra-trochlear branches. II. SUPE´RIOR MAXILLA´RIS, *sensory*: forwards through foramen rotundum from Casserian ganglion appearing on face through infra-orbital foramen. *Orbita´lis*; (*a*) temporal, (*b*) malar branches. *Spheno-palati´ni* (2). *Posterio´res denta´les* (2); (*a*) anterior branches, (*b*) posterior branches. *Ante´rior denta´lis. Palpebra´les. Nasa´les. Labia´les.* All inosculate with branches from facial. III INFE´RIOR MAXILLA´RIS: *sensor* root from Casserian ganglion, *motor* unites with it after passing through foramen ovale. ANTERIOR DIVISION: (*a*) masseteric, (*b*) deep temporal, (*c*) buccal, (*d*) pterygoid branches. POSTERIOR DIVISION: *Auric´ulo-tempora´lis*; (*a*) anterior temporal, (*b*) posterior temporal (out under cover of parotid, (*c*) communicating with facial, (*d*) inferior and superior auricular, (*e*) 2 branches to meatus, (*f*) branches to temporo-maxillary articulation, (*g*) branches to parotid gland. *Gustato´rius*, side of tongue to lip, (*a*) communicating branches, (*b*) branches of distribution to tongue, gums, etc. *Infe´rior denta´lis*, in dental canal inferior maxilla to teeth, etc.; (*a*) mylo-hyoid to do muscle, etc., (*b*) dental branches.

6th or **Abdu´cens.**—From pons. corpus pyramidale, and medulla through foramen lacerum anterius to supply motor influence to the rectus externus oculi.

7th or **Facia´lis.**—From lateral tract medulla and 6th ventricle, out through stylo-mastoid foramen to *all* the muscles of the face, ear and their integument, the platysma, buccinator, digastric, stylo-hyoid, lingualis, stapedius, laxator and tensor tympani, levator palati, and azygos uvulæ. Is essentially a motor nerve. *Tympan´icus. Chor´da tym´pani. Poste´rior auricula´ris*; (*a*) auricular branch, (*b*) occipital branch. *Stylo-hyoide´us. Digastric branch. Temporo-facia´lis*; (*a*) temporal branches, (*b*) infra-orbital, (superficial and deep branches, (*c*) malar branches. *Cervico-facia´lis*; (*a*) buccal, (*b*) supramaxillary branches, (*c*) infra- maxillary branches.

8th or **I. Glosso-pharynge´us, II. Pneumogas´tricus, III. Spina´lis Accesso´rius.**—I. and II. from floor of 4th ventricle; III. from lateral tract of cord as low as 6th cervico-spinalis, and also from medulla just below origin of I. and II. Part I. passes out through foramen lacerum posterius to supply *sensation* to mucous membrane of pharynx. Eustachian tube, tympanum, and tonsil; *motor* influence to the pharyngeal muscles; *gustation* to posterior third of tongue and its lateral papillæ. Branches of *communication* (sympathetic, facial, tympanic; *Carotid* branches. *Pharyngeal* branches. *Muscular* branches. *Tonsillar* branches. *Lingual* branches. Part II. through foramen lacerum posterius to supply *motor* and *sensor* filaments to the muscles and parts about the pharynx, larynx and trachea concerned in speech and respiration; *motor* filaments to the pharynx, heart, œsophagus, stomach, and filaments to the splenic and hepatic plexus. *Auricula´ris. Pharyngeal branch. Supe´rior laryngea´lis. Recur´rens* (or inferior) *laryngea´lis* (the motor of larynx). *Cervico-cardiac (2 or 3 in number). Thoracico-cardiac. Anterior´res pulmona´res* (2 or 3 in number). *Poste´rior pulmona´ris. Œsophagea´les. Gastric* branches. Part III. supplies *motor* filaments to sterno-mastoideus and trapezius, the accessory part arising from lateral tract of cord; the spinal portion, as low down as 6th cervical nerve, passing up in spinal foramen into skull, then out, with the accessory portion, through jugular foramen.

9th or **Hypoglos´sus.**—From floor of medulla. Is the *motor* of the tongue. Out through anterior condyloid foramen to supply the genio-hyoid, genio-hyo-glossus, hyo-glossus, stylo-glossus, thyro-hyoid, sterno-hyoid, omo-hyoid, and sterno-thyroid muscles. Is deep-seated (beneath internal carotid), but finally curves over externally to the carotid to muscles for distribution. Has *branches of communication* with pneumogastric, sympathetic, lst and 2d cervical and gustatory. *Descen´dens no´ni* (on carotid sheath), joining with 2d and 3d cervical. *Thyro-hyoid* branch. *Muscular* branches.

CERVICA´LES: each increase in size from lst to 5th; 8 pairs in all. Have anterior and posterior branches, the latter having ganglionic enlargements. The lst, or *sub-occipital,* (anterior branch) has exit between atlas and occiput; the remaining 7 between their respective vertebræ. The 4 upper (anterior branches) unite to form the cervical plexus; the 4 lower (anterior) with the lst dorsal form the brachial plexus.

Cervi´cis plex´us: SUPERF. BRS. *Superficia´lis col´li,*

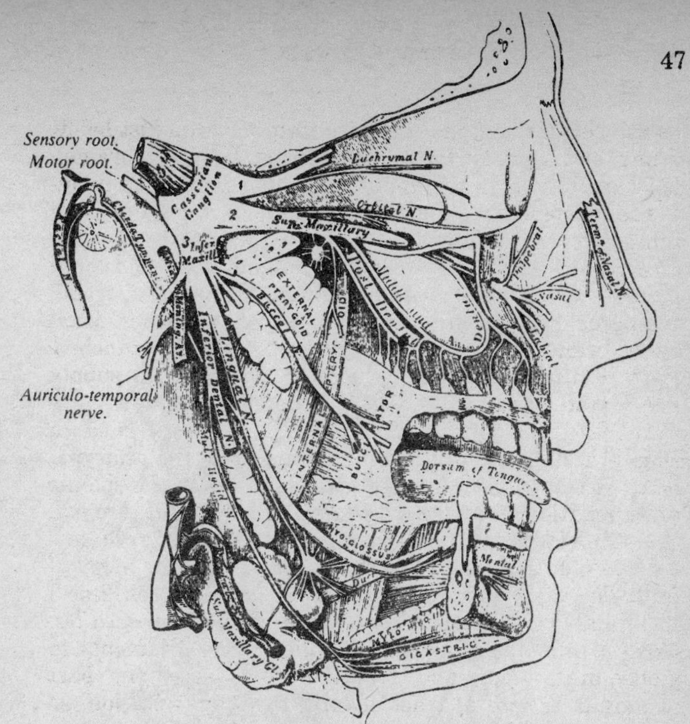

Distribution of the second and third divisions of the fifth nerve and submaxillary ganglion.

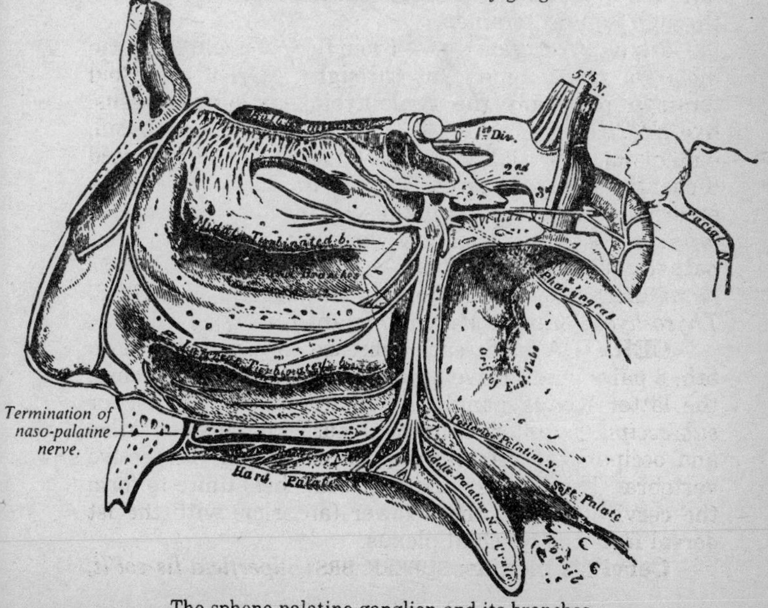

The spheno-palatine ganglion and its branches.

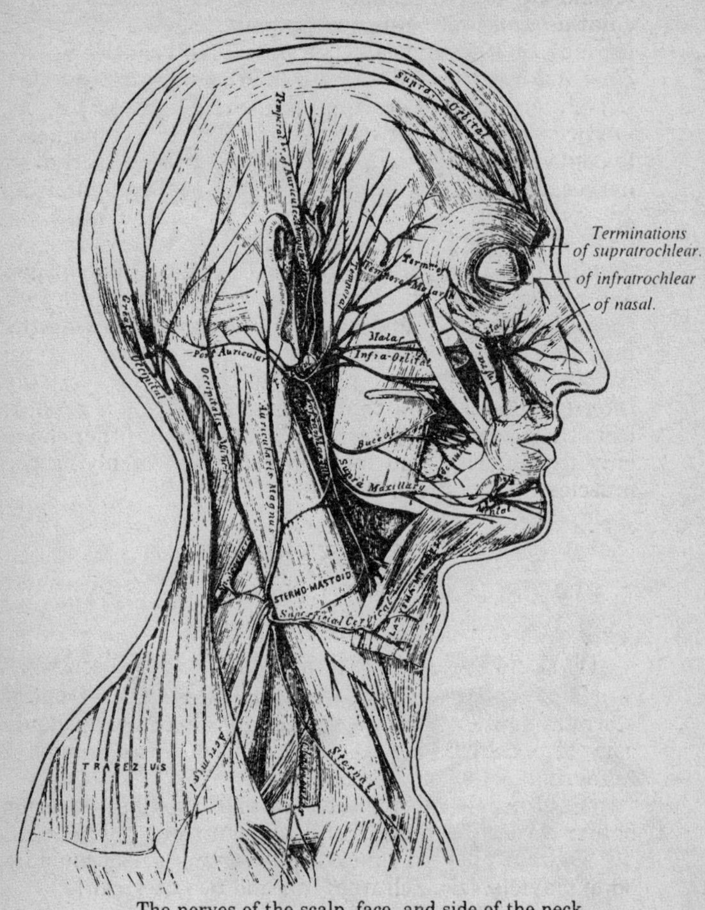

The nerves of the scalp, face, and side of the neck.

from 2d and 3d; obliquely forwards to anterior and lateral parts of neck. *Auricula´ris mag´nus*, from 2d and 3d; ascends to parotid gland, having facial, posterior auricular and mastoid branches. *Occipita´lis mi´nor*, from 2d; ascends to side of head; has auricular branch. *Supraclavicula´res*, from 3d and 4th; downwards, having sternal, clavicular, and acromial branches. DEEP BRS.: *Communican´tes*, loop between 1st and 2d, to sympathetic, hypoglossal, pneumo-gastric, and spinal accessory nerves. *Muscula´res*, from 1st. *Commu´nicans no´ni*, from 2d and 3d, uniting with descendens noni. *Phreni´cus*, from 3d, 4th and 5th; crosses subclavian artery, down to middle mediastinum, thence to pericardium, diaphragm and pleura. *The left is the longer.* POSTERIOR BRANCHES: each have external and internal divisions supplying the muscles of the back of the neck, etc. *Sub-occipita´lis*, from 1st, up to sub-occipital region. *Occipita´lis ma´jor*, the internal branch from the 2d cervical. *Occipital branch*, from the 3d cervical, internal branch. All the others have only the external and internal branches, supplying the muscles contiguous to them.

UPPER EXTREMITY.

MUSCLES OF THE UPPER EXTREMITY.

(*15 Regions, 46 Muscles.*)

(1) REGION ANTERIOR THORACIC REGION, 3 MUSCLES.

Pectora´lis ma´jor: sternal half clavicle, ½ front of sternum down to 7th rib, cartilage of true ribs, aponeurosis of external oblique—anterior bicipital ridge of humerus. [Anterior thoracic.]

Pectora´lis mi´nor: 3d, 4th and 5th ribs—anterior border coracoid process of scapula. [Anterior thoracic.]

Subcla´vius: 1st rib-cartilage—under surface middle 3d of clavicle. [Branch from 5th and 6th cervical.]

(2) LATERAL THORACIC REGION, 1.

Serra´tus mag´nus: 9 digitations from the 8 superior ribs—whole length inner margin scapula, posterior surface. [Posterior thoracic.]

(3) ACROMIAL REGION, 1.

Deltoide´us: outer 3d anterior border, upper surface, of clavicle; outer margin, upper surface acromian process; whole length lower border spine of scapula—prominence outer surface (middle) humerus. [Circumflex.]

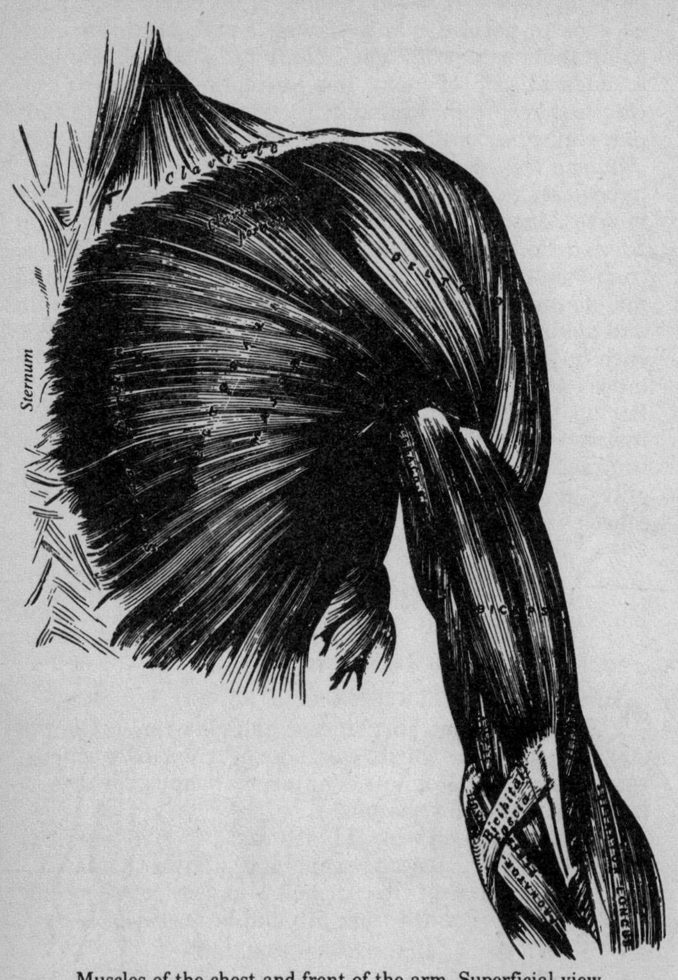

Muscles of the chest and front of the arm. Superficial view.

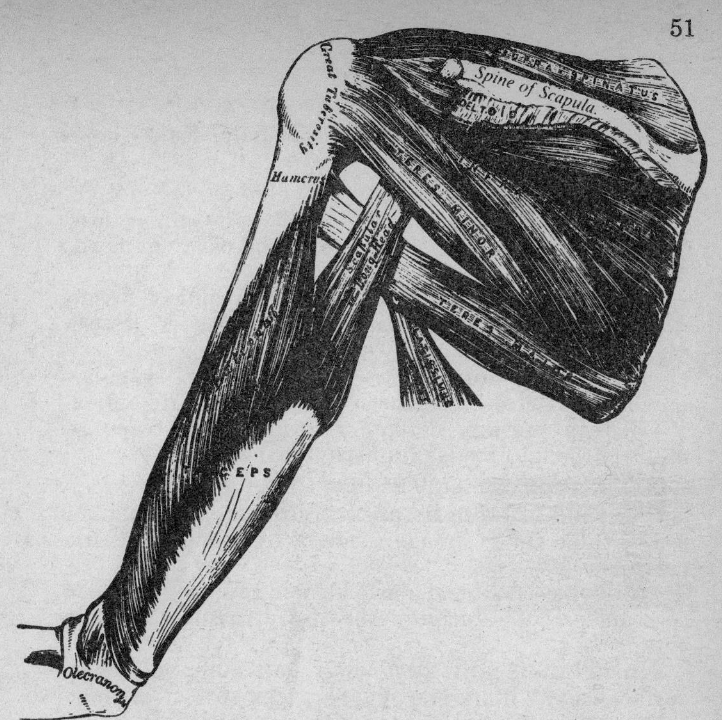

Muscles on the dorsum of the Scapula and the Triceps.

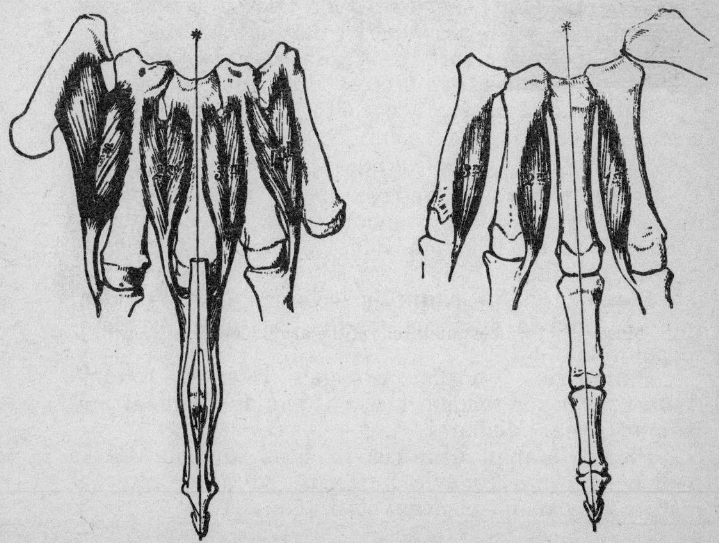

The Dorsal interossei of left hand.

The Palmar interossei of left hand.

(4) ANTERIOR SCAPULAR REGION, 1.

Subscapula´ris: inner ⅔ subscapular fossa—lesser tuberosity humerus. [Subscapular.]

(5) POSTERIOR SCAPULAR REGION, 4.

Supra-spina´tus: internal ⅔ of supra-spinous fossa of scapula—upper facet greater tuberosity humerus. [Supra-scapular.]

Infra-spina´tus: internal ⅔ of infra-spinous fossa—middle facet greater tuberosity humerus. [Supra-scapular.]

Te´res mi´nor: dorso-axillary border scapula—lowest facet greater tuberosity of humerus. [Circumflex.]

Te´res ma´jor: dorsum inferior angle scapula—posterior bicipital ridge humerus. [Subscapular.]

(6) ANTERIOR HUMERAL REGION, 3.

Coraco-brachia´lis: apex coracoid process scapula—rough ridge inner (middle) side of humerus. [Musculo-cutaneous.]

Bi´ceps: long head above glenoid cavity; short head, coracoid process—bicipital tuberosity radius. [Musculo-cutaneous.]

Brachia´lis anti´cus: lower half outer and inner surfaces shaft humerus, septa—under surface coronoid process ulna. [Musculo-cutaneous, musculo-spiral.]

(7) POSTERIOR HUMERAL REGION, 2.

Tri´ceps: long head, depression below glenoid cavity; external head, posterior superior part of humerus; internal head, posterior surface of humerus *below* musculo-spiral groove—olecranon process ulna. [Musculo-spiral.]

Subancone´us: just above olecranon fossa humerus—posterior ligament elbow-joint. [Musculo-spiral.]

(8) ANTERIOR BRACHIAL REGION, SUPERFICIAL LAYER, 5.

Prona´tor ra´dii te´res: above internal condyle humerus, common flexor tendon, fascia, inner side coronoid process ulna—rough ridge radius, outer (middle) surface. [Median.]

Flex´or car´pi radia´lis: common flexor tendon, internal condyle humerus, fascia—base of index metacarpal. [Median.]

Palma´ris lon´gus: common internal condyle (humerus) flexor tendon, fascia—annular ligament and palmar fascia. [Median.]

Flex´or car´pi ulna´ris: 1st head, common flexor tendon, internal condyle humerus; 2d head, internal margin olecranon—pisiform bone. [Ulnar.]

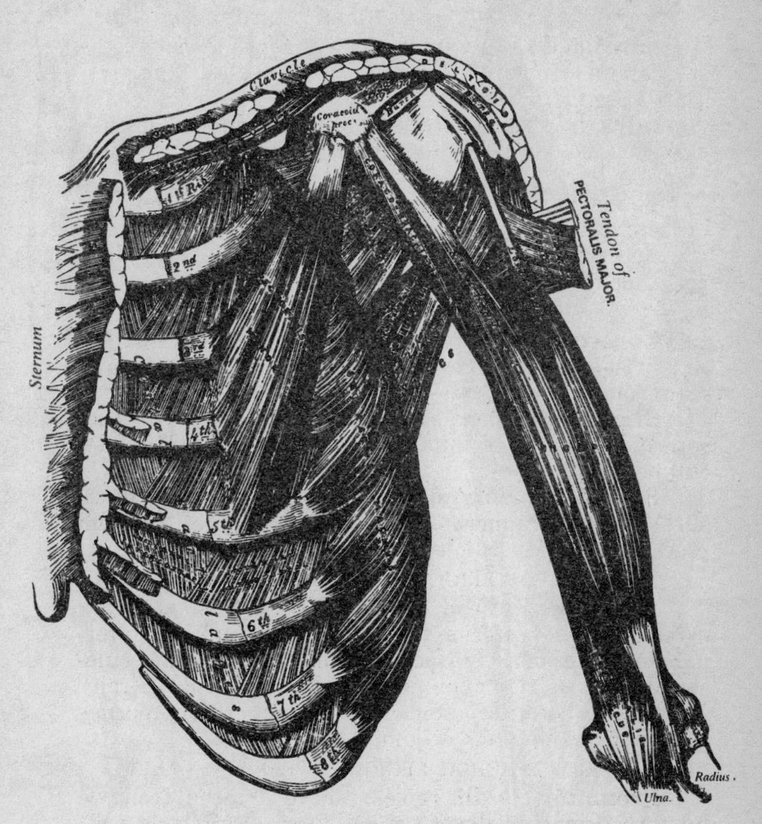

Muscles of the chest and front of the arm, with the boundaries of the axilla.

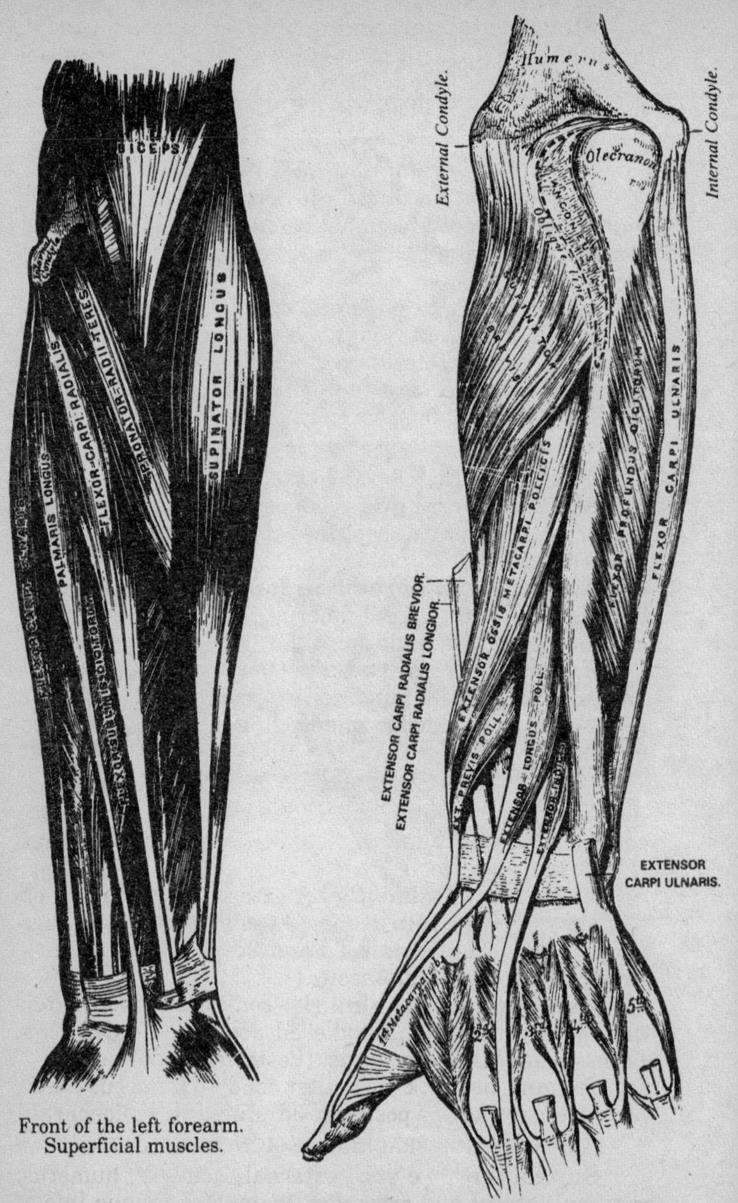

Front of the left forearm.
Superficial muscles.

Posterior surface of the forearm.
Deep muscles.

Flex´or subli´mis digito´rum: lst, internal condyle humerus (common flexor tendon); 2d head, inner side coronoid process ulna; 3d head, oblique line radius—lateral margins 2d phalanges; tendon split for passage of flexor profundus digitorum. [Median.]

(9) ANTERIOR BRACHIAL REGION, DEEP LAYER, 3

Flex´or profun´dus digito´rum: upper ⅔ anterior and inner surface ulna, inner side coronoid process, interosseous membrane—bases last phalanges. [Ulnar, anterior interosseous.]

Flex´or lon´gus pol´licis; upper ⅔ anterior surface radius, interosseous membrane—base last phalanx thumb. [Anterior interosseous.]

Prona´tor quadra´tus: oblique line and lower 4th ulna—lower 4th anterior surface and external border radius. [Anterior interosseous.]

(10) RADIAL REGION, 3.

Supina´tor lon´gus: upper ⅔ external condyloid ridge humerus, septum—styloid process radius. [Musculo-spiral.]

Exten´sor car´pi radia´lis lon´gior: lower third external condyloid ridge humerus, septum—base metacarpus indicis. [Musculo-spiral.]

Exten´sor car´pi radia´lis bre´vior: common tendon external condyle humerus, external lateral ligament, septa—base metacarpus middle finger. [Posterior interosseous.]

(11) POSTERIOR BRACHIAL REGION, SUPERFICIAL LAYER, 4.

Exten´sor commu´nis digito´rum: common tendon external condyle humerus, septa—2d and 3d phalanges. [Posterior interosseous.]

Exten´sor min´imi dig´iti: external condyle humerus, septa—unites with tendon extensor communis digitorum to be inserted into 2d and 3d phalanges of little finger. [Posterior interosseous.]

Exten´sor car´pi ulna´ris: common tendon external condyle humerus, middle 3d posterior border ulna, fascia—base 5th metacarpus. [Posterior interosseous.]

Ancone´us: back part outer condyle humerus—side, olecranon and upper posterior 3d ulna. [Musculo-spiral.]

(12) POSTERIOR BRACHIAL REGION, DEEP LAYER, 5.

Supina´tor bre´vis: external condyle humerus, external lateral and orbicular ligaments, oblique line of ulna—(surrounds radius at its upper part) back part inner surface, outer edge bicipital tuberosity; oblique line of radius. [Posterior interosseous.]

Exten´sor os´sis metacar´pi pol´licis: posterior surface shaft ulna and radius (middle 3d), interosseous membrane—base 1st metacarpus. [Posterior interosseous.]

Exten´sor pri´mi interno´dii pol´licis: posterior surface radius, interosseous membrane—base 1st phalanx of thumb. [Posterior interosseous.]

Exten´sor secun´di interno´dii pol´licis: posterior surface ulna, interosseous membrane—base 2d phalanx thumb.

Exten´sor in´dicis: posterior surface ulna, interosseous membrane—joins tendon extensor communis digitorum to 2d and 3d phalanges indicis. [Posterior interosseous.]

(13) THUMB, RADIAL REGION, 4.

Abduc´tor pol´licis: ridge trapezium and annular ligament—radial side base 1st phalanx thumb. [Median.]

Oppo´nens pol´licis: palmar surface trapezium, annular ligament—whole length 1st metacarpus, radial side. [Median.]

Flex´or bre´vis pol´licis: trapezium, outer ⅔ annular ligament, trapezoid, os magnum, base 3d metacarpus, tendon flexor carpi radialis—both sides base 1st phalanx thumb. [Median, ulnar.]

Adduc´tor pol´licis: whole length 3d metacarpus—ulnar side base 1st phalanx thumb. [Ulnar.]

(14) LITTLE FINGER, ULNAR REGION, 4.

Palma´ris bre´vis: annular ligament palmar fascia—skin inner border palm. [Ulnar.]

Abduc´tor min´imi dig´iti: pisiform bone, tendon flexor carpi ulnaris—ulnar side base 1st phalanx little finger. [Ulnar.]

Flex´or bre´vis min´imi dig´iti: tip unciform process, annular ligament—base 1st phalanx little finger. [Ulnar.]

Oppo´nens min´imi dig´iti: unciform process, annular ligament—ulnar side 5th metacarpus. [Ulnar.]

(15) MIDDLE PALMAR REGION, 3.

Lumbrica´les: (4); accessories to flexor profundus digitorum—tendon extensor communis digitorum. [Median and Ulnar.]

Interos´sei dorsa´les: (4); metacarpi—base 1st phalanges 1st, 2d, 3d fingers. [Ulnar.]

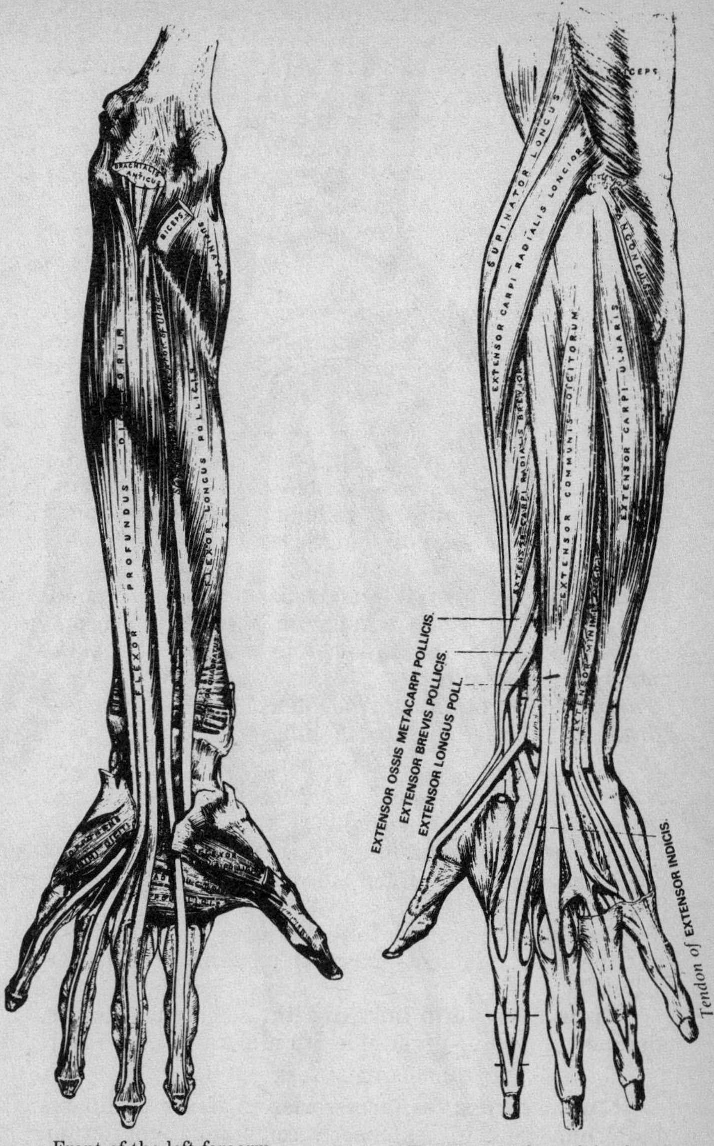

Front of the left forearm.
Deep muscles.

Posterior surface of the forearm.
Superficial muscles.

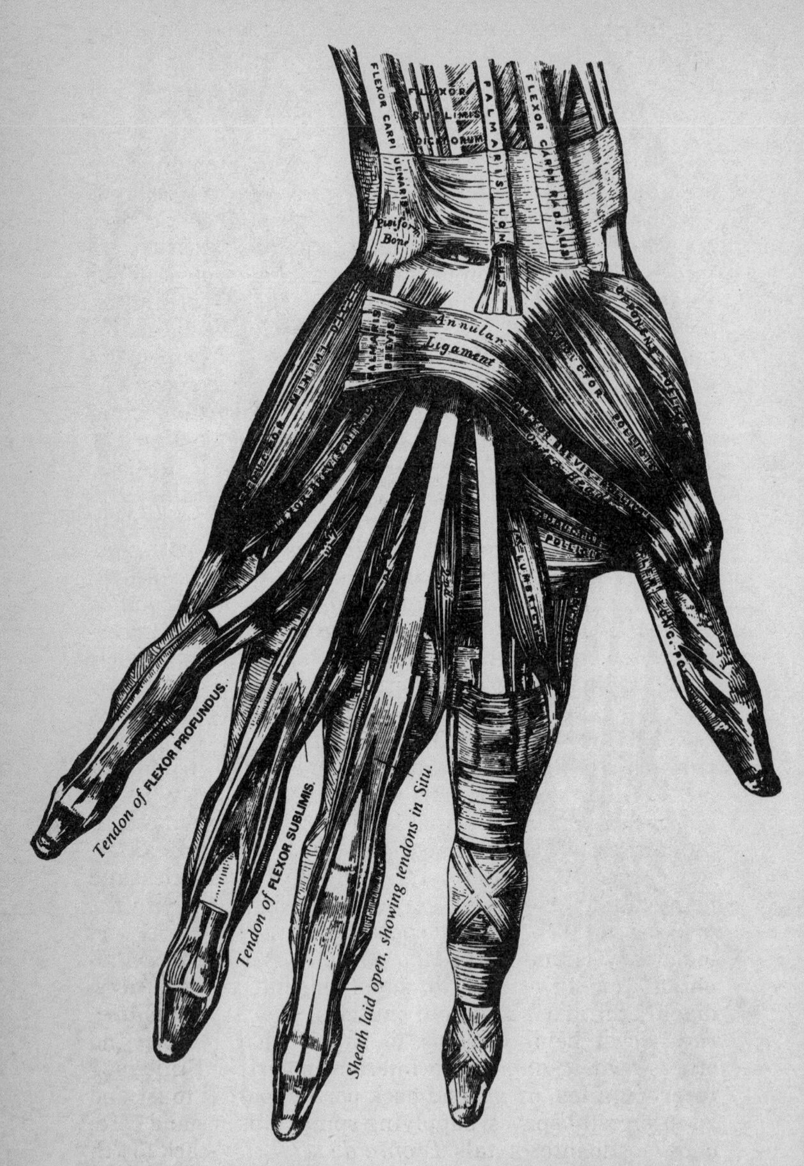

Muscles of the left hand. Palmar surface.

Interos´sei palma´res: (3); 2d, 4th and 5th metacarpi—1st phalanges of same fingers. [Ulnar.]

ARTERIES OF THE UPPER EXTREMITY.

SUB´CLAVIA: (4 brs.) *Right*, 1. from innominate at sterno-clavicular articulation to inner margin scalenus anticus. (*Left*, I. from transverse portion aortic arch opposite 2d dorsal vertebra to scalenus anticus); II. internal border scalenus anticus to outer of scalenus medius; III. from external border scalenus medius to lower border lst rib, midway along clavicle. **Vertebra´lis**, upper and back portion of part I.—enters foramen 6th cervical vertebra to be continued upwards (see page 11). **Thyroide´us axis**, anterior part of first portion, inner side scalenus anticus. *Infe´rior thyroide´a* (see page 12). *Transversa´lis col´li*, (a) superficial cervical beneath anterior margin trapezius—to trapezius and glands in that region. (b) Poste´rior scapula´ris to superior angle of scapula to anastomose at the inferior angle with subscapular. *Supra-scapula´ris—outwards and backwards, parallel with clavicle, to supra-spinous fossa; distributed to muscles in that region.* **Sterna´lis inter´na** (mammary), origin just below thyroid axis, behind clavicle along inside chest to 6th intercostal space, there dividing into musculo-phrenic and superior epigastric. *Co´mes ner´vi phren´ici*, to diaphragm; anastomoses with other phrenic branches. *Mediastina´les* to areolar of anterior mediastinum, also remains of thymus. *Pericardia´les*, to upper pericardium, triangularis sterni, *anas.* musculo-phrenic. *Anterio´res intercosta´les*, to 5 or 6 upper intercostal spaces, to intercostal and pectoral muscles and mammary gland; *anas.* aortic intercostal. *Perforan´tes*, to 5 or 6 upper intercostal spaces, to pectoral muscles and mammary gland. *Mus´culo-phren´ica*, perforates diaphragm at 8th or 9th rib, supplying intercostal spaces, diaphragm and abdominal muscles. *Epigas´trica supe´rior*—down behind rectus to supply that muscle and others near it; *anas.* with inferior epigastric. **Supe´rior intercosta´les**, upper and back portion part II to lst and 2d intercostal spaces, supplying spinal muscles and cord; *anas.* aortic intercostals. *Profun´da cervi´cis*—back to 7th cervical vertebra and between complexus and semi-spinalis colli runs to axis, supplying contiguous muscles; *anas.* anterior princeps cervicis.

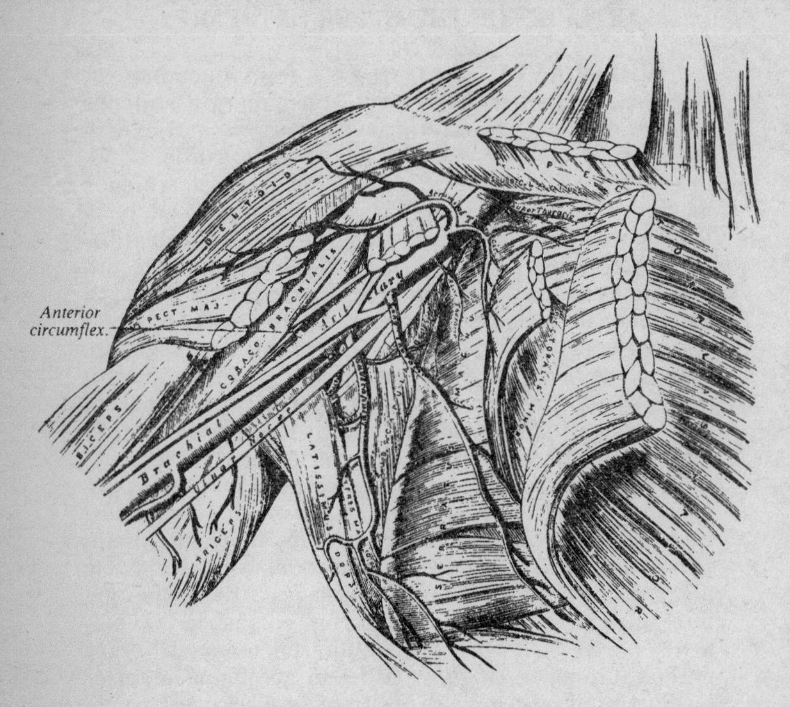

The axillary artery and its branches.

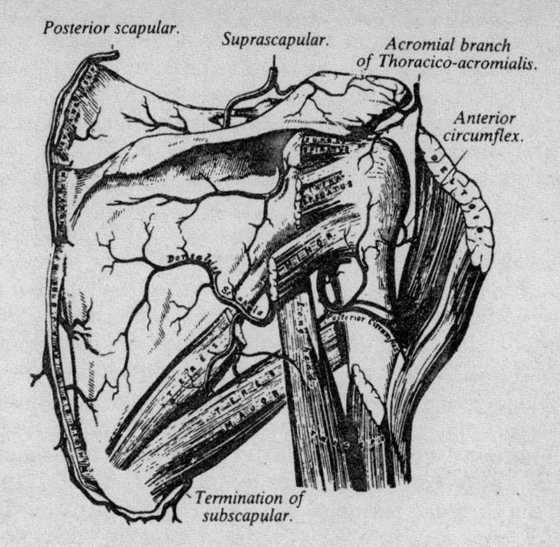

The scapular and circumflex arteries.

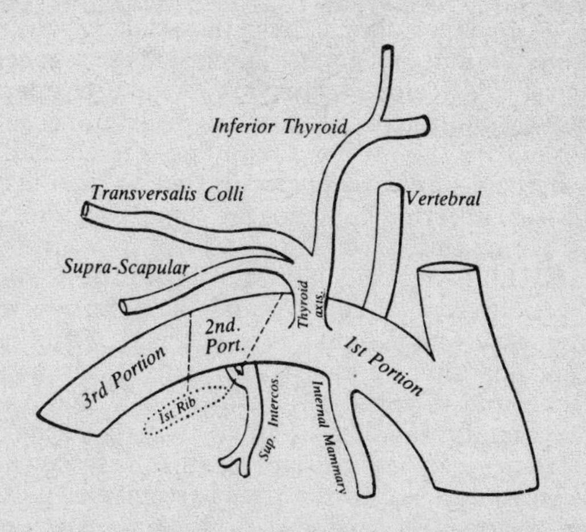

Plan of the branches of the right subclavian artery.

AXILLA´RIA: (7 brs.); lower border 1st rib to tendons latissimus dorsi and teres major; Ist part, 1st rib to pect. minor; II. from superior border pectoralis minor to inferior border same; III. from inferior border pectoralis minor to tendon latissimus dorsi. **Supe´rior Thorac´ica,** 1st part—forwards and inwards along superior border pectoralis minor, supplying pectoral muscles; *anas.* internal mammary and intercostal. **Acromia´les Thorac´ica,** 1st part to upper border pectoralis minor; *Acromia´les,* toward acromian process to deltoid; *anas.* suprascapular and posterior circumflex. *Thorac´icæ,* 2 or 3 in number, supplying serratus magnus and pectoral muscles; *anas.* intercostals of internal mammary. *Descending* branches supply pectoralis major and deltoid, as accompanying cephalic vein. **Thorac´ica lon´ga, II.** part, down- and inwards along inferior border pectoralis minor to pectoral muscles, axillary and mammary glands, serratus magnus and subscapularis; *anas.* internal mammary and intercostal. **Thorac´ica ala´ris, II.** part, to glands and areolar tissue of the axilla. **Subscapula´ris, III.,** part, opposite inferior border do muscle, down and back inferior margin do muscle to inferior angle scapula; *anas.* posterior scapula. *Dorsa´lis scap´ulæ*—dividing into 3 branches, "subscapular," "infra-spinous," and "median." Altogether they supply the scapular, latissimus dorsi and serratus magnus muscles. Make a general anastomosis. **Poste´rior circumflex´a,** opposite inferior border sub-scapularis, winds around neck humerus to supply deltoid; *anas.* anterior circumflex, suprascapular, acromio-thoracic. **Ante´rior circumflex´a,** just below above, passes anterior to humerus supplying deltoid; *anas.* post circumflex, acromio-thoracic.

BRACHIA´LIS: (5 brs.); inferior border teres major to ½ inch below bend of elbow. Runs along inner border biceps and coraco-brachialis; is superficial. **Supe´rior profun´da,** opposite inferior border trochanter major, winds backwards in spiral groove down to elbow; *anas.* recurrent radial; supplies deltoid, coraco-brachialis, triceps. *Poste´rior articula´ris,* perpendicularly down to back to elbow-joint; *anas.* interosseous recurrent, posterior ulnar recurrent, anastomotica magna. **Nutri´cia,** middle of arm to bone near insertion coraco-brachialis. **Infe´rior profun´da,** just below middle arm to *anas.* posterior ulnar recurrent and anastomotica magna at elbow; accompanied by ulnar nerve. **Anastomot´ica**

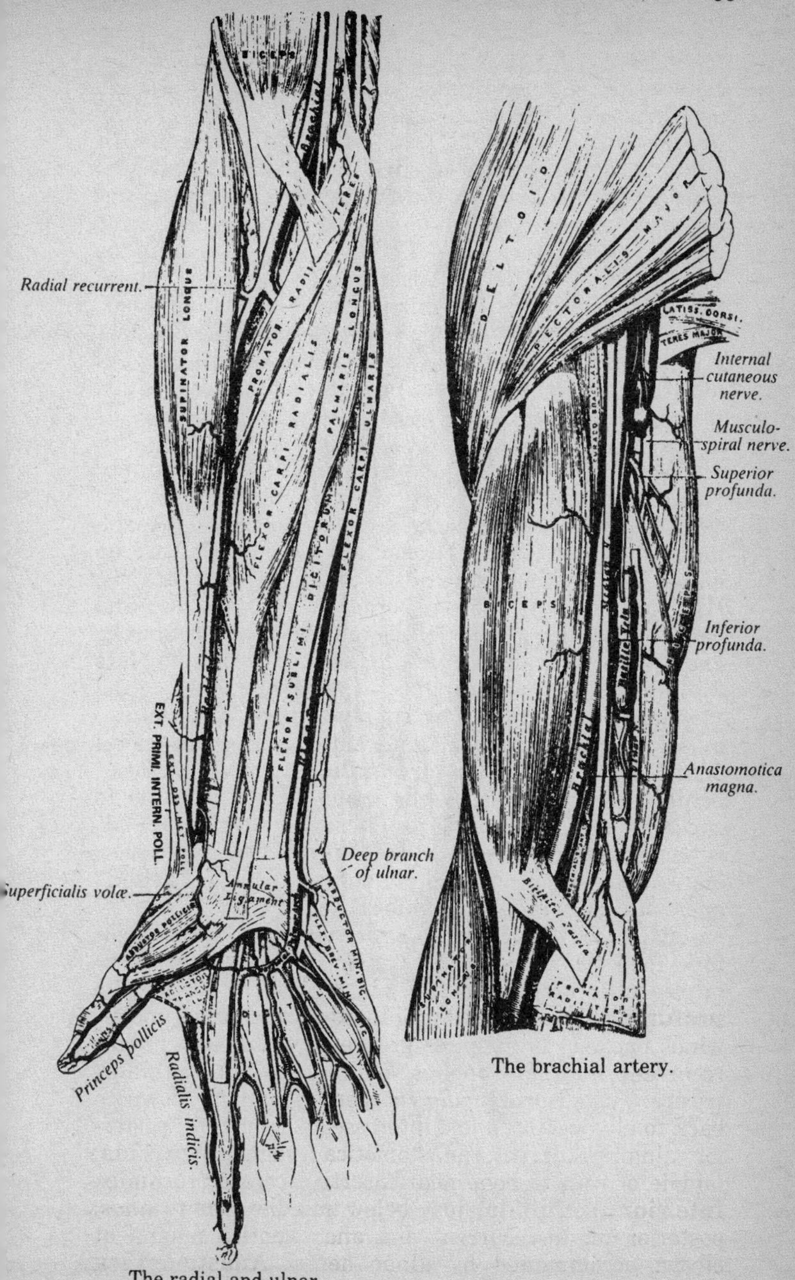

The radial and ulnar arteries.

The brachial artery.

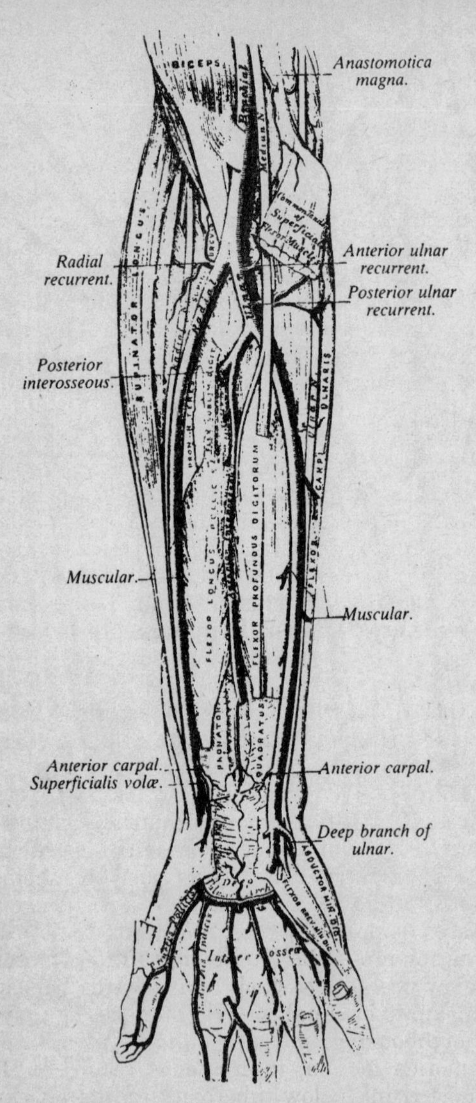

Ulnar and radial arteries. Deep view.

mag´na, 2 inches above elbow-joint, winds around and down humerus to elbow-joint; *anas.* posterior ulnar recurrent, inferior profunda, anterior ulnar recurrent. **Muscula´res**, 3 or 4, to coraco-brachialis, brachialis anticus.

RADIA´LIS: (12 brs.); end of the brachialis down radial side forearm, along inner border supinator longus to wrist; then winds around carpus beneath thumb-extensors to enter palm of hand between thumb and index finger to form "deep palmar arch;" *anas.* deep branch of ulnar. **Radia´lis recur´rens**, below elbow—up arm supplying brachialis anticus, supinator longus, supinator brevis, *anas.* superior profunda. **Muscula´res**, to radial side forearm. **Superficia´lis vo´læ**, just as artery about to wind around the carpus—to muscles in ball of thumb; *anas.* with ulnar forming "superficial palmar arch." **Ante´rior carpa´lis**, to wrist; *anas.* anterior carpalis of ulnar. **Poste´rior carpa´lis**, to wrist; *anas.* posterior carpalis of ulnar, anterior interosseous, and posterior perforating of deep palmar arch as *2 dorsal interosseous* branches. **Metacarpa´lis**, (1st dorsal interosseus) supplies adjoining sides index and middle fingers. **Dorsa´les pol´licis**, (2); along dorsum of thumb. **Dorsa´lis in´dicis**, radical side back of index. **Prin´ceps pol´licis**, beginning palmar arch to sides of palmar aspect to thumb. **Radia´lis in´dicis**, palmar arch to radial side index. **Perforan´tes**, (3); to inosculate with 3 dorsal interosseous. **Palma´res interos´seæ**, (3 or 4); from arch to *anas.*, at finger-clefts, with digital branches of superficial arch.

ULNA´RIS: (8 brs.); little below bend of elbow—along radial side flexor carpi ulnaris to palm of hand, forming "superficial palmar arch" with superficialis volæ. **Ante´-rior ulna´ris recur´rens**, just below elbow-joint, up- and inwards between brachialis anticus and pronator radii teres, supplying these; *anas.* anastomotica magna, and inferior profunda. **Poste´rior ulna´ris recur´rens**, just below preceding—back and inwards beneath flexor sublimis up to internal condyle humerus, supplying joint and neighbouring muscles; *anas.* inferior profunda, anastomotica magna, interosseous recurrent. **Interos´-sea**, short trunk below tuberosity radius—backwards to interosseous membrane, dividing into: INTEROS´SEA ANTE´RIOR, passing down forearm on interosseous membrane, piercing membrane at superior border pronator quadratus to descend to back of wrist, supplying *nutrient* (to radial and ulnar arteries) and *muscular* branches; gives off *median* branch, accompanied by do nerve. *Anas.* posterior carpal of radial and ulnar. INTEROS´SEA POSTE´-RIOR, down back forearm, between deep and superficial muscular layers to wrist, supplying these muscles; *anas.* as preceding. *Poste´rior interos´sea recur´rens*, near its origin to interval between olecranon and external condyle, beneath supinator brevis; *anas.* superior profunda, posterior ulnar recurrent. **Muscula´res**, to muscles of ulnar side of forearm. **Carpa´lis ante´rior**, beneath flexor profundus, *anas.* anterior carpal of radial. **Poste´-rior carpa´lis**, above pisiform bone, beneath flexor carpi ulnaris, giving small branch to inosculate with posterior carpal of radial forming "*Posterior carpal arch*"; continued along 5th metacarpus, forming its dorsal branch. **Commu´nicans**, from commencement palmar arch, deeply inwards, *anas.* with radial forming "*deep palmar arch.*" **Digita´les**, (4); from convexity of superficial palmar arch, supplying ulnar side 4th and adjoining sides 3d, 2d and 1st fingers.

VEINS OF THE UPPER EXTREMITY.

Ulna´ris ante´rior: from anterior carpus and ulnar side hand, up along side forearm to elbow-joint, to form basilica. Communicates with median and posterior ulnar.

Ulna´ris poste´rior: posterior ulnar border hand and vein of little finger (*v. salvatel´la*)—unites with preceding just below elbow-joint.

Basil´ica: coalescence of anterior and posterior ulnares; receives median-basilic at elbow; ascends inner side arm to venæ comites of brachial artery, or axillary vein.

Radial´is: dorsum thumb, radial side index and hand—at bend elbow receives median-cephalic to become the cephalic.

Cephal´ica: up between deltoid and pectoralis major to axillary veins.

Me´dia: palmar surface of hand and middle of forearm (communicates with ulnar and radial), to median-cephalic and median-basilic at elbow.

Cephal´ica me´dia: obliquely outwards from bend elbow, between supinator longus and biceps; empties into cephalic as a formative branch.

Basil´ica me´dia: obliquely inwards behind biceps and pronator radii teres; empties into basilic as a formative branch.

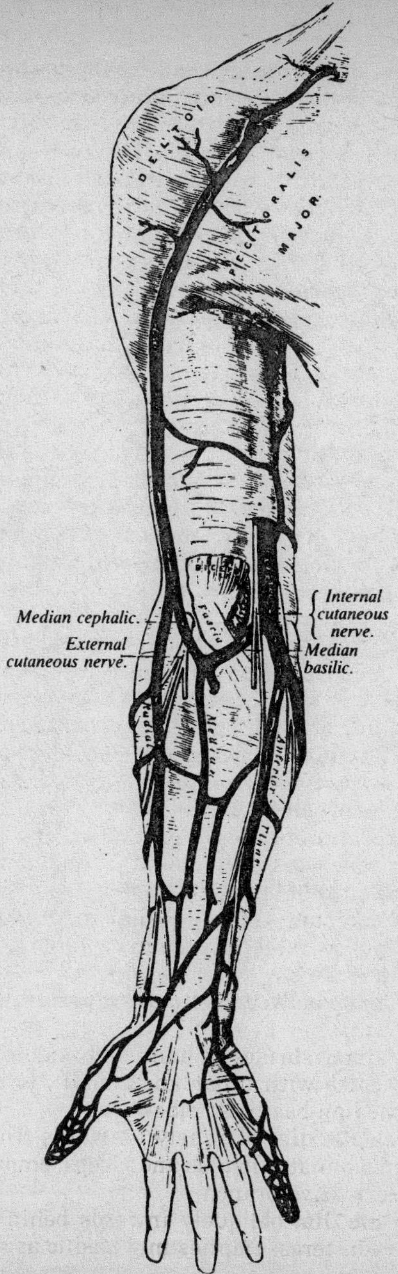

The superficial veins of the upper extremity.

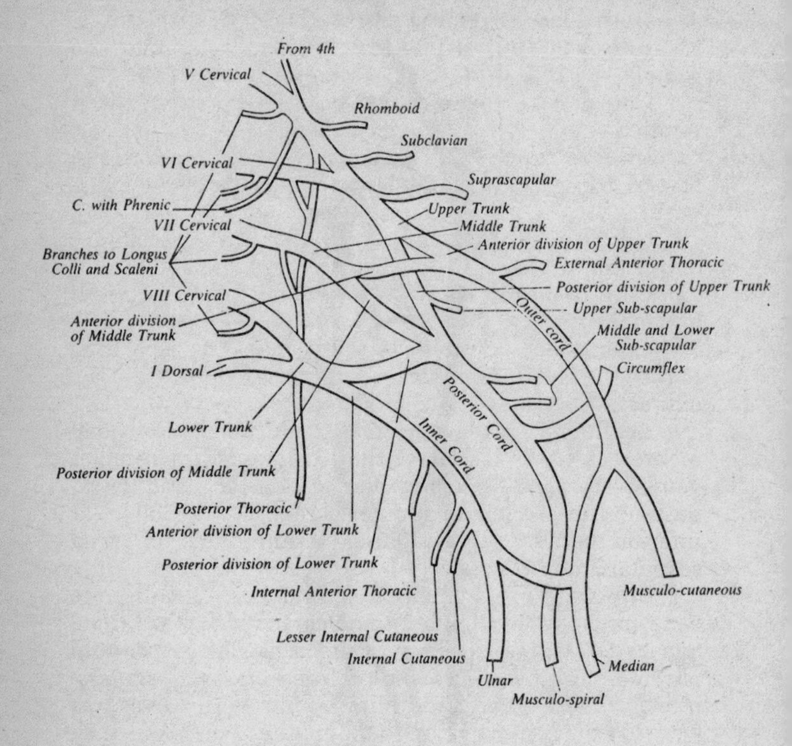

Plan of the brachial plexus.

The following are deep veins and accompany their respective arteries as *ve´næ com´ites*, intercommunicating with each other, and the superficial veins, frequently.

Digita´les: (2); empty into the superficial palmar.

Palma´res superficia´les: (2) empty into ulnar and radial.

Palma´res profun´dæ: empty into the radial venæ comites.

Interos´seæ: (2); accompany the anterior and posterior interosseous arteries, commencing at the wrist, terminating in venæ comites of the ulnar.

Com´ites radia´lis: form, with the ulnar, the comites of brachial.

Com´ites ulna´ris: with the radial, form comites of brachial.

Com´ites brachia´lis: receiving veins corresponding to the branches of the brachial artery, and empty into the axillary vein.

Axilla´ria: is the continuation of the basilic. Commences at lower border of the axillary space; receives veins corresponding to branches of its artery, and terminates in the subclavian at outer border lst rib. [Valves at inferior border subscapularis, terminations of vena scapularis and cephalica.]

Subcla´via: continuation of axillary, emptying into vena innominata at right sterno clavicular articulation. Separated from its artery by scalenus anticus muscle and phrenic nerve. Receives external and anterior jugulars, branch from cephalic, and internal jugular. [Valves just external to entrance of external jugular, or about 1 inch from its termination.]

NERVES OF THE UPPER EXTREMITY.

PLEX´US BRACHIA´LIS: formed by anterior roots 4 lower cervical and lst dorsal nerves. 5th and 6th cervical unite, then are joined by 7th to form *upper trunk*. 8th cervical and 1st dorsal unite to form *lower trunk*. Both trunks accompany the subclavian artery to the axilla, lying upon its outer side. Opposite clavicle, each of the trunks gives off a fasciculus, which, uniting, form a *third trunk*; in the centre of the axilla the original *upper* cord lies to the *outside* of the artery; the original *lower* cord to the *inside*; the cord formed from fascicular union, *posteriorly*. The plexus lies between the anterior and middle scaleni, beneath the clavicle upon lst serration of the serratus magnus and the subscapular muscles. (Has 4 brs. above, 9 below the clavicle.) Receives communicating branches from cervical plexus, phrenic, inferior cervical, sympathetic ganglia. **Commu´nicans**, 5th cervical to phrenic on scalenus anticus. **Muscula´res**, to longus colli, scaleni, rhomboidei and subclavius. **Poste´rior thorac´icus**, from 5th and 6th cervical to serratus magnus. Passes behind brachial plexus. **Supra-scapula´ris**, from "outer cord" obliquely outwards beneath trapezius, to supra-spinous fossa through supra-scapular notch, here giving 2 branches to supra-spinatus muscle and 1 to joint; in infra-spinous fossa, 2 branches to muscle, 1 to joint, all of these are given off ABOVE, the clavicle. Those BELOW the clavicle are: **Exter´nus ante´rior thorac´icus**, "outer cord" inwardly across axillary vessels to pectoralis major. **Inter´nus ante´rior thorac´icus**, "inner cord" passing up between axillary artery and vein (*sometimes perforating the vein*) to pectorales major and minor. **Subscapula´res**, (3); "posterior cord" the *Upper* to subscapular muscles; the *longer* to latissimus dorsi; the *lower* to teres major. **Circumflex´us**, "posterior cord," down behind axillary vessels to lower border subscapularis, dividing into *upper* branch winding round neck of humerus, supplying deltoideus and integument; *lower* branch to teres minor deltoideus and integument over posterior surface deltoid. *Articula´ris*, given off before division, to joint. **Mus´culo-cuta´neus**, continuation of outer cord, *perforates coraco-brachialis*, obliquely outwards between biceps and brachialis anticus to these muscles, integument to elbow, and to the joint. *Anterior* branch, down radial border of forearm from elbow, to wrist, supplying integument to ball of thumb; communicates with radial. *Posterior* branch given off middle of forearm, supplying integument to wrist, on radial side; communicates with radial and external cutaneus. **Inter´nus cuta´neus**, "inner cord," down in company with brachial artery, becoming cutaneous at middle of arm, then dividing into *anterior* branch, supplying integument of ulnar side of arm to wrist, communicating with branch from ulnar; *posterior* branch down, on inner side of basilic vein, over internal condyle, on posterior ulnar side of forearm to wrist, communicating at wrist with dorsal branch of ulnar, at elbow, with lesser internal cutaneous. **Cuta´neus mi´nor inter´nus**, from "inner cord" to integument inner side of arm. **Me´dius**, (4

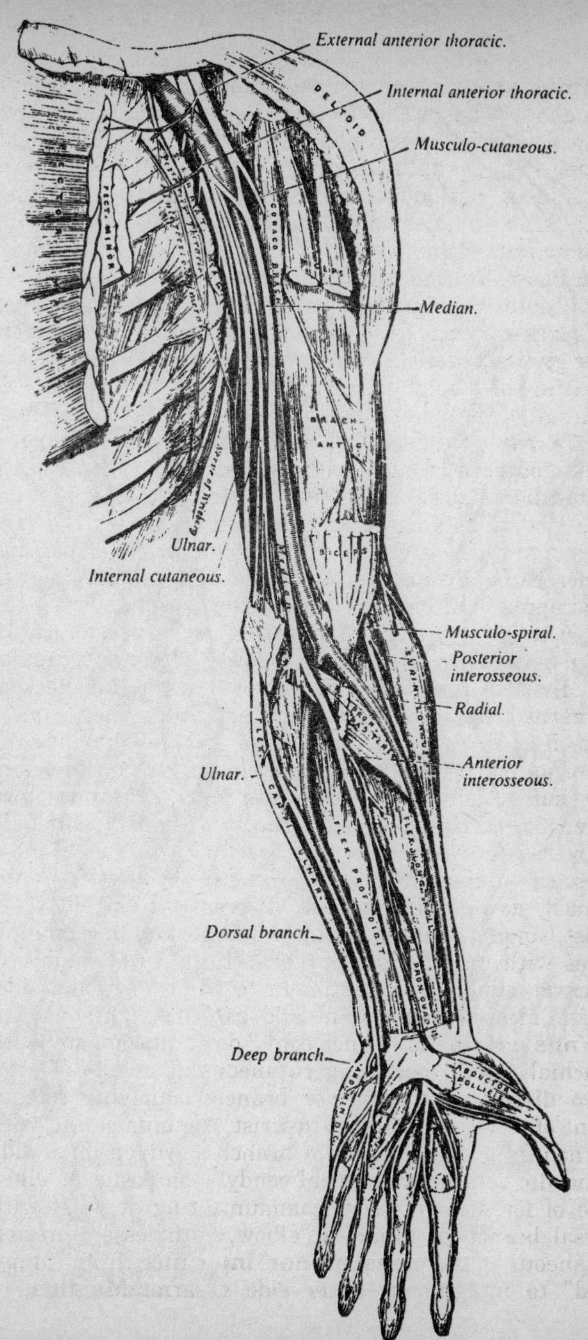

Nerves of the left upper extremity.

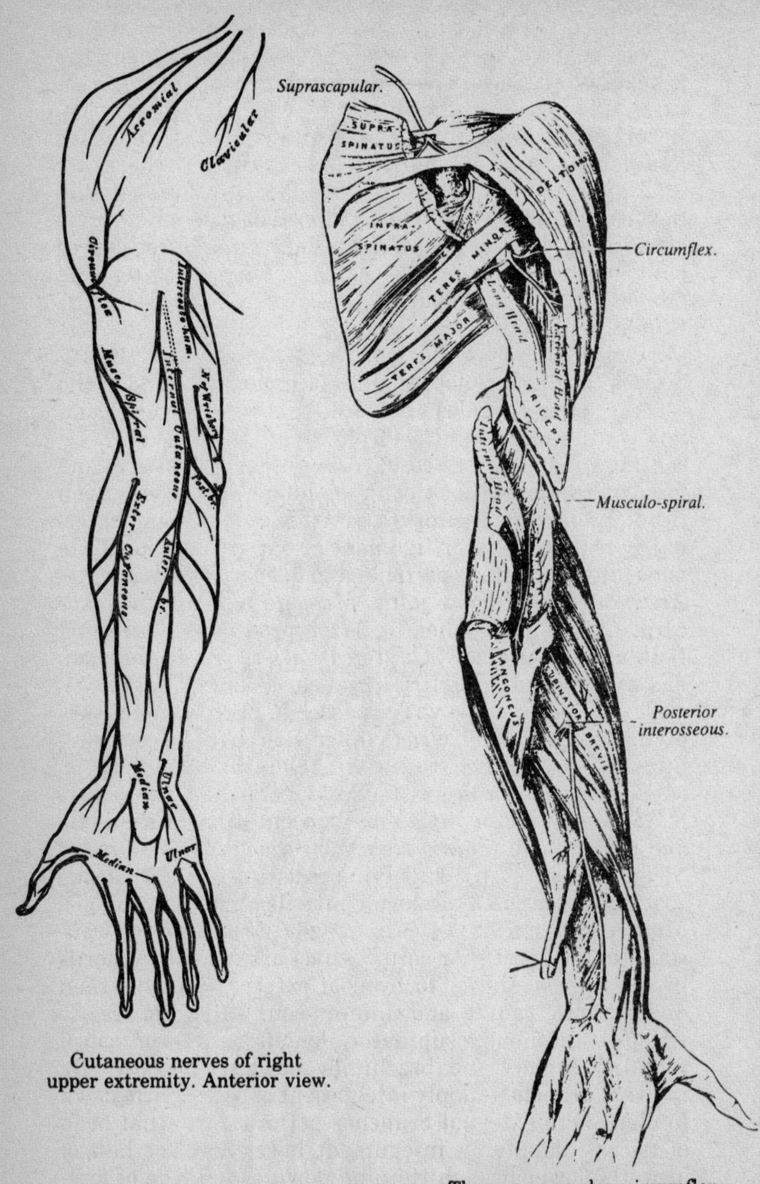

Cutaneous nerves of right upper extremity. Anterior view.

The suprascapular, circumflex, and musculo-spiral nerves.

branches) arises by 2 roots, one from "outer" and one from "inner" cord; at first lies to outer side of the artery, crosses it at middle of arm; at forearm runs between the 2 heads of the pronator radii teres, beneath flexor sublimis till near annular ligament, when it lies between flexor sublimis and flexor carpi radialis; it passes beneath annular ligament to hand. (No branches in the arm.) *Muscula´res*, from near elbow, to forearm muscles save flexor carpi ulnaris. *Ante´rior interos´seus* follows course of the artery, to flexor profundus digitorum, flexor longus pollicis, and pronator quadratus. *Cuta´neus palma´ris*, crosses annular ligament, the outer branch supplying the thumb region; the inner branch, the palmar. *Digita´les*, (5); two go to thumb, the 3d to radial side of index; the 4th divides to supply adjacent sides of index and middle; the 5th the adjacent sides of middle and ring fingers, communicating with branches from ulnar. **Ulna´ris**, (7 brs.); continuation of "inner cord," down ulnar side of arm and forearm (over the back of inner humeral condyle) upon flexor profundus digitorum, having ulnar artery externally, crosses annular ligament at outer side of pisiform bone, dividing into superficial and deep palmar branches. *Articula´res*, to elbow joint. *Muscula´res*, one to flexor carpi ulnaris, the other to flexor profundus digitorum. Both arise near elbow. *Cuta´neus*, arises middle forearm, has a deep and superficial branch. *Dorsa´lis cuta´neus*, arises 2 inches above wrist, passes to back of hand, supplying ulnar side of wrist, inner side of little and ring fingers. *Articula´res*, to wrist. *Palma´ris superficia´lis*, supplies palmaris brevis, and integument inner side of hand, ulnar side of little and adjacent sides of the little and ring fingers. *Palma´ris profun´dus*, follows course of "deep palmar arch," supplying muscles of interosseous spaces, lumbricales, abductor and flexor brevis pollicis. **Mus´culo-spira´lis**, (4 brs.; *largest br. of plexus*) continuation of "posterior cord;" winds around the humerus in spiral groove, etc., to front of external condyle, then divides into radial and interosseous. *Muscula´res*, to triceps, anconeus, supinator longus, extensor carpi radialis longior, and brachialis anticus. *Cuta´nei*, (3); internal branches supply integument of back of arm down to olecranon: external branches perforate external head of triceps, supplying integument lower anterior half of arm, the lower branch running down radial side of forearm (posteriorly) to wrist, supplying contiguous integument. *Radia´lis*, down by outer side of radial artery, just concealed by supinator longus till within 3 inches of wrist, where pierces deep fascia of outer side forearm; divides to supply radial side of ball of thumb (communicating with external cutaneous nerve), and on back of hand forms an arch with ulnar, giving off 4 *digital* nerves; the 1st to ulnar side of thumb; the 2d to radial side of index; the 3d, adjoining sides of index and middle; the 4th, adjoining sides of middle and ring fingers. *Interos´seus poste´rior* pierces supinator brevis, winding to back of forearm, passing down to wrist, there having ganglionic enlargement. Supplies carpus, and all muscles on back of forearm *except* anconeus, supinator longus and extensor carpi radialis longior.

BODY.

MUSCLES OF THE BODY. (*10 Regions, 48 Muscles.*)

(REGION 1) BACK, FIRST LAYER, 2 MUSCLES.

Trape´zius: inner 3d superior curved occipital line, ligamentum nuchæ, spinous processes of 7th cervical and all the dorsal vertebræ—posterior border clavicle, superior margin and acromian process and superior border spine scapula. [Spinal accessory, cervical plexus.]

Latis´simus dor´si: aponeurosis from spinal processes 6 lower dorsal, all lumbar and sacral vertebræ, external lip iliac crest—twisting upon itself so as to be inserted into bicipital groove of humerus. [Subscapular.]

(2) BACK, SECOND LAYER, 3.

Leva´tor an´guli scap´ulæ: transverse processes of 3 or 4 superior cervical vertebræ—posterior border scapula. [5th cervical, cervical plexus.]

Rhomboide´us mi´nor: ligamentum nuchæ, spinal processes 7th cervical and 1st dorsal vertebræ—down- and outwards to root scapular spine. [5th cervical.]

Rhomboide´us ma´jor: spinal processes superior dorsal vertebræ—tendinous arch along vertebral border scapula. [5th cervical.]

(3) BACK, THIRD LAYER, 4.

Serra´tus posti´cus supe´rior: ligamentum nuchæ, spinal processes 7th cervical and 2 or 3 superior dorsal vertebræ—superior border 2d, 3d, 4th 5th ribs. [Posterior external brs. cervical.]

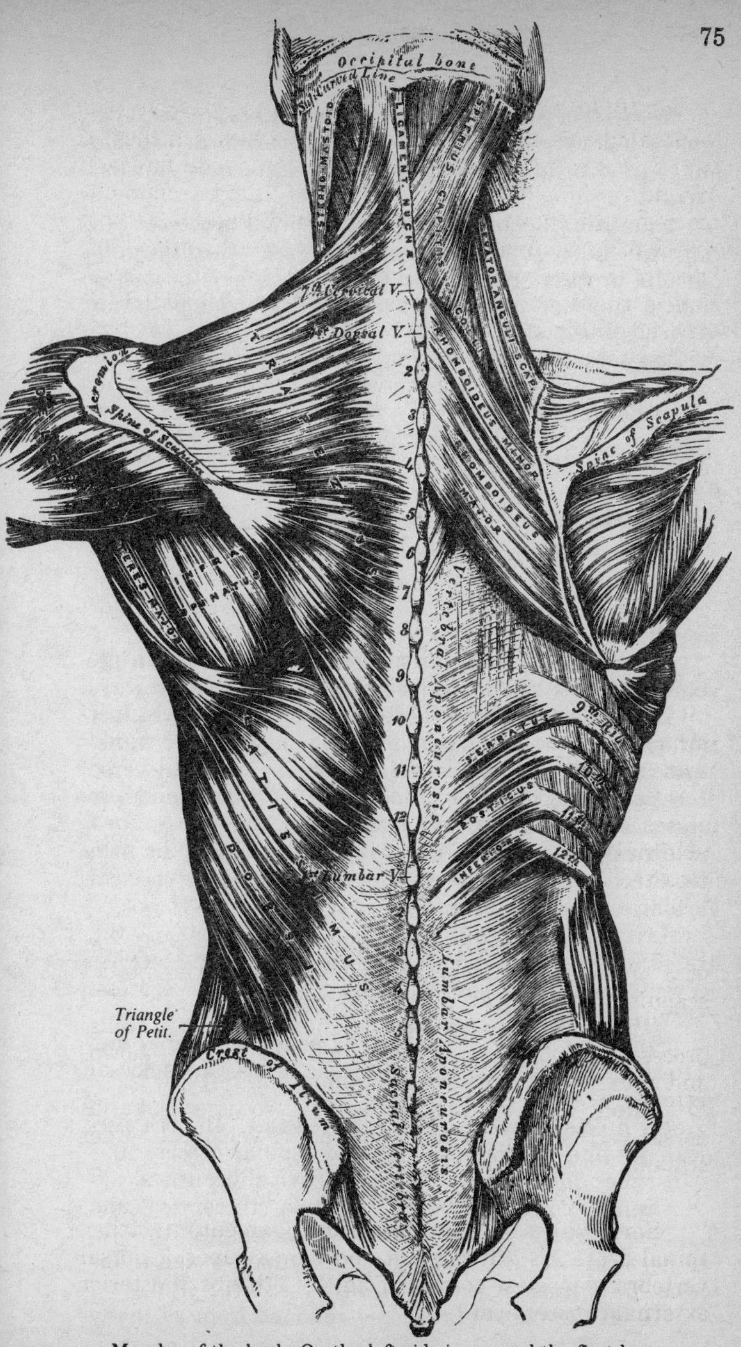

Muscles of the back. On the left side is exposed the first layer; on the right side, the second layer and part of the third.

Serra´tus posti´cus infe´rior: spinal processes 11th and 12th dorsal, 1st, 2d and 3d lumbar vertebræ—up and out to inferior border 4 inferior ribs. [External branches dorsal.]

Sple´nius: ligamentum nuchæ, spinal processes 7th cervical and 6 superior dorsal vertebræ—**Capitis**, into mastoid process and occiput; **Colli**, transverse processes 3 or 4 superior cervical vertebræ. [External posterior branches cervical.]

(4) BACK, FOURTH LAYER, SACRAL AND LUMBAR REGION, 1.

Erec´tor spi´næ: sacro-iliac groove, lumbo-sacral tendon, iliac crest, transverse processes sacrum—sacro-lumbalis, longissimus dorsi. [External posterior branches lumbar.]

(5) BACK, FOURTH LAYER, DORSAL AND CERVICAL REGION, 10.

Sa´cro-lumba´lis: (see above)—angles inferior ribs. [Dorsal.]

Accesso´rius: angles 6 lower—angles 6 superior ribs. [Dorsal.]

Cervica´lis ascen´dens: 4 or 5 superior ribs—transverse processes 4th, 5th, 6th cervical vertebræ. [Cervical.]

Longis´simus dor´si: see erector spinæ, of which it is the larger portion; inserted (*lumbar* region) into transverse processes lumbar vertebræ: *dorsal*, tips transverse processes of all vertebræ, and 7 to 11 ribs, between their tubercles and angles. [Lumbar dorsal.]

Transversa´lis col´li: transverse processes 3d, 4th, 5th, 6th dorsal—transverse processes 5 inferior cervical vertebræ. [Cervical branches.]

Trache´lo-mastoide´us: transverse processes 3d, 4th, 5th, 6th dorsal, and articular processes 3 or 4 inferior cervical vertebræ—posterior margin mastoid process. [Cervical branches.

Spina´lis dor´si: spinal processes 1st, 2d lumbar and 11th and 12th dorsal vertebræ—spinal processes of dorsal vertebræ. [Dorsal branches.]

Spina´lis cervi´cis: spinal processes 5th, 6th cervical (1st and 2d dorsal), vertebræ—spinal process axis (sometimes 3d and 4th cervical). [Cervical branches.]

Complex´us: transverse processes 7th cervical and 3 superior dorsal vertebræ, articular processes 4th, 5th, 6th cervical—occipital bone between superior and inferior occipital lines. [Cervical branches, sub-occipital.]

Biven´ter cervi´cis: 2 or 4 tendons from as many

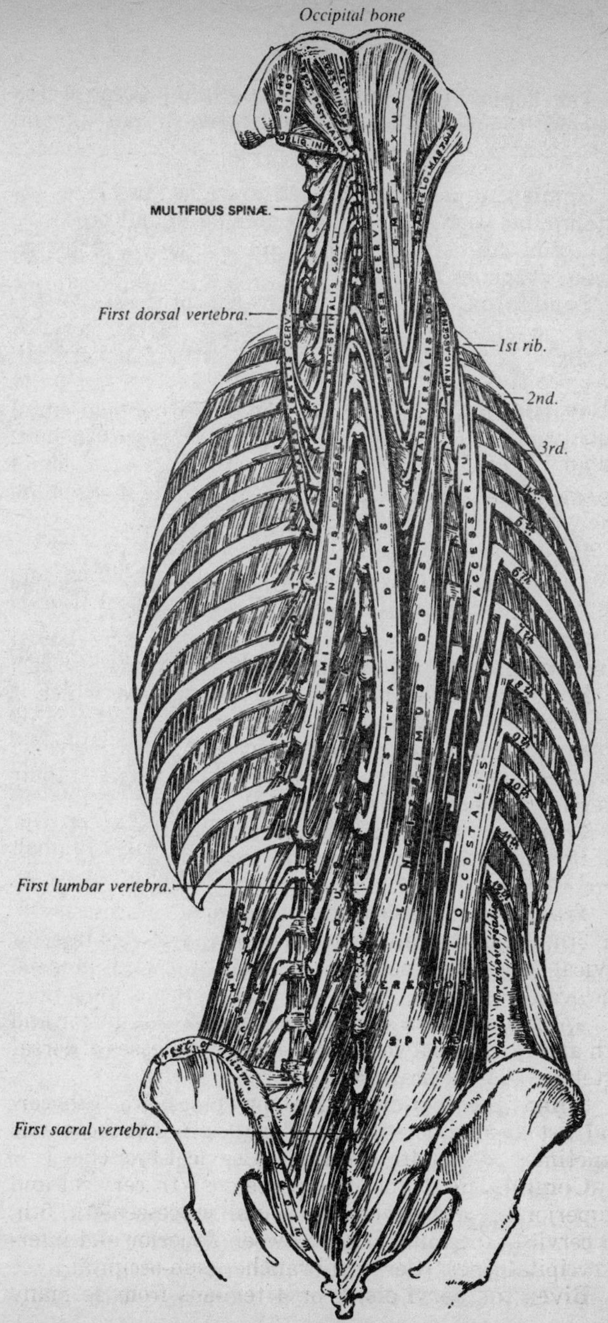

Muscles of the back. Deep layers.

superior dorsal vertebræ—superior curved occipital line of occiput to inside of complexus. [Cervical branches.]

(6) BACK, FIFTH LAYER, 8.

Semispina´lis dor´si: transverse processes of vertebræ between 11th and 5th dorsal—spinal processes of 6th and 7th cervical and 4 superior dorsal vertebræ. [Dorsal branches.]

Semispina´lis col´li: transverse processes 4 superior dorsal and articular processes 4 inferior cervical vertebræ—spinal processes 2d, 3d, 4th, 5th cervical. [Cervical branches.]

Multif´idus spi´næ: fills groove on either side spinal processes back part sacrum, articular processes in lumbar and cervical region, transverse processes in dorsal region—spinal processes and laminæ of the 4 vertebræ above. [Posterior spinal branches.]

Rotato´res spi´næ: (11); upper and back part transverse processes of dorsal vertebræ—inferior border and outer surface of laminæ of vertebræ above. [Dorsal branches.]

Supra-spina´les: on spinal processes of cervical vertebræ. [Cervical branches.]

Inter-spina´les: in pairs between spinal processes of adjacent vertebræ; 6 cervical, 3 dorsal (lst to 4th, and 11th to 12th), 4 lumbar. [Spinal branches.]

Exten´sor Coccy´gis: last bone sacrum—inferior part coccyx, lying on posterior surface.

Inter-transversa´les: 7 cervical, 1 dorsal, 4 lumbar, lying between transverse processes. [Spinal branches.]

(7) ABDOMINAL REGION, 6.

Obli´quus abdom´inis exter´nus: 8 digitations from inferior borders 8 lower ribs—*down* to anterior outer iliac crest, pubic spine and symphysis, linea alba. *Poupart's ligament* formed by its aponeurosis. [Inferior intercostal, ilio-hypogastric, ilio-inguinal nerves supply this and the 5 following muscles.]

Obli´quus inter´nus: outer ½ Poupart's ligament, anterior ⅔ middle lip iliac crest, lumbar fascia—pectineal line, linea alba, pubic crest, inferior edges cartilages of 4 inferior ribs.

Transversa´lis: outer ⅓ Poupart's, anterior ⅔ internal lip ilium, internal surfaces cartilages of 6 inferior ribs, aponeurosis from spinal and transverse processes lumbar vertebræ—pubic crest (forming with above

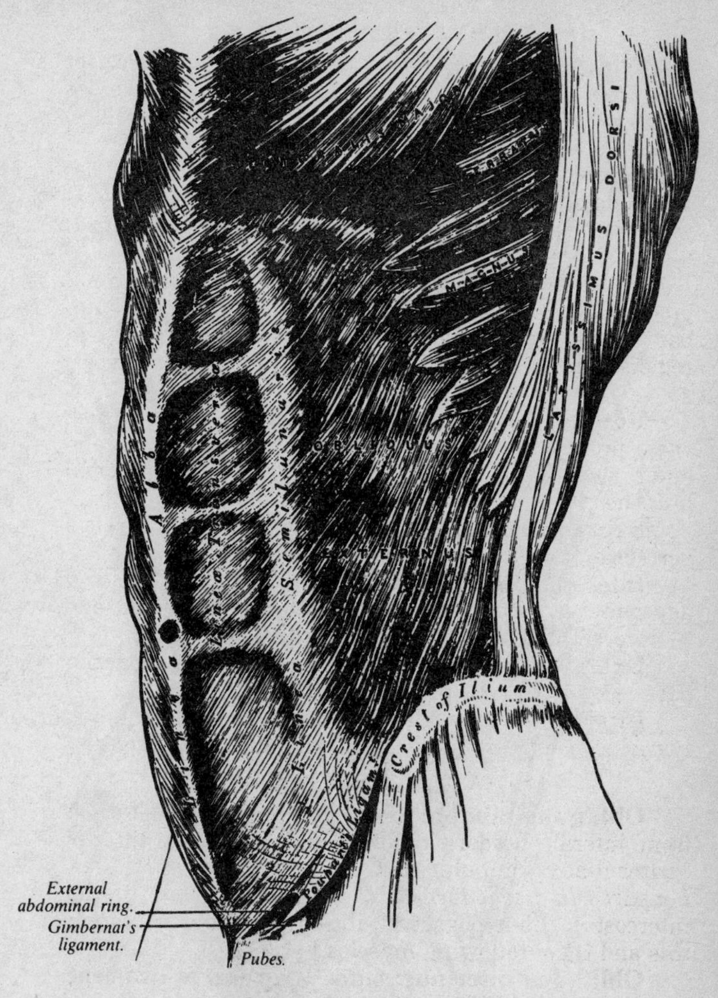

The External oblique muscle.

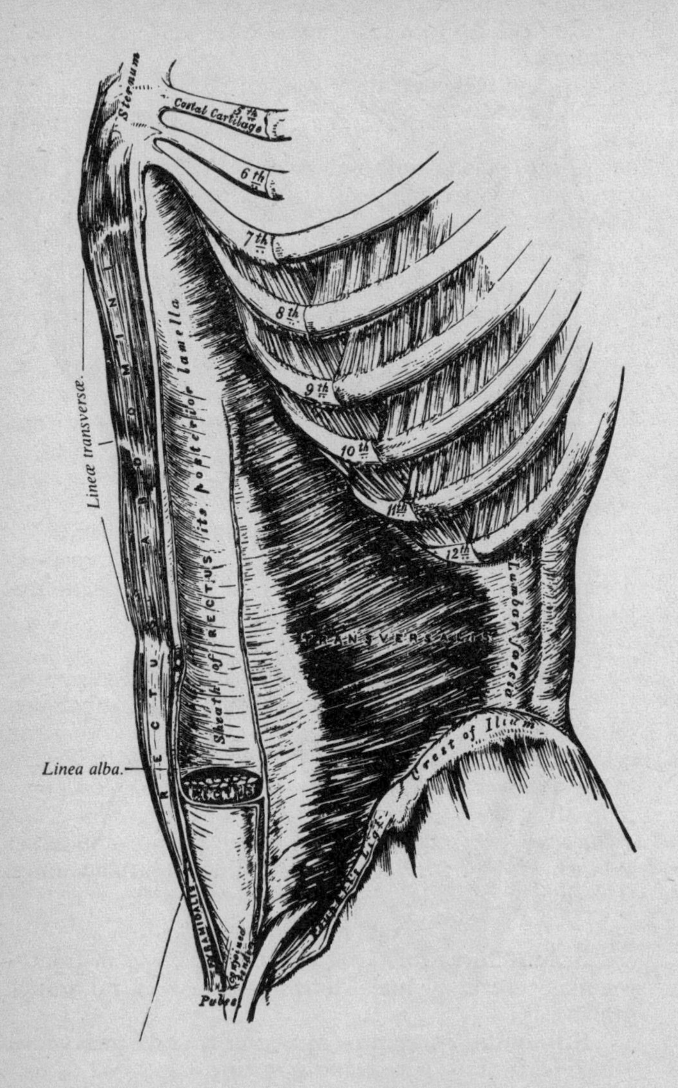

The Transversalis, Rectus, and Pyramidalis muscles.

"conjoined tendon"), lineæ ilio-pectinea and alba.

Rec´tus abdom´inis: pubic crest and symphysis—cartilages 5th, 6th, 7th ribs. (In sheath formed by internal oblique and transversalis aponeuroses.)

Pyramida´lis: pubes—linea alba midway to umbili´-cus.

Quadra´tus lumbo´rum: posterior fourth of iliac crest, ilio-lumbar ligament, transverse processes 3d, 4th, 5th lumbar vertebræ and last rib.

(8) THORACIC REGION, 5.

Intercosta´les exter´ni: (11); outer lip of groove in inferior borders of ribs—down and *forwards* to superior border rib below. [Intercostal.]

Intercosta´les inter´ni: (11); inner lip of groove—down and *backwards* to rib below. [Intercostal.]

Infracosta´les: inferior surface of one rib—internal surface lst, 2d or 3d rib below. [Intercostal.]

Triangula´ris ster´ni: side of gladi´olus, internal surface ensiform appendix, cartilage of 3 or 4 lower true ribs—cartilages of 2d, 3d, 4th, 5th ribs. [Intercostal.]

Levato´res costa´rum: (12); transverse processes dorsal vertebræ—superior border rib below, near angle. [Intercostal.]

(9) DIAPHRAGMATIC REGION, 1.

Diaphrag´ma: internal surfaces of 6 or 7 lower ribs, ligamenta arcuata, crures from 2d, 3d, 4th lumbar vertebræ, ensiform cartilage—converge forming common central tendon. *Aortic* opening for aorta, vena azygos major thoracic duet; *œsophageal*, œsophagus and pneumogastric nerves; *vena cava* for inferior vena cava; *right crus* transmits sympathetic and greater and lesser splanchnics; *left crus*, vena azygos minor and splanchnics. [Phrenic.]

(10) PERINÆAL REGION, 8.

Sphine´ter a´ni: tip of coccyx and fascia in front—common central perinæal tendon. (Hemorrhoidal branch 4th sacral.]

Sphine´ter inter´nus: muscular ring (½ inch wide), 1 inch from anus, surrounding rectum.

Accelera´tor uri´næ: central perinæal tendon and raphé—covers bulb corpus cavernosum, and corpus spongiosum, and dorsal vessels.

Erec´tor pe´nis: internal surface tuber ischii—sides and inferior surface crus.

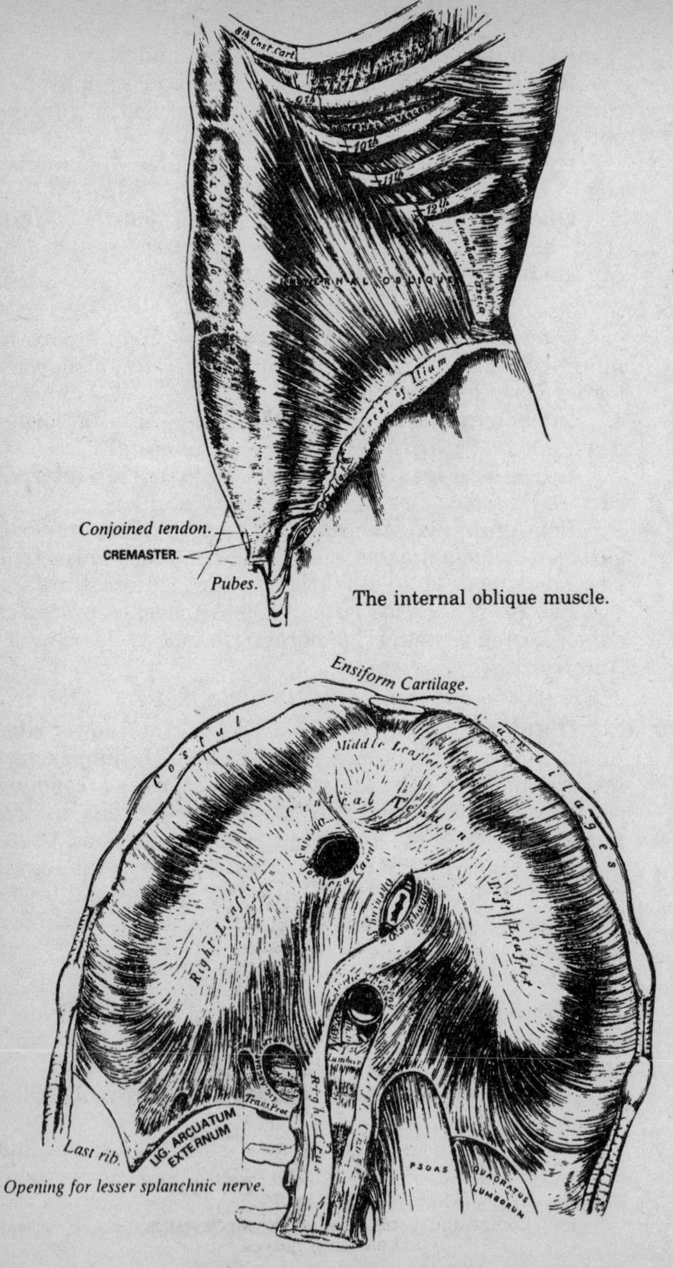

The internal oblique muscle.

The Diaphragm. Under surface.

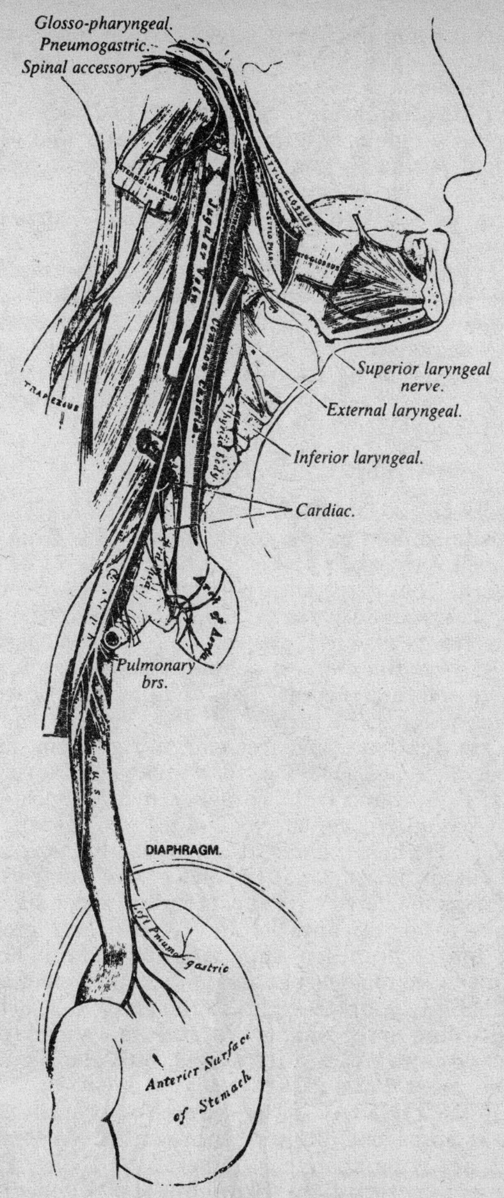

Course and distribution of the ninth, tenth, and eleventh nerves.

Transver´sus perinæ´i: internal surface ascending ischic ramus—obliquely for- and inwards to central perinæal tendon.

Leva´tor a´ni: inside of pubic ramus and body, ischic spine, fascia (angle of division into obturator and vesical)—central perinæal tendon, rectum, coccyx; assists to form floor of pelvic cavity.

Compres´sor ure´thræ: pubic ramus—surrounds membranous portion.

Coccyge´us: ischic spine and lesser sacro-sciatic ligament—side of coccyx and last sacral segment.

(In the female the above perinæal muscles are essentially the same; the *erec´tor clitori´dis* takes the place of erector penis, being let into the sides of the clitoris; *sphinc´ter vagi´nx* represents somewhat the accelerator urinæ of the male, surrounding the vagina.)

ARTERIES OF THE BODY.

ARCH OF AORTA: (5 branches); from left ventricle, opposite middle of sternum, upwards for 2 inches, arching backwards over root of left lung (on level 2d dorsal vertebra); the "descending portion," runs down on the left side of 2d and 3d vertebræ, there becoming thoracic aorta. In front, are left pleura, lung, pneumogastric, phrenic and cardiac nerves; behind, trachea, right pulmonary vessels and nerves, root of right lung, cardiac plexus, œsophagus, thoracic duct, left recurrent nerve. **Corona´ria dex´tra:** above free margin right semi-lunar valve, between pulmonary artery and right auricular appendix; runs round right border of heart to posterior interventricular groove, there dividing into 2 branches, supplying right heart; *anas.* at apex with left coronary. **Corona´ria sin´istra:** (smaller); above semi-lunar valve, passes forwards between left auricular appendix and pulmonary artery to anterior inter-ventricular groove, dividing into 2 branches, supplying left side of heart. **Innomina´ta:** commencement transverse portion of arch, ascends obliquely up to right sterno-clavicular articulation, dividing into common carotid and sub-clavian. **Caro´tis commu´nis sin´istra** and **Sub-cla´via sin´istra:** (see pages 6 and 21.)

AOR´TA THORAC´ICA: (see arch); 5 branches; terminates at aortic opening in diaphragm as "Abdominal Aorta," there lying upon front of vertebral bodies. **Pericardi´acæ:** to pericardium. **Bronchia´les:** (3 generally); to the left bronchus. Œsophage´æ: (4 or 5); front of aorta, obliquely down to œsophagus, *anas.* with inferior thyroid,

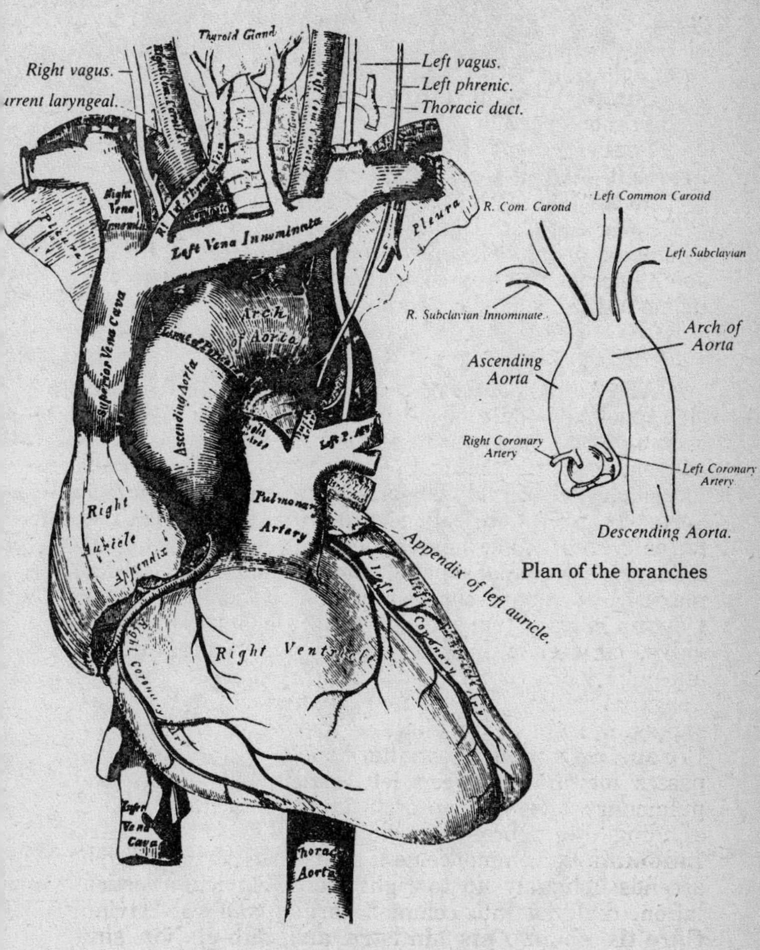

Plan of the branches

The arch of the aorta and its branches.

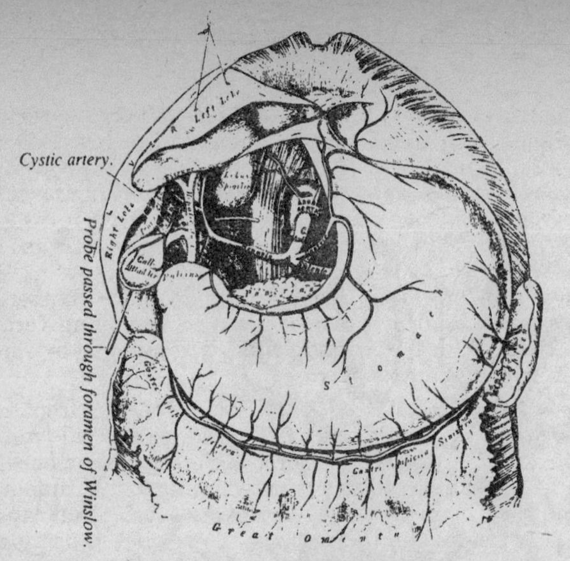

The cœliac axis and its branches, the liver having been raised and the lesser omentum removed.

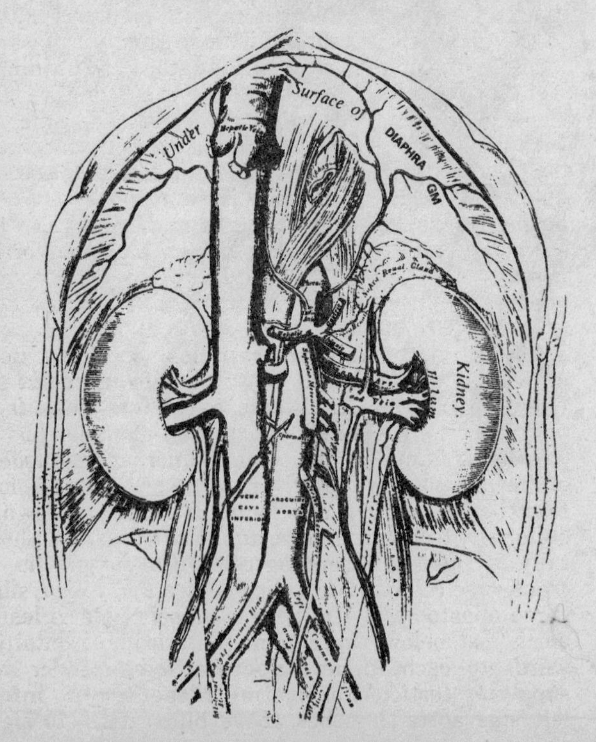

The abdominal aorta and its branches.

gastric and phrenic. **Mediastina´les posterio´res:** glands and areolar tissue therein. **Intercosta´les:** (10 pairs); right longer than left; pass out to do spaces, there dividing into *anterior* branches ascending to inferior border rib above, the smaller branch of it on the superior border rib below, running towards sternum, *anas.* with internal mammary, thoracic branches of axillary, superior intercostal, epigastric, phrenic, lumbar, etc. *Posterior* branch passes backwards, supplying vertebræ, cord, and muscles of back. (lst space supplied by superior intercostal of subclavian.)

AOR´TA ABDOM´INIS: (9 brs.); from aortic opening of diaphragm, in front last dorsal vertebra, terminates on body 4th lumbar, in the "Common Iliacs." **Phren´icæ:** (2); obliquely outwards to supply diaphragm, inferior vena cava, œsophagus and supra renal capsules; *anas.* freely. **Cœli´aca:** (axis ½ inch long); horizontally forwards, dividing into CORONA´RIA VENTRIC´ULI (gastric), which passes round lesser curvature stomach from cardiac end to pylorus, there *inosc.* with hepatic. HEPAT´ICA, to the transverse fissure of liver to supply right and left lobes, giving off *pylo´ric* branch to stomach, running from right to left; *gas´tro duodena´lis* that supplies greater curve of stomach (gas´tro epiplo´ica dex´tra which *inosc.* with gas´tro-epiplo´ica sin´istra of splenic), pancreas and duode´num (pancreat´ico-duodena´lis, with *inosc.* which duodenal branch of superior mesenteric); *cys´tica*, small branch to gall-bladder. SPLEN´ICA, horizontally left to spleen: *pancreat´icæ* (*mag´na* and *par´væ*) small branches to pancreas; *va´sa bre´via*, 5 to 7 small branches to cardiac end of stomach; *gas´tro-epiplo´ica sin´istra*, around greater curve stomach from left to right, *anas.* gas´tro epiplo´ica dex´tra. **Supra-rena´les:** obliquely up- and outwards to supra-renal capsules. **Mesenter´ica supe´rior:** ¼ inch below cœlic axis, to the intestines. *Infe´rior pancreat´ico-duodena´lis*, up to head pancreas and lower %(duode´num, *anas.* with pancreat´ico-duodena´lis of hepatic. *Va´sa intesti´na ten´uis*, 12 to 15 looping branches to jejunum and ileum. *Il´io-col´ica*, down right obliquely, to ileum and cæcum. *Col´ica dex´tra*, horizontally to right to ascending colon. *Col´ica me´dia*, up to transverse colon, *inosc.* colica dextra and colica, sinistra. (Free anastomosis of all these vessels.) **Rena´les:** sides aorta just below superior mesenteric, horizontally outwards to each kidney. **Spermat´icæ:** slender vessels supplying testicles, or ovaries. **Mesenter´ica infe´rior:** left side aorta 2 inches above bifurcation, to sigmoid-

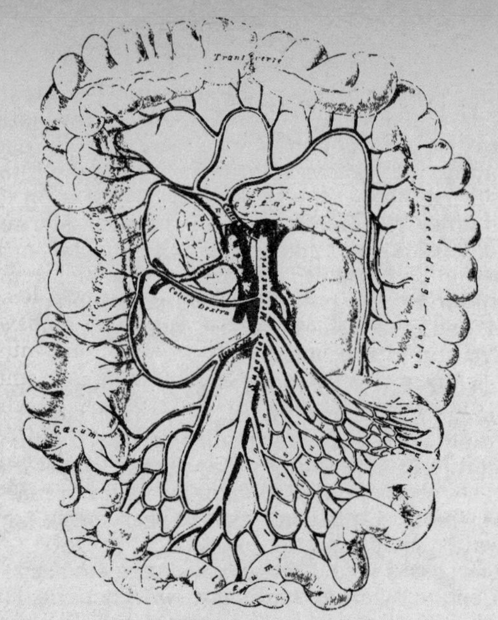

The superior mesenteric artery and its branches.

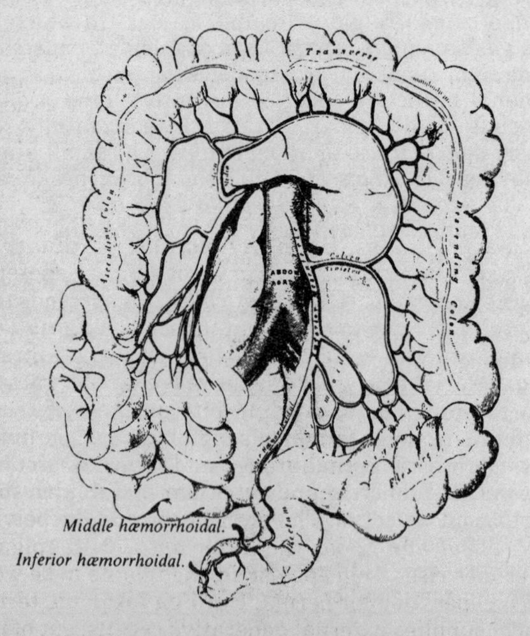

The inferior mesenteric and its branches.

flexure of colon, and rectum. *Col´ica sin´istra*, horizontally to left to descending colon. *Sigmoide´æ*, branches passing obliquely downwards to sigmoid flexure. *Hæmorrhoida´lis supe´rior*, termination of inferior mesenteric, supplying superior part of rectum, *anas.* with middle hæmorrhoidal of internal iliac, and inferior hæmorrhoidal of internal pudic. The branches of both mesenteric arteries are in free anastomosis. **Lumba´les:** 4 pairs arising from back aorta, dividing near transverse processes into *abdominal* branches (supplying muscles and *anas.* with epigastric, internal mammary, intercostal, ilio-lumbar and circumflex iliac branches) and *dorsal* branches (supplying back muscles, etc., with a spinal branch to meninges and cord), *anas.* intercostal. **Sa´cra me´dia:** back of aorta at its bifurcation, down median line to coccyx, there *anas.* with lateral sacral, supplying adjacent parts.

ILI´ACÆ COMMU´NES: from bifurcation of aorta, obliquely out- and downwards to intervertebral substance between sacrum and last lumbar vertebra, there dividing into Internal and External Iliac; each about 2 inches long. Give small branches to peritoneum, ureters, psoæ, etc. *The left is the larger.*

ILI´ACA INTER´NA: (see above); 1½ inches long, dividing at greater sacro-sciatic foramen into anterior and posterior trunks. Branches from the ANTERIOR trunk are: **Vesica´lis supe´rior:** part of fœtal-hypogastric that remains pervious; to fundus of bladder, and vas deferens. **Vesica´lis me´dia:** base of bladder and vesiculæ seminales. **Vesica´lis infe´rior:** base bladder, prostate, and vesiculæ seminales. **Hæmorrhoida´lis me´dia:** rectum, *anas.* with hemorrhoidal brarch of inferior mesenteric and internal pudic. (**Uterine:** to neck, and ascends to fundus, giving branch to ovary and tube, etc. **Vaginal:** corresponds to inferior vesicle, supplying vagina, urethra, etc.) **Obturato´ria:** forwards to superior border obturator foramen, escaping there, dividing into an *internal* (curving round inner border foramen, supplying adjacent muscles, etc. *anas.* with external branch and internal circumflex) and *external* branch (round outer margin foramen supplying adjacent muscles). The branches inside the pelvis are *iliac, resical* and *pubic*; the latter *anas.* with epigastric. Sometimes rises from epigastric, then liable to be wounded in an operation for hernia. **Pudi´ca inter´na:** terminal branch; supplies external generative organs; out of pelvis

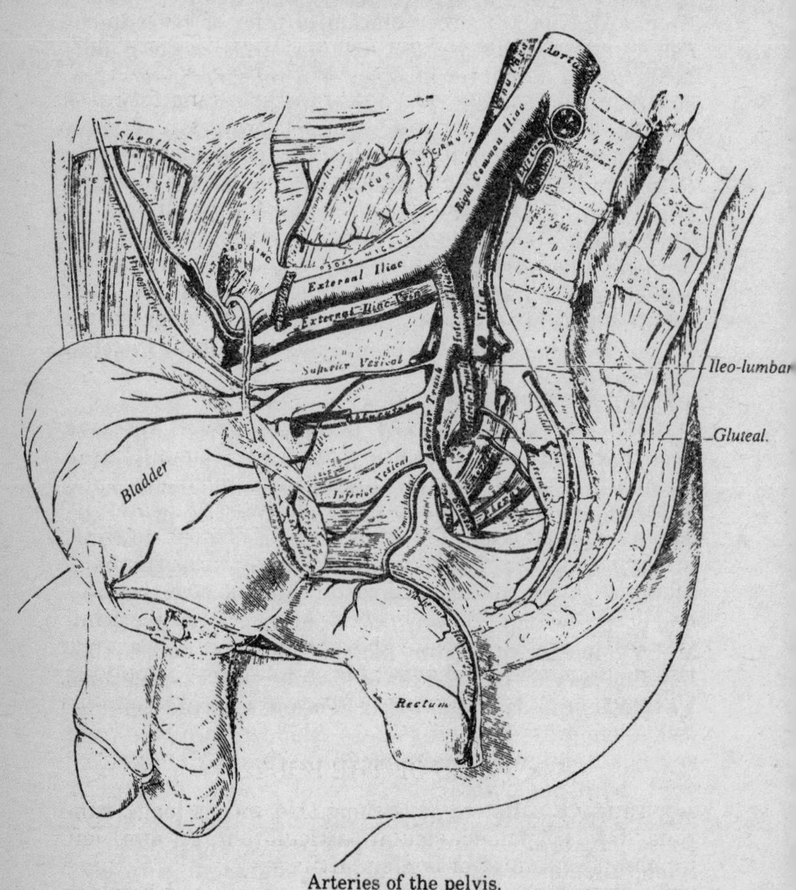

Arteries of the pelvis.

in front of pyriformis (great sacro-sciatic foramen) crosses ischic spine, re-enters pelvis, through lesser sacro-sciatic foramen, ascends ischic ramus up to pubes. *Hæmorrhoida´les, inferio´res*, 2 or 3 to rectum, etc. *Superficia´lis perinæ´i*, to scrotum and perinæum. *Transver´sa perinæ´i. A. corpo´ris bulbo´si*, to bulb and Cowper's gland. *A. corpo´-ris caverno´si*, terminal branch running forwards in this structure. *Dorsa´lis pe´nis*, forwards to glands. **Sciat´ica:** terminal branch (see lower extremity). Branches from the POSTERIOR trunk are: **Glute´a supe´-rior:** (see lower extremity.) **Ilio-lumba´lis:** divides at upper part iliac fossa into *lumbar* (to psoas and quadratus muscles, branches to spinal canal) and *iliac* branch (to iliacus internus, *anas.* with gluteal, epigastric, etc.) **Sacra´les latera´les:** (2); *superior* enters lst or sacral foramen, *anas.* with fellow and middle sacral; *inferior*, descends on sacrum, *anas.* over coccyx with middle sacral and opposite fellow.

ILI´ACA EXTER´NA: from bifurcation common iliac to femoral arch. Line drawn from left of umbili´cus to a point on Poupart's ligament midway between pubes and anterior superior spinal process of ilium, indicates its course. **Epigas´trica:** few lines above Poupart's upwards and inwards to umbili´cus, there *anas.* with internal mammary and inferior intercostal. *Spermat´ica externa´-lis*, to cremaster. *Pudic* branch. *Muscula´res.* **Circum-flex´a il´ii:** origin opposite above, from outer side artery, runs obliquely up- and outwards on iliac crest, supplying adjacent muscles, and *anas.* with gluteal, epigastric and lumbar arteries.

VEINS OF THE BODY.

Innomina´tæ: *right* is short (1½ inches long), running from sterno-clavicular articulation to join left innominate at inferior border cartilage of lst rib, forming vena cava superior. Is external to artery, and receives right lymphatic duct, right vertebral, right internal mammary, right inferior thyroid and right superior intercostal veins. *Left* is 3 inches long, runs in front of the three large arterial branches of aorta; receives corresponding venous branches as right. Neither have valves.

Mamma´ria inter´na: 2 to each artery; unite in single trunk, emptying into innominate.

Thyroidea inferior; (sometimes 3 or 4) from thyroid venous plexus, emptying into right and left innominate.

Intercosta´les superio´res: from 2 or 3 superior intercostal spaces, emptying into innominatæ. Left bronchial empties into left intercostal.

Ve´na ca´va supe´rior: 2½ to 3 inches long formed of venæ innominatæ, emptying into right auricle; receives vena azygos major, and pericardial veins. No valves.

Az´ygos ma´jor: opposite lst or 2d lumbar vertebra, from right lumbar veins, up through aortic diaphragmatic opening to right side 3d dorsal vertebra, arching over root right lung, emptying into vena cava. Receives the 10 lower right intercostal veins, vena azygos minor, several œsophageal, mediastinal, vertebral, and right bronchial veins. Imperfect valves, though its branches have complete ones.

Az´ygos mi´nor infe´rior: lumbar regions of left side from lumbar veins, or branches of renal, through left crus of diaphragm to 6th or 7th dorsal vertebra, there crossing to terminate in azygos major. Receives 4 or 5 lower intercostal, and some œsophageal and mediastinal veins.

Az´ygos mi´nor supe´rior: from branches intercostal and azygos minor inferior veins; empties into one of the other azygos veins.

Bronchia´les: from lungs; the right terminating in azygos major; the left in the left superior intercostal.

Spina´les: *dorsi-spina´lis*, whole length of back of spine, forming network, terminating in the vertebral (of neck), the intercostal (of thorax), lumbar and sacral veins. *Longitudina´les spina´les anterio´res*, whole length vertebral foramen; anterior surface terminating at dorsi-spinal, etc. *Longitudina´les spina´les posterio´res*, whole length vertebral foramen, posterior surface, terminating in dorsi-spinal. *V. ba´sis vertebra´rum*, from bodies of vertebræ, terminating in anterior longitudinal. *Medul´li-spina´les*, cover cord, between pia and arachnoid, from sacrum to occiput; *anas.* freely with those contiguous. No valves in any of the spinal veins.

Ili´aca exter´na, inter´na and **commu´nis**, see lower extremity.

Ve´na ca´va infe´rior: junction of the 2 common iliacs, up on right side of aorta, terminating in lower and back part of right auricle. It receives: the *lumbar* branches (3 or 4 in No.) from muscles and integument of loins; the *right spermatic* (the left emptying into left renal), both having valves; *ovarian*, have same termi-

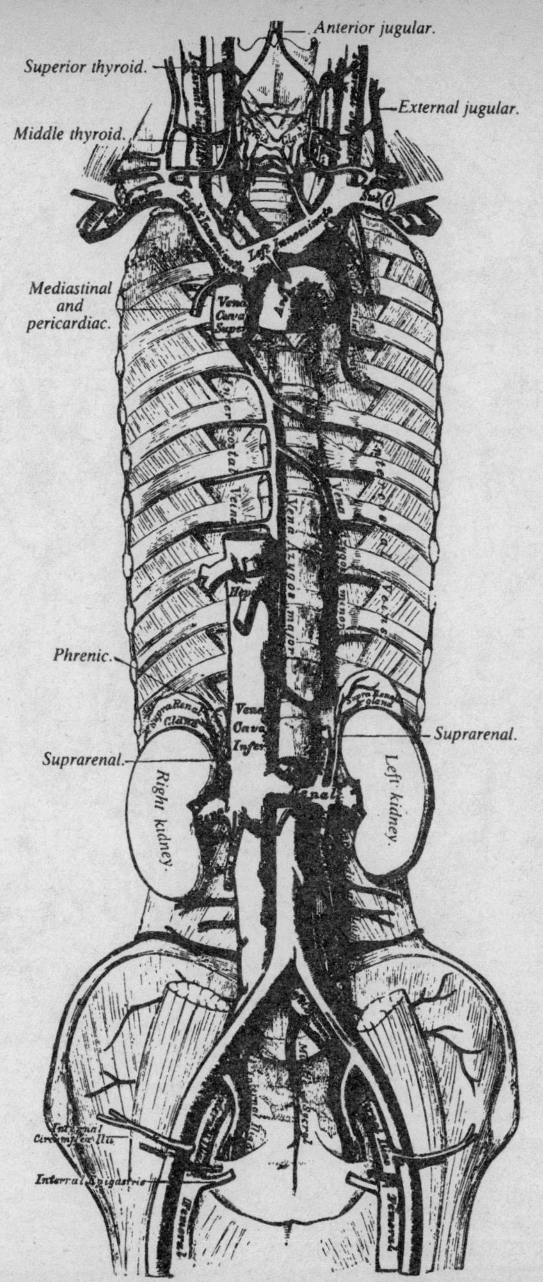

The venæ cavæ and azygos veins, and their formative branches.

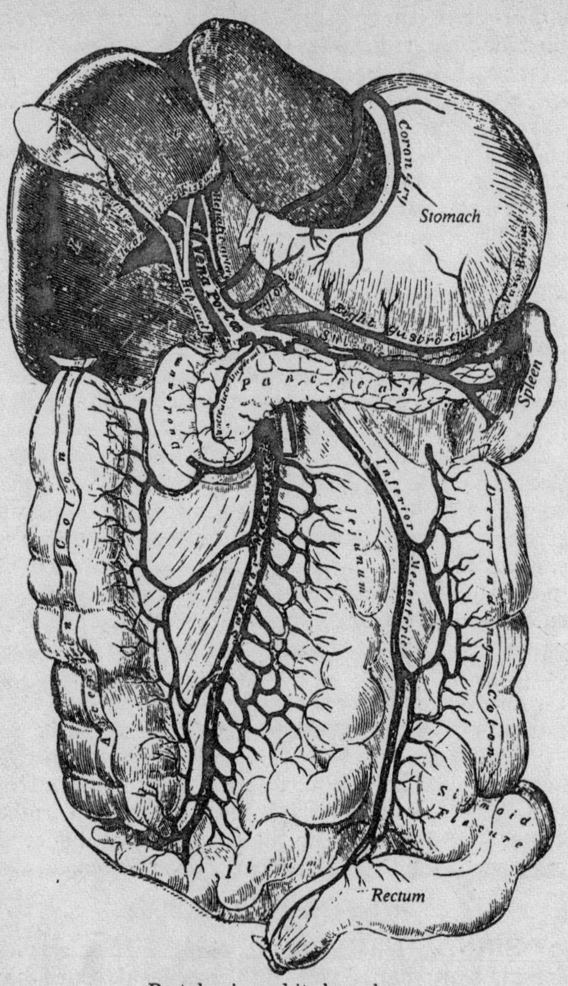

Portal vein and its branches.

nation; the *renal*, the left being the longer; the right *supra-renal* (the left terminating in the left renal, or phrenic); the right *phrenics* (the left superior emptying into superior intercostal or internal mammary, and the inferior into the left renal): the *hepatic*, 3 branches (no valves), these commencing as the *intra*-lobular veins (in the centre of the lobule), forming the *sub*-lobular, and these last finally the large hepatic trunks.

Ve´na por´ta: 4 inches long; no valves in it or its branches; formed by *mesenter´ica infe´rior*, (draining rectum, sigmoid flexure, and descending colon; its branches *inosc.* with internal iliac); *mesenter´ica supe´rior* (draining small intestines, cæcum, ascending and transverse colon); *splen´icæ* (5 or 6 branches from spleen; receiving branches of va´sa bre´via, left gas´tro-epiplo´ica, pancreat´ica and pancreat´ico-duodena´lis veins); *gas´trica*, from lesser stomachic curvature.

Cardi´acæ: *ve´na cor´dis mag´na*, from apex, up anterior interventricular groove to base ventricles, curving to left side to back part of heart, emptying into coronary sinus, guarded by 2 valves; receives posterior cardiac and left cardiac veins. *Ve´na cor´dis me´dia*, (posterior. cardiac) from apex, up posterior interventricular groove, terminating in coronary sinus, guarded by valve. *Ve´næ par´væ*, (anterior veins), 3 or 4 small branches from anterior surface of right ventricle, emptying into lower part right auricle. *Ve´næ thebe´sii* drain muscular substance, opening into right auricle.

Pulmona´les: 4 in number; commence in capillary network upon bronchial cells, uniting to form a trunk for each lobe; the one of the middle lobe of the right lung unites with the one from the superior lobe, hence 2 veins from each side. No valves; *carry arterial blood.*

NERVES OF THE BODY.

SPINAL NERVES: 31 pairs, viz.: 8 cervical, 12 dorsal, 5 lumbar, 5 sacral, 1 coccygeal. Each have an anterior and posterior root, hence have moto-sensor functions.

Cervica´les: (see pages 16 and 17). **Dorsales:** 1st from between 1st and 2d dorsal vertebræ, the last from between 12th dorsal and 1st lumbar. The POSTERIOR branches have *external* and *internal* branches. The *cutaneous* branches are the 6 upper from the internal branches, the 6 lower from the external branches. These nerves supply the structures of the back. ANTERIOR branches supply walls of the chest and abdomen, each having branches from the sympathetic. *Superio´res Inter-costa´les*, pass forwards with the arteries, giving off numerous branches, the chief being the lateral cutaneous, which have anterior and posterior branches. The 1st intercostal has no lateral branches; the 2d has a large one (the intercos´to-humera´lis,) which supplies the integument of upper inner half of arm. *Intercosta´les inferio´res*, having nearly the same course as the superior, supplying the anterior cutaneous nerves to abdomen, and having lateral branches.

Lumba´les: have largest roots of all; have *anterior* and *posterior* branches; the latter having external and internal branches; the *anterior* branches unite to form the lumbar plexus. Supply muscles and integument in their region. The *anterior* branches communicate with sympathetic. **Sacra´les** and **Coccygea´les:** (see nerves of lower extremity.)

LOWER EXTREMITY.

MUSCLES OF THE LOWER EXTREMITY.

(*14 Regions, 57 Muscles.*)

(REGION 1) ILIAC REGION, 3 MUSCLES.

Pso´as mag´nus: last dorsal and all lumbar vertebræ (transverse processes)—lesser trochanter, in union with iliacus. [Anterior branches lumbar.]

Pso´as par´vus: sides of bodies last dorsal and last lumbar vertebræ—ilio-pectineal eminence. [Anterior branches lumbar.]

Ili´acus: iliac fossa, crest and anterior spinous processes of ilium, base sacrum—outer side tendon psoas magnus. [Anterior crural.]

(2) ANTERIOR FEMORAL REGION, 7.

Ten´sor vag´inæ fem´oris: outer crest ilium, anterior superior spinous process—fascia lata ¼ way (laterally) down the thigh. [Superior gluteal.]

Sarto´rius: (*longest muscle of body*); anterior superior spinal process ilium, part of notch below—upper, inner side of tibial shaft, having crossed the anterior surface of the thigh obliquely. [Anterior crural.]

Quad´riceps exten´sor: (vas´ti inter´nus and exter´-nus, rec´tus and crure´us); the *Rec´tus* from anterior inferior spinal process ilium and groove above acetabulum–

Muscles of the iliac and anterior femoral region.

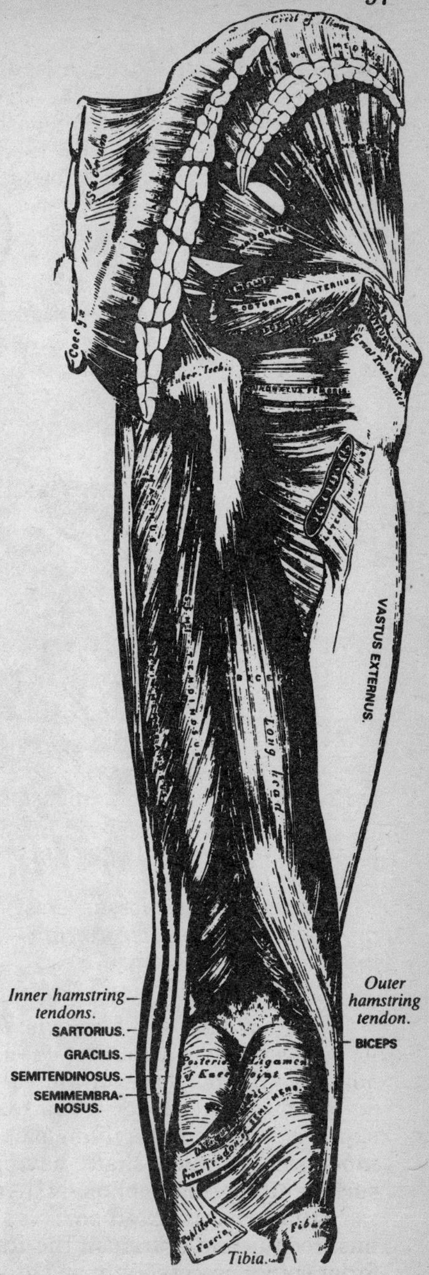

Muscles of the hip and thigh.

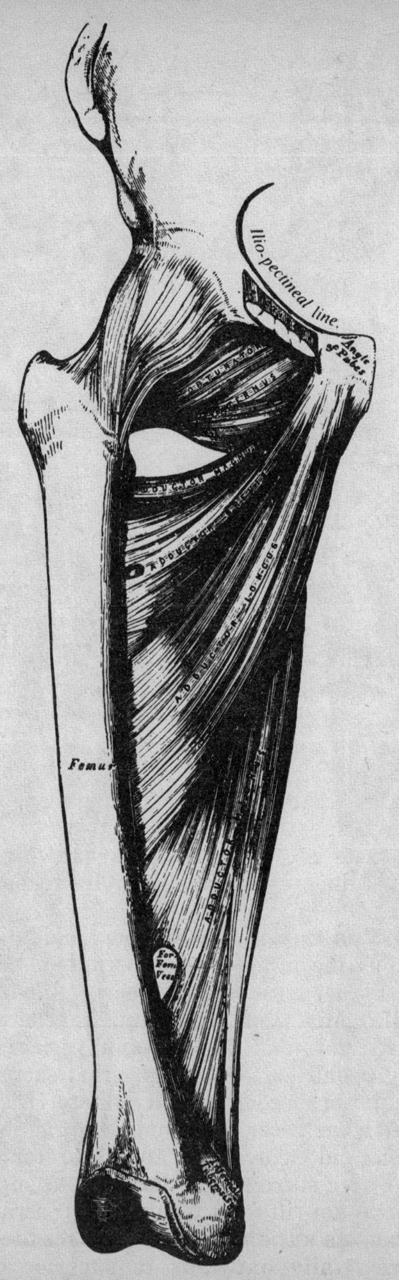

Deep muscles of the internal femoral region.

the *Vas´tus Exter´nus* from anterior border great trochanter, linea aspera,—the *Vas´tus Inter´nus* and *Crure´us* from inner lip of linea aspera and nearly all internal, anterior and external surface of femur-shaft—all joining into a common tendon to be inserted into patella. [Anterior crural.]

Subcrure´us: lower anterior surface of femur—synovial pouch behind patella. [Anterior crural.]

(3) INTERNAL FEMORAL REGION, 5.

Grac´ilis: inner margin rami of pubes and ischium—inner side upper part tibia above insertion semi-tendinosus and beneath sartorius. [Obturator.]

Pectine´us: Gimbernat's ligament, linea ilio-pectinea—rough line between trochanter minor and linea aspera. [Obturators and anterior crural.]

Adduc´tor lon´gus: front of pubes—middle 3d of linea aspera. [Obturators.]

Adduc´tor bre´vis: descending ramus of pubes between gracilis and obturator—upper part linea aspera, behind pectineus. [Obturators.]

Adduc´tor mag´nus: ramus of pubes and ischium, and tuber ischii—from great trochanter to inner condyle. [Obturator and great sciatic.]

(4) GLUTEAL REGION, 9.

Glutæ´us max´imus: superior curved line of ilium down to coccyx and sacro-sciatic ligaments—rough line between great trochanter and linea aspera. [Inferior gluteal branch sacral plexus.]

Glutæ´us me´dius: between superior and middle iliac curved lines, crest, fascia—great trochanter. [Superior gluteal.]

Glutæ´us min´imus: between middle and inferior curved lines, margin great sacro-sciatic notch-impression anterior border trochanter major. [Superior gluteal.]

Pyrifor´mis: front of sacrum, anterior margin great sacro-sciatic foramen and anterior surface great sacro-sciatic ligament, etc.— through great sacro-sciatic foramen to superior, border great trochanter. [Sacral plexus.]

Obtura´tor inter´nus: inner margin obturator foramen, pubic and ischic rami, and obturator membrane—through lesser sacro-sciatic foramen to superior border great trochanter, in front pyriformis. [Sacral plexus.]

Gemel´lus supe´rior: outer surface of spine of ischium—horizontally outwards to superior border great trochanter, in company with obturator internus. [Sacral plexus.]

Gemel´lus infe´rior: superior outer border tuber ischii—superior border great trochanter with obturator internus. [Sacral plexus.]

Obtura´tor exter´nus: inner side obturator foramen, pubic and ischic rami, internal ⅔ of external surface obturator membrane—out- and backwards to digital fossa of femur. [Obturator.]

Quadra´tus fem´oris: outer border tuber ischii—horizontally outwards to linea quadrati of posterior surface of trochanter. [Sacral plexus.]

(5) POSTERIOR FEMORAL REGION, 3.

Bi´ceps: *long* head from tuber ischii, *short* head from linea aspera—outer side head fibula, covering external lateral ligament. *Forms outer "ham-string."* [Great sciatic.]

Semi-tendino´sus: tuber ischii in company with biceps, and the aponeurosis—tendon (inner side popliteal space) curves round internal tibial tuberosity to inner surface of shaft (external and beneath sartorius.) [Great sciatic.]

Semi-membrano´sus: tuber ischii, above and external to biceps and semi-tendinosus—back of tibial tuberosity in 3 digitations, beneath internal lateral ligament. The preceding, with this, and gracilis and sartorius, *form inner "ham-string."* [Great sciatic.]

(6) ANTERIOR TIBIO-FIBULAR REGION, 4.

Tibia´lis anti´cus: outer tibial tuberosity and superior shaft, external surface—inner under surface internal cuneiform and base 1st metatarsus. [Anterior tibial.]

Exten´sor pro´prius pol´licis: middle anterior surface fibula and interosseous membrane—base last phalanx great toe. [Anterior tibial.]

Exten´sor lon´gus digito´rum: external tuberosity tibia, upper ¾ anterior surface shaft of fibula, interosseous membrane—3 tendons distributed to 4 lesser toes. [Anterior tibial.]

Perone´us terti´us: (part of above); lower outer fourth fibula—base 5th metatarsus. [Anterior tibial.]

(7) POSTERIOR TIBIO-FIBULAR REGION, SUPERFICIAL LAYER, 3.

Gastrocne´mius: 2 heads, one from each femuric condyle—unites with soleus to form *tendo Achillis*, insert-

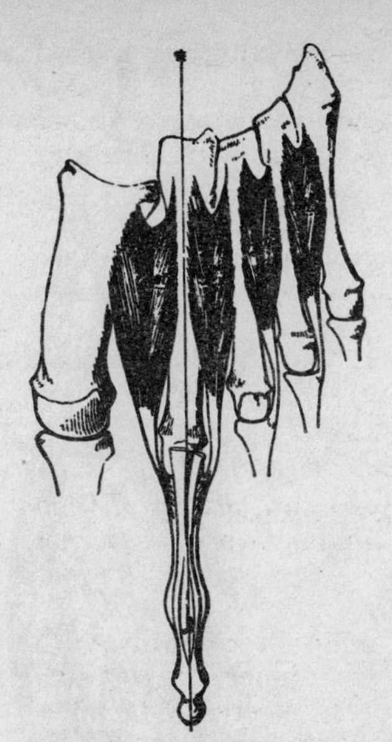

The Dorsal interossei. Left foot.

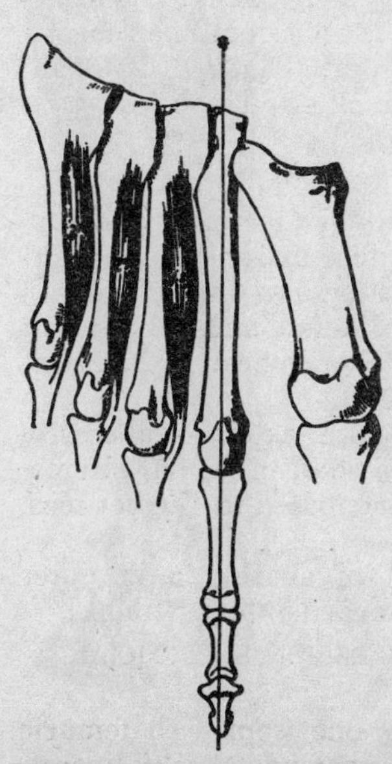

The Plantar interossei. Left foot.

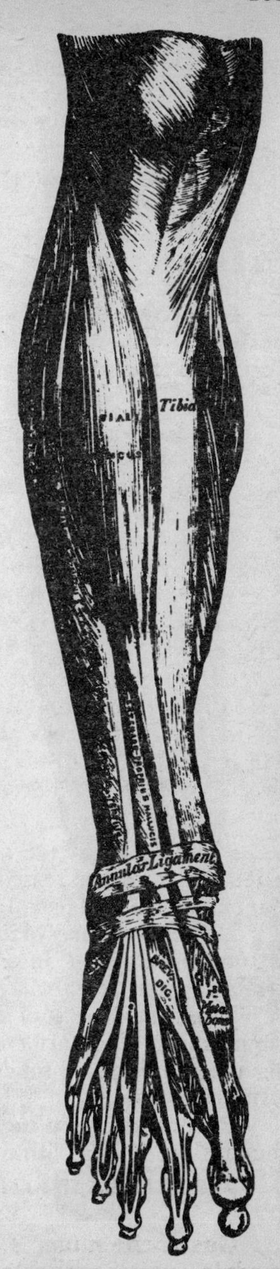

Muscles of the front of the leg.

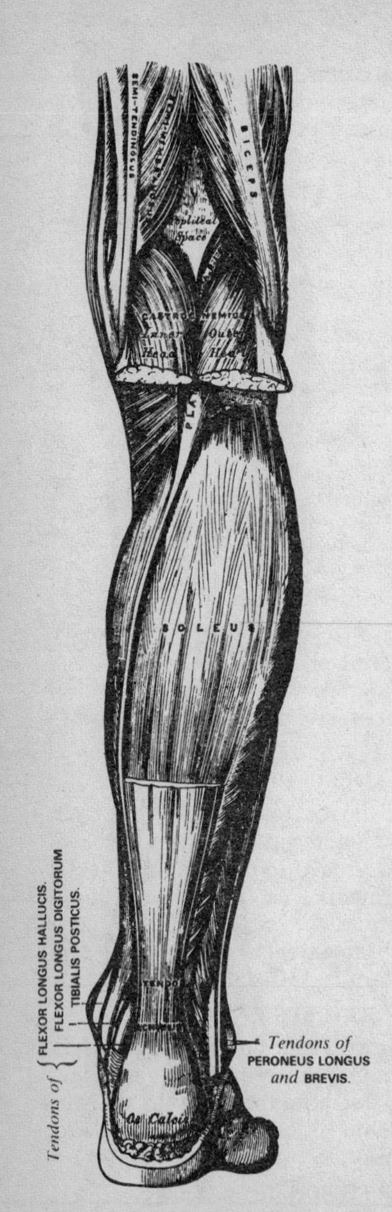

Muscles of the back of the leg.
Superficial layer.

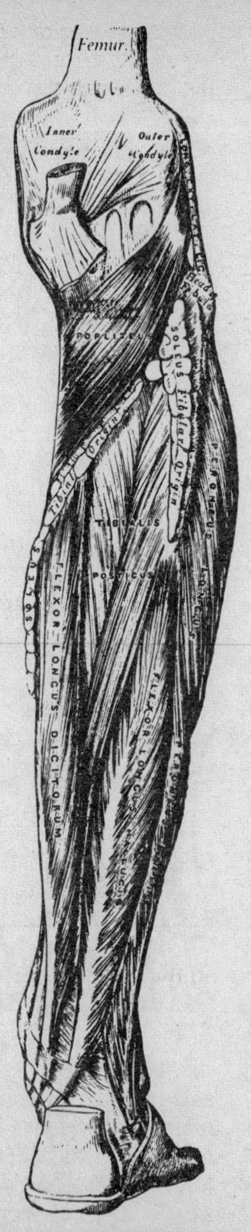

Muscles of the back of the leg.
Deep layer.

ed into posterior surface os calcis. [Internal popliteal.]

Sole´us: oblique line tibia, back of head and superior portion of fibular shaft—os calcis. [Internal popliteal.]

Planta´ris: outer surface external femuric condyle and posterior ligament knee-joint—os calcis, posterior surface. *Noted for long, slim tendon.* [Internal popliteal.]

(8) POSTERIOR TIBIO-FIBULAR REGION, DEEP LAYER, 4.

Poplitæ´us: (forms floor popliteal space); depression below tuberosity of external femuric condyle—inner ⅔ triangular space above oblique line on posterior surface tibia. [Internal popliteal.]

Flex´or lon´gus pol´licis: lower internal ⅔ fibular shaft, interosseous membrane, muscular septum and fascia—through grooves in tibia, astragalus, and calcis to base last phalanx big toe. [Posterior tibial.]

Flex´or lon´gus digito´rum: posterior surface tibia below oblique line, intermuscular septum—behind inner malleolus, calcic arch, joined by tendon flexor accessorius, divides into 4 tendons which pass through slits in the tendons of flexor brevis digitorum to be inserted into bases of last phalanges of the 4 outer toes. [Posterior tibial.]

Tibia´lis posti´cus: interosseous membrane, superior ⅓ posterior surface tibial shaft, superior ⅔ fibula, inner surface—behind inner malleolus, beneath calcaneo-scaphoid articulation to tuberosity scaphoid and internal cuneiform. [Posterior tibial.]

(9) FIBULAR REGION, 2.

Peronæ´us lon´gus: head, and upper, outer ⅔ fibular shaft, muscular fascia and septa—behind external mal-leolus, through cuboid groove to outer side base 1st metatarsus. [Musculo-cutaneous.]

Peronæ´us bre´vis: middle ⅓ outer surface fibular shaft, muscular septa—behind external malleolus to dorsal surface base 5th metatarsus. [Musculo-cutaneous.]

(10) FOOT, DORSAL REGION, 1.

Exten´sor bre´vis digito´rum: outer side os calcis, astragalo-calcanean ligament, anterior annular ligament—4 tendons; 1st into 1st phalanx of great toe, the rest into outer sides of tendons of long extensor to 2d, 3d, and 4th toes. [Anterior tibial.]

(11) FOOT, PLANTAR REGION, 1ST LAYER, 3.

Abduc´tor pol´licis: inner tuberosity os calcis,

internal annular ligament, plantar fascia—inner side base 1st phalanx great toe. [Internal plantar.]

Flex´or bre´vis digito´rum: internal tuberosity os calcis, plantar fascia, muscular septa—4 tendons, sides 2d phalanges of outer toes. [Internal plantar.]

Abduc´tor min´imi dig´iti: outer tuberosity os calcis plantar fascia, muscular septum—outer side base 1st phalanx little toe; joins tendon of short flexor. [External plantar.]

(12) FOOT, PLANTAR REGION, 2D LAYER, 2.

Flex´or accesso´rius: *inner* head from inner surface os calcis and calcaneo-scaphoid ligament; *outer* head, inferior surface os calcis and plantar ligament—tendon flexor longus digitorum. [External plantar.]

Lumbrica´les: (4); tendon of long flexor—inner sides bases of 2d phalanges of 4 outer toes. [Internal plantar to 1st and 2d, external plantar to 3d and 4th.]

(13) FOOT, PLANTAR REGION, 3D LAYER, 4.

Flex´or bre´vis pol´licis: internal border of the cuboid and contiguous surface of external cuneiform, tendon of tibialis posticus—outer and inner sides base first phalanx big toe. [Internal plantar.]

Adduc´tor pol´licis: tarsal extremity of 2d, 3d and 4th metatarsi and sheath of peroneus longus—outer side base 1st phalanx big toe. [External plantar.]

Flex´or bre´vis min´imi dig´iti: base of 5th metatarsus and sheath peroneus longus—outer side base 1st phalanx of little toe. [External plantar.]

Transver´sus pe´dis: under surface head 5th metatarsus, transverse ligament of metatarsus—outer side 1st phalanx of big toe. [External plantar.]

(14) FOOT, PLANTAR AND DORSAL INTEROSSEOUS REGIONS, 7.

Interos´sei dorsa´les: (4); bipenniform, from adjacent sides of metatarsi—bases of 1st phalanges, outer (except the 1st) side of the 4 outer toes. **Planta´res:** (3); arise from the shafts of the 3d, 4th and 5th metatarsi, inner side—inner sides of the bases of the 1st phalanges of the same toes, and common extensor tendon.

ARTERIES OF THE LOWER EXTREMITY.

SCIAT´ICA: (5 branches); larger terminus of anterior trunk of internal iliac; out through lower part of the great sacro-sciatic foramen resting on pyriformis,

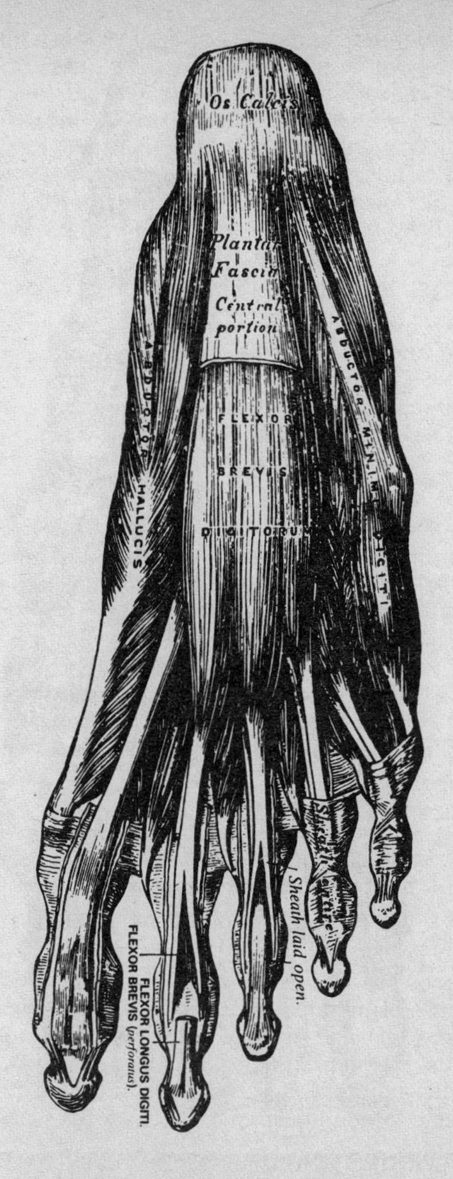

Muscles of the sole of the foot. First Layer.

Muscles of the sole of the foot. Third layer.

descending between tuber ischii and great trochanter, to supply muscles of the thigh. **Coccygea´lis:** inwards, *piercing* great sacro-sciatic ligament, supplying glutæus maximus and integument. **Glutæ´æ inferio´res:** 3 or 4 supplying glutæus maximus. **Co´mes ner´vi ischiad´ici:** accompanying great sciatic nerve, and finally *pierces* it and is lost in its substance. **Muscula´res:** to back part of hip, *anas.* with gluteal, superficial perforating—external and internal circumflex. **Articula´res:** to hip-joint capsule.

GLUTÆ´A SUPE´RIOR: largest branch of internal iliac; out above pyriformis, dividing into deep and superficial branches; supplies iliacus, obturator internus, pyriform. *Superficial* branch, beneath glutæus maximus, supplying it; *anas.* with posterior branch sacral. *Deep* branch, between glutæus medius and glutæus minimus, the superior division *anas.* at anterior superior spinous process of ilium with circumflex iliac and external circumflex; the inferior division goes to great trochanter, *anas.* with external circumflex. Branches supply all muscles in this region, also joint.

FEMORA´LIS: (7 branches): from Poupart's ligament to opening in adductor magnus. A line drawn from the middle of said ligament to internal femuric condyle lies over its course. Vein lies on inside; anterior crural nerve on the outside of artery. **Superficia´lis epigas´trica:** ½ inch below Poupart's ligament, through saphenous opening upwards to umbili´cus in the fascia covering the external oblique abdominis; *anas.* deep epigastric and internal mammary. **Superficia´lis circumflex´a ili´aca:** arises close to above, outwards to iliac crest, supplying glands fascia and integument; *anas.* circumflex iliac, gluteal, external circumflex. **Superficia´lis exter´na pudi´ca:** inner side, ½ inch below Poupart's ligament, pierces fascia lata, crosses spermatic cord, supplies integument of lower part of abdomen, penis, scrotum (or labia); *anas.* internal pudic branches. **Profun´da exter´na pudi´ca:** passes inwards on pectineus, piercing fascia at pubes, supplies integument of perinæum, scrotum (or labia); *anas.* superficial perinæal. **Profun´da fem´oris:** outer and back part, 1 to 2 inches below Poupart's ligament, passing back of artery and the femoral vein to inner side femur, terminating in adductor magnus, lower 3d; *anas.* with popliteal and inferior perforating. *Circumflex´a exter´na,* having ascending, descending and trans-

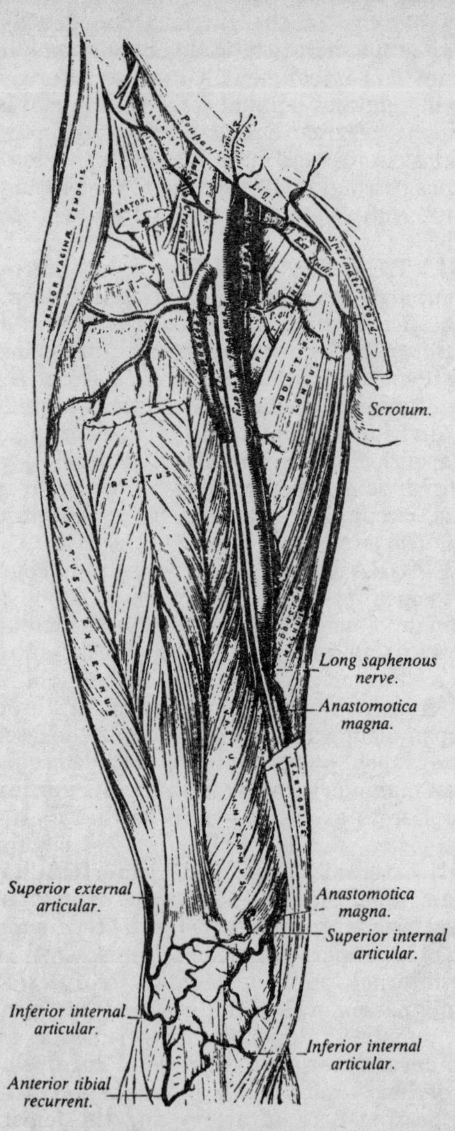

The femoral artery.

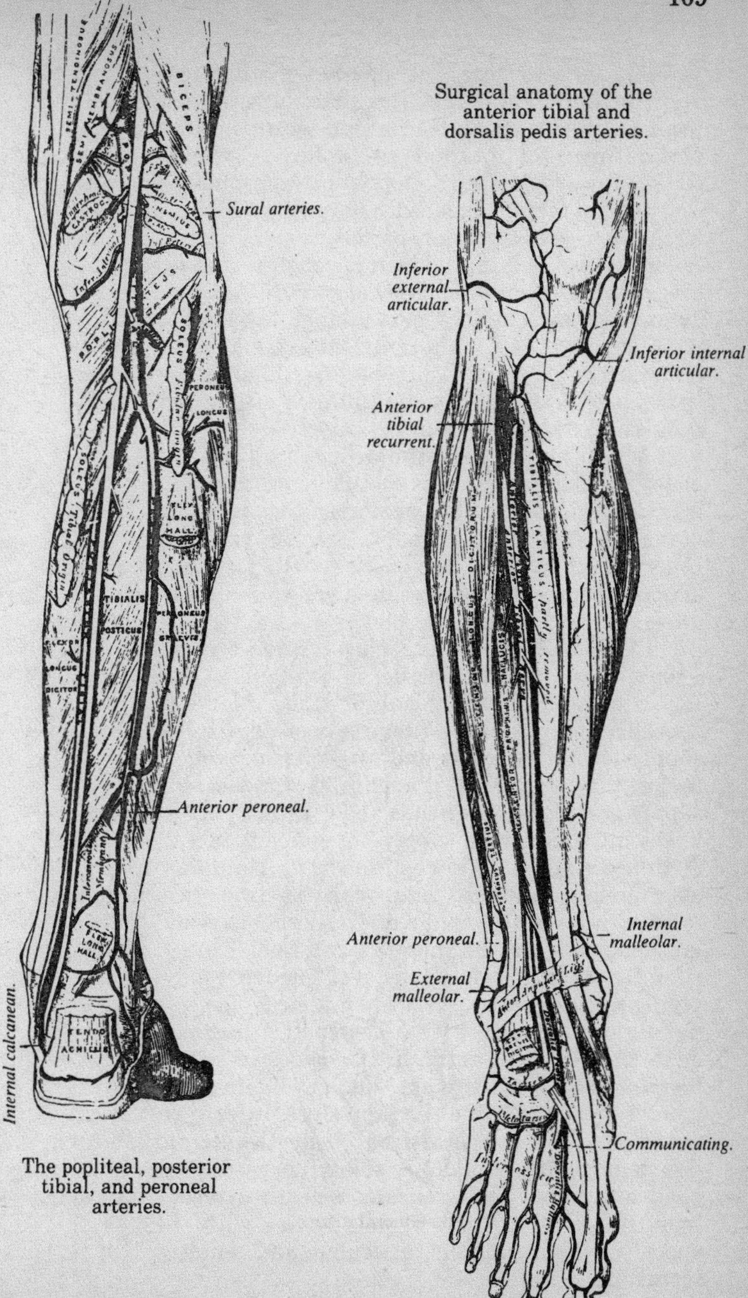

Surgical anatomy of the anterior tibial and dorsalis pedis arteries.

The popliteal, posterior tibial, and peroneal arteries.

verse branches, supplying muscles in that region, and *anas.* with gluteal, circumflex iliac, superior articular of popliteal, near great trochanter with sciatic, superior perforating and internal circumflex. *Circumflex´a in-ter´na*, inwards to joint, supplying contiguous muscles, and head of femur; *anas.* with obturator, sciatic, external circumflex and superior perforating. *Perforan´tes*, the "superior," supplying adductors magnus and brevis, biceps, glutæus maximus and *anas.* with sciatic, internal circumflex and middle perforating; "middle" one supplies flexors of thigh and nutrient artery, *anas.* with its fellows; the "inferior" supplies the thigh flexors, *anas.* with its fellows and terminal branch of profunda. **Muscula´res:** 2 to 7 in number, supplying sartorious and vastus internus. **Anastomot´ica mag´na:** arises just before the femoral, pierces the adductor magnus, dividing into *superficial* branch, accompanying long saphenous nerve, to supply integument; *deep* branch descends to inner side of knee, where it *anas.* with superior internal articular and recurrent of anterior tibial, and supplies knee-joint and contiguous parts.

POPLITÆ´A: (7 branches), from termination of femoral down to lower border of popliteus muscle, dividing into anterior and posterior tibial. Nerve and vein superficial to artery. **Muscula´res:** *superior* (2 or 3), supply vastus externus and thigh flexors; *anas.* inferior perforating, terminal branches profunda. *Inferior* (2), supply gastrocnemius heads and plantaris; arise opposite kneejoint. **Cuta´nei:** supply integument of calf of leg. **Articula´res superio´res:** *internal*, running inwards over femuric condyles, *anas.* with anastromotica magna, inferior internal articular and superior external articular, supplying vastus internus and knee-joint. *External*, running circularly outwards over femuric condyles, supplying vastus externus, knee-joint, etc.; *anas.* with external circumflex, and with anastomotica magna, forming an arch. **Az´ygos articula´ris:** opposite bend of joint, piercing posterior ligament, supplies ligaments, synovial membranes and joint. **Articula´res inferio´res:** wind round tibial head; the *internal*, beneath internal lateral ligament, to front and inner side of joint, supplying tibial head and joint. The *external*, beneath external lateral ligament, etc., to front of joint, *anas.* with the one of opposite side, superior articular and anterior tibial recurrent.

TIBIA´LIS ANTE´RIOR: (3 branches); forward through interosseous membrane and 2 heads of tibialis posticus, lying upon anterior surface of interosseous membrane down to front of ankle, there becoming dorsalis pedis. A line drawn from inner fibular head to midway between the 2 malleoli indicates its course. Has venæ comites; the anterior tibial nerve lies a little superficial and to its outer side. **Recur´rens:** arises just as artery passes through interosseous membrane, running up in tibialis anticus muscle to front of joint, *anas.* with the articulares. **Muscula´res:** numerous, supplying integument and muscles throughout the course, *anas.* with branches from posterior tibial and peroneal. **Malleola´res:** *internal* arises 2 inches above articulation, inwards, beneath tendons ramifying upon inner malleolus, *anas.* with branches from posterior tibia and internal plantar. *External*, outwards beneath tendons, supplying outer malleolus, *anas.* with anterior peroneal, and tarsea branch of dorsalis pedis.

DORSA´LIS PE´DIS: (4 branches); from bend of ankle to 1st interosseous space, there dividing into communicating and dorsalis hallucis. Has venæ comites; anterior tibial nerve lies on outer side. **Tar´sea:** arises over scaphoid, passing outwards beneath extensor brevis digitorum, supplying that muscle and tarsal articulations; *anas.* metatarsal, external malleolar, peroneal, and external plantar. **Metatar´sea:** outwards over metatarsal heads, giving off 3 *interos´seæ* branches which pass forwards to clefts of the 3 outer toes, there dividing to supply adjacent sides of the toes, and outer side of little toe. *Anas.* with tarsea and external plantar; the 3 interosseous, each, receive a posterior perforating branch from plantar arch near their origin, and each a branch from anterior perforating of digital near the toe-clefts. **Dorsa´lis hallu´cis:** forwards along outer border 1st metatarsus to 1st toe-cleft, there dividing to supply inner side of big toe, and the adjacent sides of big and 2d toes. **Commu´nicans:** dips down into sole, *anas.* with external plantar to form plantar arch, there dividing to supply toes same as dorsalis hallucis.

TIBIA´LIS POSTE´RIOR: (5 brs.); from lower border popliteus, parallel inner border tendo Achillis, to fossa between inner ankle and heel, there dividing into the plantar arteries. Has venæ comites; nerve to the outside for the lower ¾ of its course. **Peronæ´a:** from 1 inch below popliteus, obliquely outwards to fibula, descending along inner border of it to outer ankle, supplying contiguous structures, *anas.* with external malleolar, tarsal and external plantar. *Ante´rior peronæ´a,* given off 2 inches above ankle, pierces interosseous membrane, passes down to front of outer ankle and tarsus, supplying adjacent structures, *anas.* with tarsal and external malleolar. *Nutri´tia,* to fibula. *Muscula´res,* to fibular muscles. **Nutri´tia:** near origin of posterior tibia, being largest of its kind in the body, enters tibia just below oblique line. **Muscula´res:** to soleus and deep muscles. **Commu´nicans:** transversely across tibia 2 inches above its inferior extremity to *anas.* with peroneal. **Calca´neæ interna´les:** several branches arising just before division of posterior tibial, supplying fat and integument about heel and muscles of inner side of foot; *anas.* with peroneal, internal malleolar.

PLANTA´RIS INTER´NA: forwards along inner side of foot to big toe, *anas.* with digital branches, supplies adductor pollicis, flexor brevis digitorum, etc.

PLANTA´RIS EXTERNA: (2 brs.); out- and forwards to base 5th metatarsus, then turning obliquely inwards to 1st interosseous space, *inosc.* with communicating branch from dorsalis pedis, forming *plantar arch.* **Perforan´tes posterio´res:** (3); ascend through back part of the 3 outer interosseous spaces; *anas.* with interosseous branches of metatarsal. **Digita´les:** (4); arise from arch and supply both sides of the 3 outer toes and outer side of the 2d toe, bifurcating at the respective toe-clefts to do this. At each bifurcation a branch (the *anterior perforating*) is sent upwards through the interosseous space; *anas.*with interosseous branches of the metatarsal.

VEINS OF THE LOWER EXTREMITY.

Saphe´na inter´na: or *long* saphenous; from plexus at dorsum and inner side of foot, ascends, in front of inner ankle, behind inner margin of tibia, bends behind inner femuric condyle, empties into femoral through saphenous opening, 1½ inches below Poupart's ligament, where it receives *superficia´lis circumflex´a ili´aca, superficia´lis epigas´trica,* and *superficia´lis exter´na pudi´ca.* Communicates with internal plantar, tibial, etc. 2 to 6 valves.

Saphe´na exter´na: plexus at dorsum and outer side of foot, up behind outer ankle to median line of leg,

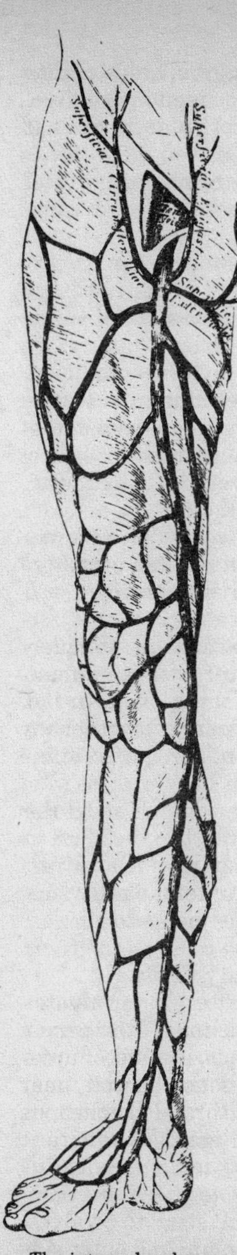

The internal or long saphenous vein and its branches.

The plantar arteries. Deep view.

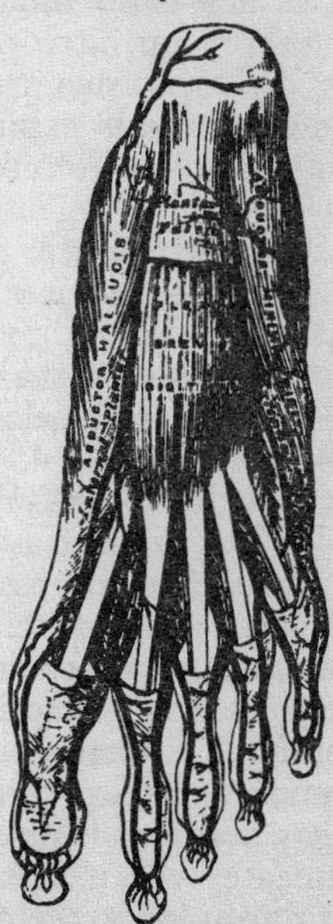

The plantar arteries. Superficial view.

accompanied by external saphenous nerve; empties into popliteal vein, between heads of gastrocnemius; 2 valves, one near termination. *Communicates* with deep veins of foot.

Tibia´les posterio´res: formed from *external* and *internal plantar* joining with the *peroneal*. Course same as artery.

Tibia´les anterio´res: continuation of *ve´næ dorsa´les pe´dis*, pierce interosseous membrane at upper part of leg, and form, by junction with the *posterior tibial* veins, the popliteal.

Poplitæ´a: (see tibial anterior); up to tendinous aperture of adductor magnus, there becoming the femoral; receives *sural, articular*, and *external saphenous* veins. 4 valves. Crosses artery from within outwards.

Femora´lis: (see above); up to Poupart's ligament, there becoming external iliac. Lies (below) to outside, but crosses beneath the artery to its inside. Receives *muscular* branches, and *profun´da fem´oris*, and *internal saphenous*, at 1½ inches below Poupart's ligament. 4 or 5 valves.

Ili´aca exter´na: (see above), to sacro-iliac symphysis, there uniting with internal iliac to form common iliac. On right side, lies to inside of artery at first, but gradually passes behind it. On left side, altogether on inside of artery. Receives *epigastric* and *circumflex iliac*. No valves.

Ili´aca inter´na: formed by venæ comites of *all* the branches of the iliac artery, but the umbilical; lies first to inside, but finally gets behind the artery. No valves, though the plexus that help form it are abundantly supplied. 1. *Hæmorrhoidal* plexus; 2. *vesico-prostatic* plexus; 3. *vaginal* plexus; 4. *uterine* plexus; 5. *dorsalis penis* plexus; these all intercommunicate very freely.

Ili´aca commu´nis: (see iliaca externa); terminates at intervertebral substance between 4th and 5th lumbar vertebræ, there, with its fellow of opposite side, forms vena cava inferior. On the right it is the shorter, and nearly vertical. Receives *ilio-lumbar*, and sometimes *lateral sacral* veins. *Middle sacral* empties into left common iliac. No valves.

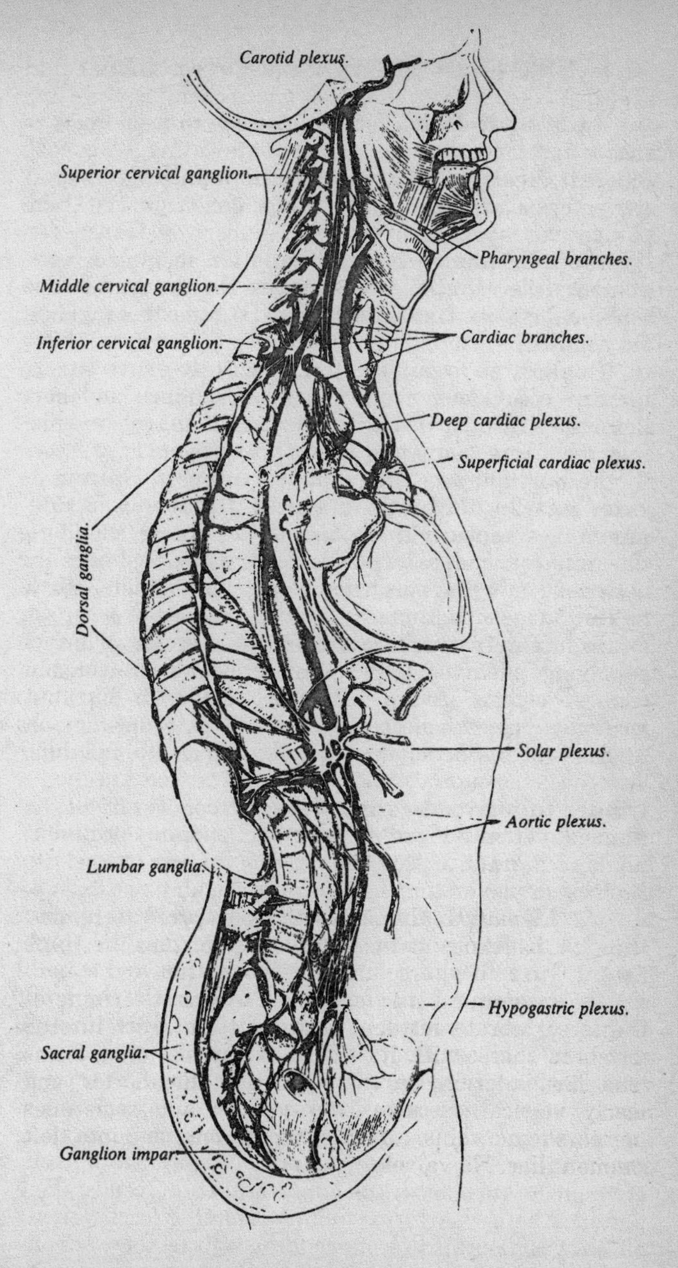

The sympathetic nerve.

NERVES OF THE LOWER EXTREMITY.

LUM´BAR PLEX´US: formed by anterior roots of the 4 upper lumbar nerves communicating with each other. It furnishes different nervous branches to supply the inferior extremities. **Ilio-hypogas´trica:** 1st lumbar, outwards to iliac crest, piercing there the transversalis, the *iliac* branch being distributed to gluteal integument; the *hypogas´tric* supplies the integument in umbilical region. **Ilio-inguina´lis:** 1st lumbar: escapes at external ring, supplying inner thigh, scrotum (labia in female) and inguinal region. **Genito-crura´lis:** 2d lumbar and branch from 1st; pierces psoas, and near Poupart's ligament divides; the *genital* branch to genitals, the *crural* to integument on anterior upper aspect of thigh; *communicates* with middle cutaneous. **Cuta´neus exter´nus: 2d lumbar; perforates psoas, and at Poupart's ligament divides; the** *anterior* branch supplying the anterior and external part of thigh to knee; the *posterior* supplying surface of thigh to its middle. **Obtura´tor:** 3d and 4th lumbar, and at upper part of obturator foramen enters thigh, dividing into: *anterior* branch supplying adductor longus and brevis, pectineus and femoral artery, giving articular branch to hip-joint; *posterior* branch pierces obturator externus, passes to front of adductor magnus, dividing into muscular branches; *articular* branch is given off for knee-joint. **Obtura´tor accesso´rius:** either from obturator, or filaments from 3d and 4th lumbar; supplies pectineus, hip-joint, and a cutaneous branch to leg. *Sometimes wanting.*

ANTE´RIOR CRURA´LIS: 3d and 4th lumbar, through psoas beneath Poupart's ligament to thigh, external to artery in pelvis, supplies iliacus, and femoral artery; without, all the muscles on front of the thigh but the tensor vaginæ femoris. **Cuta´neus me´dius**, through fascia lata below Poupart's ligament, dividing into 2 branches, supplying sartorius and integument in front as low as knee. **Cuta´neus inter´nus**, obliquely across upper part femoral sheath, the *anterior* branch perforating fascia at lower 3d of thigh, supplies integument of inside of thigh to knee-joint; the *inner* branch descends along posterior border sartorius to knee, piercing fascia, giving off numerous branches, descending still farther, supplying integument of inner side of leg. **Saphe´nus internus**,

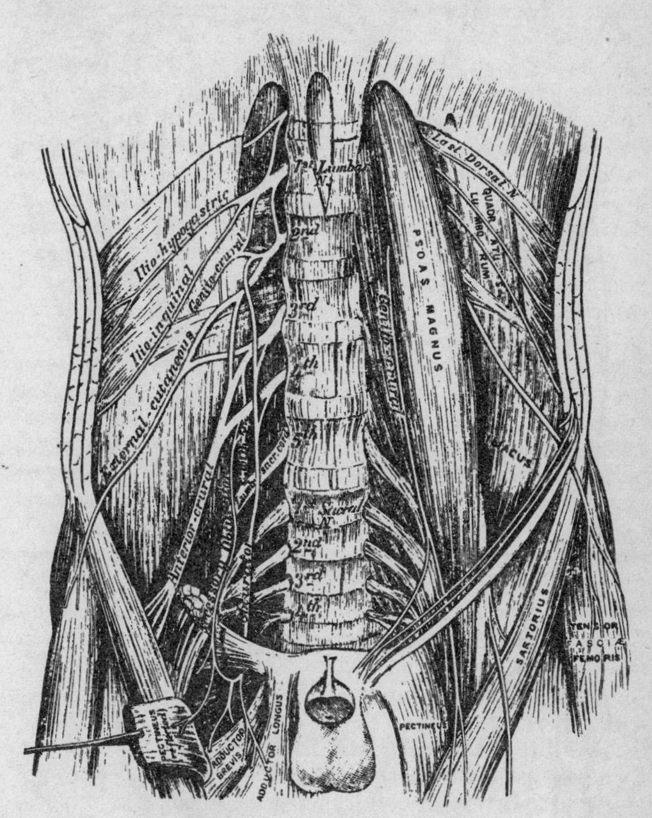

The lumbar plexus and its branches.

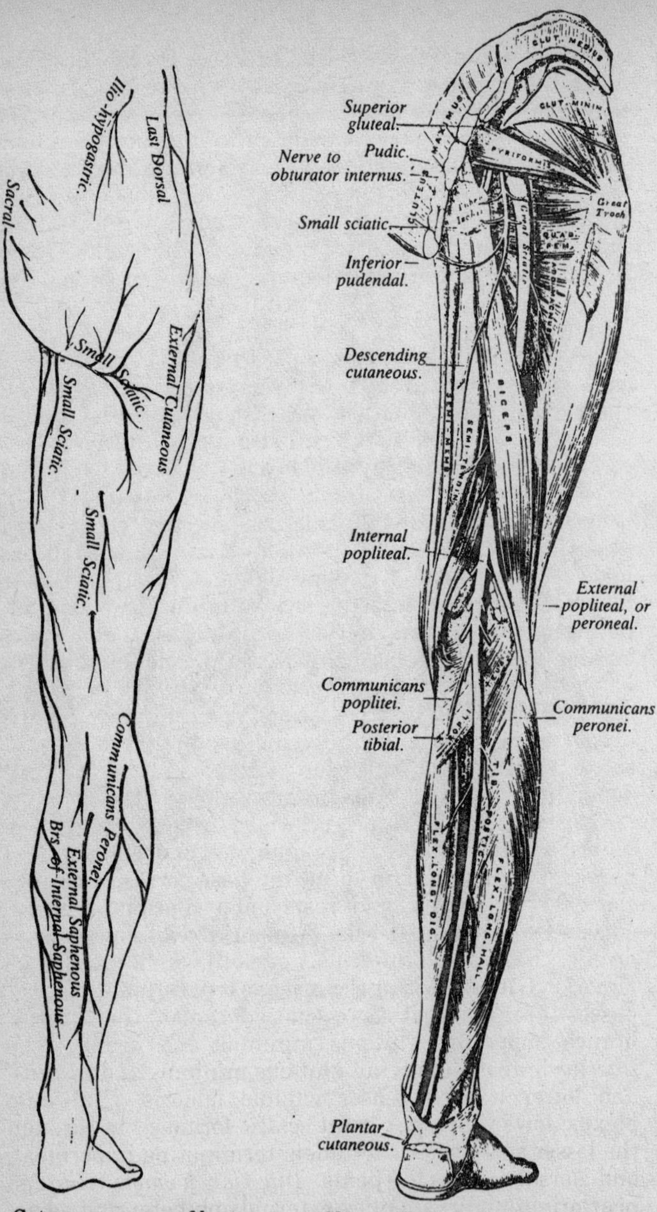

Cutaneous nerves of lower extremity. Posterior view.

Nerves of the lower extremity. Posterior view.

downwards beneath sartorius to knee, inner side, then along inner side of leg in company with internal saphenous vein, dividing into 2 branches, one terminating at inner ankle, the other distributed to integument of dorsum of foot. Supplies muscles and integument in its course, giving off branches *communicating* with internal cutaneous and obturator nerves; another to patellar integument and forms a "plexus patellæ" with other branches. **Muscula´res**, all muscles of front of leg but tensor vaginæ femoris. **Articula´res,** 2 to knee-joint ligaments.

SACRA´LES: 5; the 4 upper through anterior sacral canals; the 5th through the sacro-coccygeal foramen; the posterior are smaller and through posterior sacral canals, except the 5th, which is through posterior sacro-coccygeal foramen. Have *long roots. Posterior internal* branches supply multifidus spinæ. *Posterior external* branches supply integument over sacrum, coccyx and posterior gluteal region, forming many anastomosing loops. *Anterior*, the 4 upper supplying rectum, bladder (vaginal), and pelvic viscera (*communicating* with sympathetic); with their muscular branches they supply levator ani, coccygeus, sphineter ani, and integument between anus and coccyx, *communicating* with coccygeal.

COCCYGEA´LIS: posterior branch receives branch of *com.* from posterior sacral and is lost in fibrous cover of coccyx. Anterior branch pierces sacro-sciatic ligament, supplying integument about coccyx. *Anas*, 5th sacral.

SA´CRAL PLEX´US: is formed by lumbo-sacral, the anterior branches of 3 upper (and part of the 4th) sacral nerves. Is triangular in form, the base corresponding to the exits of nerves, and rests on pyriformis, anterior surface, covered by fascia. **Muscula´res**, supply pyriformis, obturator internus, gemelli, and quadratus femoris. **Glutæ´us supe´rior:** back part lumbo-sacral, passes through great sacro-sciatic foramen, the *superior* branch supplying glutæus minimus and medius, the *inferior* branch supplying glutæus minimus and medius, and lower portion tensor vaginæ femoris. **Pudi´cus:** plexus, lower part; out great sciatic foramen, in through the lesser sacro-sciatic foramen, terminating in perineal, and dorsal nerves of penis. *Infe´rior hæmorrhoida´lis*, near origin pudic, supplies external sphineter and adjacent integument; *communicates* with interior pudendal and superficial perineal. *Perinæ´us*, terminal branch.

accompanies perineal artery; the *anterior cutaneous* branches supply scrotum and under part of penis, (labia), and levator ani; the *posterior* branches supply sphineter ani and integument in front of anus, and back part scrotum. The *muscular* branches supply transversus perinæi, accelerator urinæ, erector penis, compressor urethræ, and bulb. *Dorsa'lis pe'nis*, along ramus ischii, with pudic artery, follows it and its branches to the glans penis, which it supplies. *Anas.* with sympathetic, and supplies integument of prepuce and of penis, and corpus cavernosum. (In female, to the analogous parts.)

SCIAT'ICUS PAR'VUS: supplies integument of perinæum, back part of thigh and leg, and glutæus maximus. Two branches from sacral plexus unite to form it; follows course of sciatic artery in distribution, piercing fascia in popliteal region, accompanies external saphenous vein to middle of leg. *Inferio'res glutæ'i*, to glutæus maximus, several large branches. *Inter'nal cuta'nei*, to skin of upper and inner side of thigh, posterior aspect; scrotum by inferior pudendal that curves around tuber ischii. *Ascenden'tes cuta'nei*, run upwards and supply integument of gluteal region, and muscles. Branches to integument of thigh, popliteal region and upper part of leg.

SCIAT'ICUS MAG'NUS: 3/4 inch wide, and continuation of lower part sacral plexus, passing out of great sacro-sciatica foramen below pyriformis, down between great trochanter and tuber ischii to lower 3d of thigh, there dividing into internal and external poplitæus. *Articula'res*, to hip-joint and capsule. *Muscula'res*, to flexors of the leg, adductor magnus; integument of thigh also supplied by this nerve.

POPLITÆ'US INTER'NUS: (see above); largest terminal branch; down through middle of popliteal space, beneath soleac arch, becoming posterior tibial. *Articula'res* (3); knee-joint, accompanying superior internal articular, and azygos arteries. *Muscula'res*, (4 or 5) to gastrocnemius, plantaris, soleus and popliteus. *Saphe'nous exter'nus*, down, between gastrocnemius heads, to middle of leg, there piercing fascia and *anas.* with communicans peronæi, then down along outer margin of tendo Achillis, in company with vein, supplying integument of outer side of foot and little toe; *communicates* with musculo-cutaneus.

TIBIA'LIS POSTE'RIOR: from lower border poplitæus, passes down leg with posterior tibial artery,

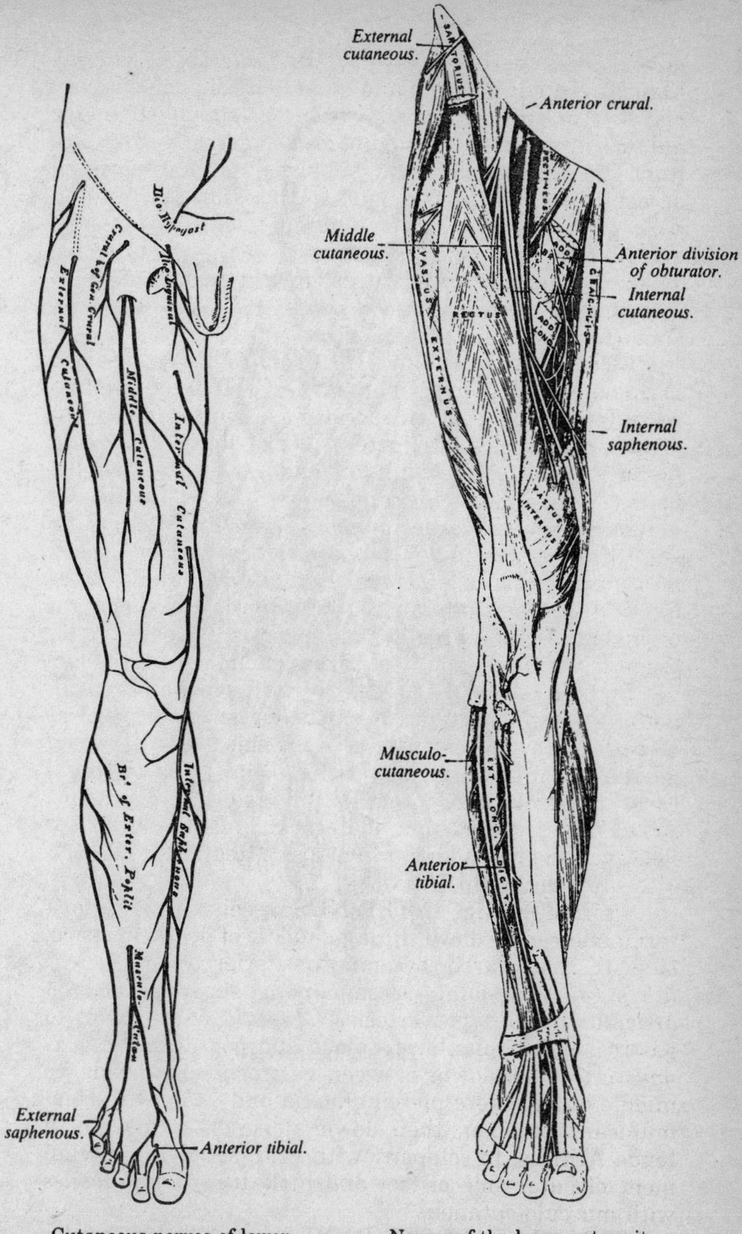

Cutaneous nerves of lower extremity. Front view.

Nerves of the lower extremity Front view.

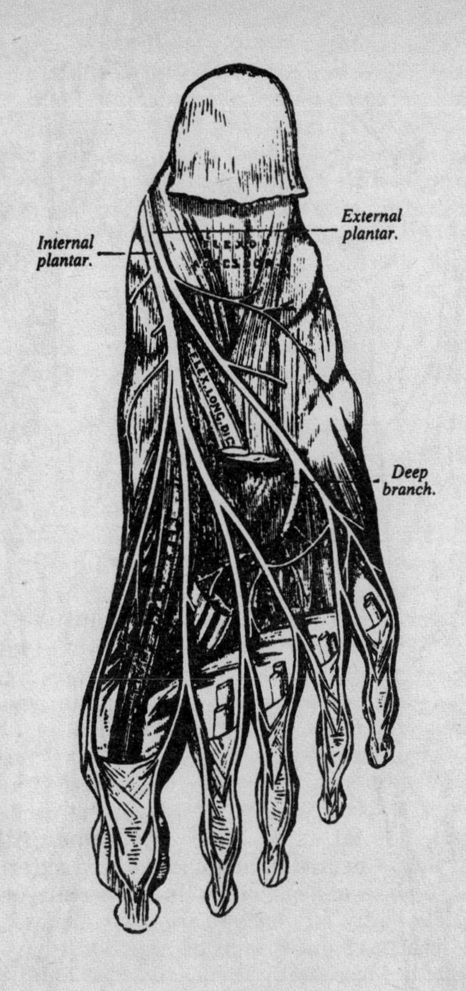

The plantar nerves.

between heel and internal ankle, there dividing into external and internal plantar; above, lies to inside of artery; below to outer side. **Muscula´res:** to tibialis posticus, flexor longus digitorum, and pollicis. **Cuta´neus planta´ris:** perforates internal annular ligament, supplying integument of heel and inner side of sole of foot. **Planta´ris inter´nus:** (see above); largest terminal branch; accompanies internal plantar artery along inner side of foot. *Cuta´nei*, to sole of foot. *Muscula´res*, to flexor brevis digitorum, and abductor pollicis. *Articula´res*, to tarsus and metatarsus. *Digita´les*, (4); supplying the first 3 toes (both sides) and inner margin of the 4th toe, integument, articulations, nails, etc., and 1st and 2d lumbricales. **Planta´ris exter´nus:** (see tibial posterior); follows course of its artery to outer side of foot, supplying little toe and outer half of 4th toe, and structures adjacent, flexor accessorius, and abductor minimi digiti. *Superficia´lis* branch goes to outer side of 5th and adjacent sides of 4th and 5th toes, flexor brevis minimi digiti, and the 2 interossei of 4th metatarsal space. *Deep* branch supplies remaining interossei, 2 outer lumbricales, adductor pollicis, transversus pedis.

POPLITÆ´US EXTER´NUS (or peronæ´us): ½ size of internus poplitæus (see great sciatic); descends along outer margin of popliteal space to fibula, and about 1 inch below its head divides into anterior tibial and musculo-cutaneus. **Articula´res:** (2); accompany external articular arteries to outer side of knee. Sometimes a 3d is given off as a recurrent, which supplies front of knee. **Cuta´nei:** (2 or 3); supply integument of back and outer side of leg as far as its lower 3d. **Commu´nicans peronæ´i:** arises near fibular head, joining external saphenus at middle of leg. **Tibia´lis ante´rior:** (see above); passes obliquely forwards to front of interosseous membrane, reaching outer side of anterior tibial artery at middle of leg descending thence to front of ankle it divides into external and internal branches. *Muscula´res*, to tibialis anticus, extensor longus digitorum, extensor proprius pollicis. *Exter´nal* or *tar´seus*, outwards across tarsus, supplies external brevis digitorum, and articulations of tarsus and metatarsus; becomes ganglionic. *Internal* branch accompanies dorsalis pedis artery, supplying 1st interosseous space and adjacent sides 1st and 2d toes; *communicates* with internal division of musculo-cutaneus. **Musculo-cuta´neus:** supplies muscles of fibular side of leg and dorsal integument of foot (see poplitæ´us exter´nus). At lower 3d of leg (its front and outer side) divides into internal and external branches. *Muscula´res*, to fibular muscles and integuments. *Internal* branch, down in front of ankle to supply inside of great toe and adjacent sides of 2d and 3d toes, integuments of inner ankle and inside of foot; communicates with internal saphenus and interior tibial. *External* branch, down from outer side dorsum of foot to supply adjacent sides of 3d, 4th and 5th toes, integument of outer ankle and outer side of foot; *communicates* with external saphenus.

THE SPINAL CORD.

THE SPINAL CORD AND ITS MEMBRANES.

The du´ra ma´ter is the most external membrane, and is continuous with that investing the brain, but it does not form the endosteum of the vertebræ, nor has it any sinuses, but it is separated from the bones by areolar tissue and a plexus of veins. It is connected above with the edge of the foramen magnum; *at the top of the sacrum it becomes impervious* and is continued as a slender cord to blend with the periosteum of the coccyx. This membrane gives sheaths to all the spinal nerves.

The arach´noid is a thin serous membrane investing the outer surface of the cord and the inner surface of the dura mater. (Some now hold that the inner surface of the dura mater is not covered by the arachnoid.)

The cavity between the arachnoid and the chord is termed the **subarach´noid** space, and contains the subarachnoid fluid.

The pi´a ma´ter is the most internal coat, and covers the entire surface of the cord. It is more fibrous and less vascular than the pia mater of the brain. A process, the *linea splendens*, is sent into the anterior median fissure at the first lumbar vertebræ; it ends in a slender cord, the *filum terminale*, which is within the prolongation of the dura mater.

The ligamen´tum denticula´tum is found between the anterior and posterior roots of the nerves; it consists of a number of serrations of the pia mater attached externally to the dura mater, and serves to support the cord.

Contents of the Neural Canal.

Venous Plexus between bone and dura mater.

Membranes.
- Dura mater.
- Arachnoid.
  - Parietal layer.
  - Visceral layer.
- Pia mater, with ligamenta denticulata.

Cerebro-spinal fluid

Spinal vessels.
- Anterior spinal artery and vein.
- Two posterior spinal arteries and veins.

Spinal cord, with anterior and posterior roots of nerves.

THE SPINAL CORD is contained in the spinal canal, occupying, *in adults*, about 2/3 the length of it; but *in the fœtus*, before the 3rd month, it occupies the whole of the canal.

Extent.—The spinal cord extends from lower border of the foramen magnum to the lower border of the 1st lumbar vertebra, there terminating in a slender filament of gray matter, extending for some distance, called the *filum terminale.* Usual length, 16 to 17 inches; weight, 1½ ounces, or 1/35 as much as encephalon.

Shape.—A transverse section would be oval, being elongated from side to side.

Enlargements.—Presents two enlargements upon its surface. The upper or *brachial* is the larger of the two, corresponding to the origin of the brachial plexus, enlarged laterally. The lower or *crural* corresponds to the origin of the lumbar and sacral plexuses, which form the *cauda equina*; is more bulbous than the upper one.

Fissures. Anterior median fissure: in longitudinal direction along the middle line, extending into the substance of the cord for about one-third its thickness, but deeper below than in the upper part; lined with pia mater.

Posterior median fissure: narrower than the preceding, but extends into the cord for nearly half its thickness; contains a septum of pia mater.

Columns: the cord being thus divided into two lateral halves, may again be subdivided into *anterior, lateral* and *posterior* columns.

The *posterior* and *lateral* columns are divided by a groove or lateral sulcus, to which the posterior nerve roots are attached.

The *anterior* and *lateral* columns are separated by the anterior roots of the nerves.

Posterior median column is formed by a groove a little outside the posterior median fissure, dividing the posterior column into two parts: a posterior median column and posterior column proper.

Central Canal: in the interior of the cord, upper part for a short distance, is a central canal, lined with spheroidal, ciliated epithelium, and opening into the cavity of the 4th ventricle. In the fœtus, up to 6th month, this canal exists throughout the cord.

THE BRAIN AND ITS MEMBRANES.

THE DU´RA MA´TER: the most external; is a dense fibrous membrane, outer surface rough and forms the endosteum of the bones of the skull. The inner surface is smooth, and covered by the arachnoid. It is continuous with the dura mater of the spinal cord through the foramen magnum. In certain parts the fibrous layers of this membrane separate to form the *sinuses* of the dura mater. On the upper surface are the *Pacchionian* bodies. There are certain processes of the dura mater, viz:

The falx cere´bri: placed vertically between the two hemispheres of the cerebrum, attached in front to the crista galli, and behind, to the internal occipital protuberance and the tentorium.

The tento´rium cerebel´li is placed horizontally between the cerebrum and the cerebellum. It is attached in front to the anterior and posterior clinoid processes, superior edge of the petrous bone, and behind to the upper margin of the lateral sinus.

The falx cerebel´li reaches vertically from tentorium to the foramen magnum, dividing the two hemispheres of the cerebellum. It is attached posteriorly to the vertical crest of the occiput, and below on either side of the magnum.

THE ARACH´NOID resembles that of the spinal cord (serous membrane), and consists of a *parietal* and *visceral* layer.

The *visceral* layer invests the brain, covering the pia mater. It is thicker at the base, and dips down into the great longitudinal fissure. It stretches across between the two middle lobes, forming the *anterior subarachnoid space*, which is just anterior to the pons. Beneath the cerebellum it forms, in a like manner, by stretching from the cerebellum to the medulla, the *posterior subarachnoid space*.

THE PI´A MA´TER is very vascular, invests the entire brain surface, and dips down between the convolutions, and gives off processes to the interior of the brain. Nervous supply is from the sympathetic, 3d, 6th, 7th, 8th and spinal accessory nerve, which accompany the arterial branches.

BRAIN: The brain is contained in, and nearly corresponds to the cranial cavity. It is divided into four parts:

(1) **Cere´brum:** the highest and largest, being nearly 7/8 of the whole, occupies vault, middle and anterior fossæ of the skull; divided into two hemispheres by the longitudinal fissure down which the falx cerebri dips.

(2) **Cerebel´lum:** contained in the posterior fossa, lying under part of the base of the cerebrum, but separated from it by tentorium cerebelli, divided into two lateral halves by the falx cerebelli.

(3) **Medul´la Oblonga´ta:** extends from the cord to the Pons Varolii, lying just above the foramen magnum.

(4) **Pons Varo´lii:** forms a process to connect all the other three parts together.

Weight: average brain weight in the male 50 oz., in the female 45 oz. At birth the brain is relatively to the weight of the body five or six times heavier than in adults. The weight may be thus distributed: cerebrum 44 oz., cerebellum 5 oz., pons and medulla 1/2 oz. Increases in weight rapidly up to the 7th year, slowly up to the 16th, and reaches the maximum at 40; after this it slowly declines, at the rate of about one ounce with each ten years.

THE MEDUL´LA OBLONGA´TA, or bulb. Extent: from the lower border of the foramen magnum to the lower border of the pons.

Dimensions: 1 1/4 inch long, 3/4 inch wide.

Shape: pyramidal, with base to the pons and apex to spinal cord.

Surfaces: anterior lies in basilar groove; posterior triangular in shape, forms lower half of the 4th ventricle.

Fissures: anterior and posterior median fissures, continuous with those of the cord.

Anterior median fissure: terminates just below the pons in the foramen cæcum, and at the upper part the fibres of one side cross over to the other, forming the decussation of the pyramids.

Posterior median fissure: reaches only half way up, widening out and gradually getting lost in the 4th ventricle.

Eminences: each lateral half of the bulb is sub-

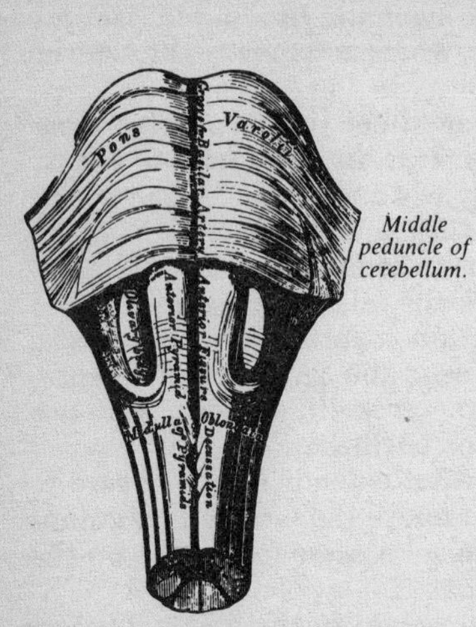

Medulla oblongata and pons Varolii.
Anterior surface.

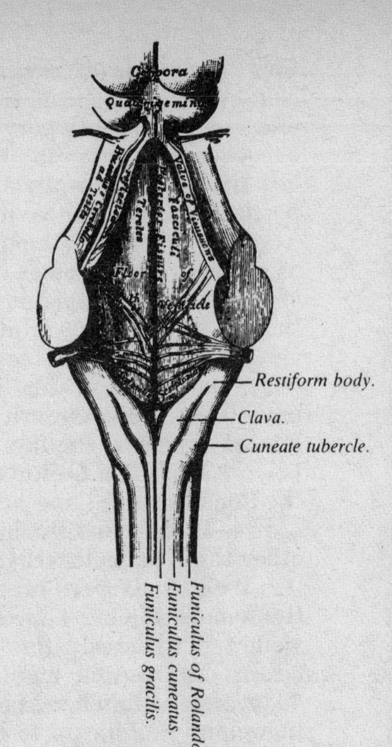

Posterior surface
of the medulla oblongata.

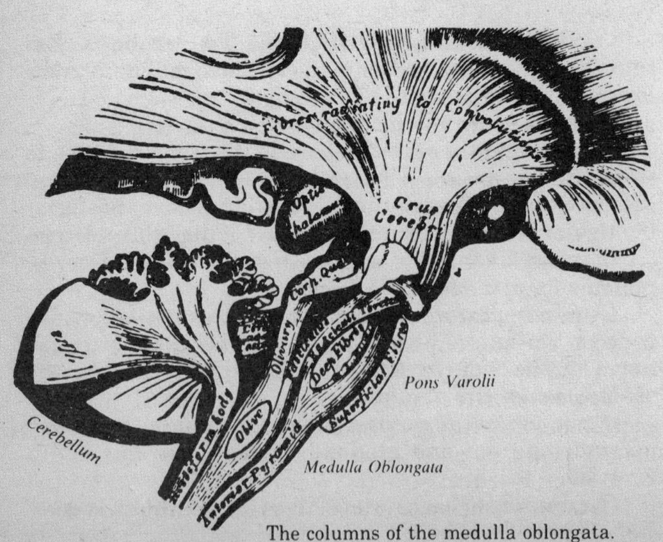

The columns of the medulla oblongata.

divided into four columns, named from within out: *anterior pyramid, olivary body, restiform body* and *posterior pyramid.*

Anterior pyramid: (one on either side) is the continuation of the anterior column of the cord; it is external to the median fissure and internal to the olivary body. It enters the pons, enlarging as it ascends, but before it disappears becomes constricted. The innermost fibres decussate with one another. Contains no gray matter.

Ol´ivary body: an oval body, half inch long, lying between the anterior pyramid and restiform body, but separated on either side by a slight groove, as it is also from the pons above.

Lateral tract: continuous with the lateral column of the cord, lying between olivary and restiform bodies.

Res´tiform body: largest column of the medulla, continuous from the posterior column of the cord, and separated from both the posterior pyramid and lateral tract by a slight groove. It diverges from its fellow in the upper half, forming lateral borders of the 4th ventricle, and enters the cerebellum, receiving the name of *inferior peduncle of the cerebellum.*

Posterior pyramids: bound the apex of 4th ventricle, and then dip down to form floor.

Gray matter is a continuation of that in the interior of the spinal cord.

THE PONS VARO´LII: (See also cut on page 38) is the band of union of the various segments of the encephalon, connecting the cerebrum above, the medulla below, and the cerebellum behind. It is situated *above* the medulla, *below* the crura cerebri, and *between* the hemispheres of the cerebellum.

Dimensions: about 1 inch or 1¼ inch long; measures little more transversely.

Under Surface: is convex, grooved along the centre for the basilar artery; this surface is marked with openings for the entrance of vessels.

Upper Surface: smaller than the under, and continous with the posterior surface of bulb; forms the upper part of the floor of the 4th ventricle.

Upper Border: longer than the inferior, with a notch in median line corresponding to groove on anterior surface. It arches over the cerebral peduncles.

Lower Border: straight, overlays bulb.

Laterally: the pons is continued backwards and outwards, and is continued as the *middle peduncle of the cerebellum.*

Structure: consists of alternate layers of transverse and longitudinal fibres intermixed with gray matter.

THE CERE´BRUM: the largest portion of the encephalon consists of two lateral halves or hemispheres, partly separated by *longitudinal fissure*, which lodges the *falx cerebri*, and which runs from before backwards; in front it entirely divides the hemispheres, but in the middle line they are connected by the *corpus callosum.* The inferior surface is divided into two parts transversely by the *fissure of Sylvius.*

The fissure of Syl´vius separates the anterior and middle lobes at the base of the brain, and as it ascends it divides into a *horizontal* part, which separates the temporal and frontal lobes, and a *vertical*, which loses itself between the convolutions of the frontal lobe. The *sulci* vary from being 1 inch in depth.

Convolutions: on removal of the pia mater the whole surface of each hemisphere presents numerous convoluted eminences separated from each other by *sulci* of various depths. Each convolution is made up of white matter centrally, gray matter outwardly. Are not regular in conformation as regards size and shape in different individuals.

Their number and extent and depth have some general relation to the intellectual power. Those which are largest and most generally present are the *convolution of the corpus callosum*; that of the *longitudinal fissure*; the *supra-orbital convolution*, and those of the outer surface of the hemispheres.

THE BASE.

The following objects are seen from before backwards: anterior or frontal lobe; fissure of Sylvius; middle or parietal lobe; occipital lobe; cerebellar lobe.

In the median line: longitudinal fissure; olfactory bulbs and nerve; corpus callosum; pituitary body; optic nerves; optic commissure; infundibulum; anterior perforated space; optic tract; tuber cinereum; corpora albicantia; post. perforated space; 3d pair nerves; crura cerebri; 4th pair nerves; pons varolii; 5th pair nerves; 6th pair nerves; 7th pair nerves; medulla oblongata; 8th pair nerves; 9th pair nerves; cerebellar vermiform process.

Cru´ra cere´bri: extend from pons varolii to optic

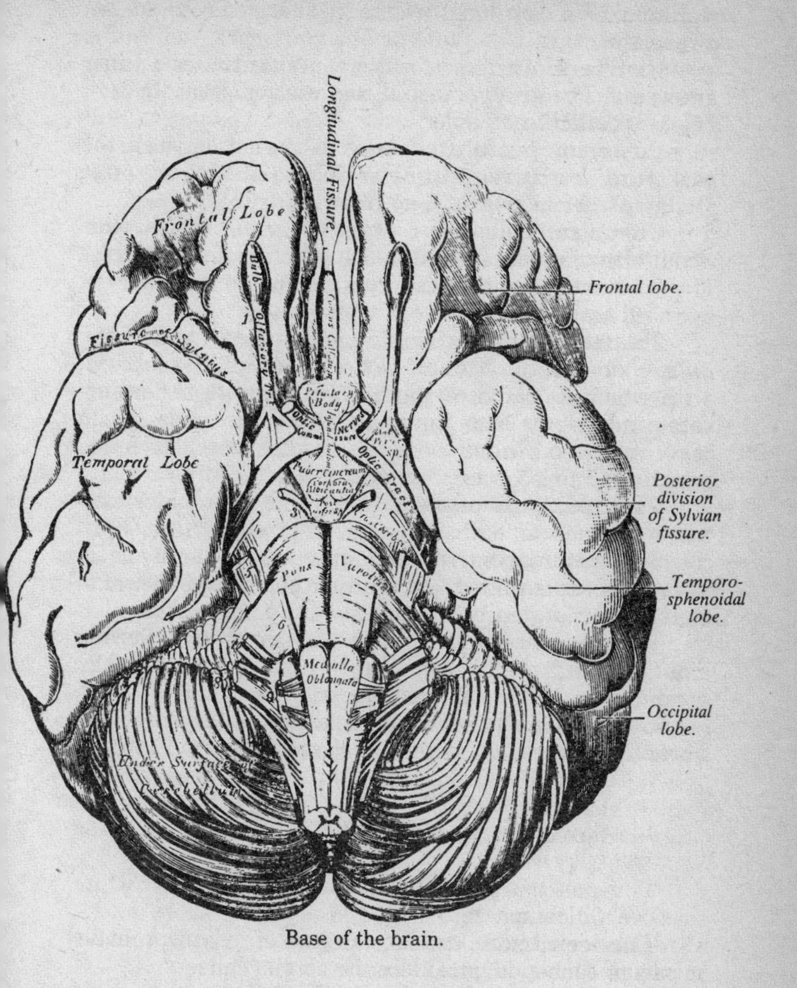

Base of the brain.

thalamus; ¾ inch long, widening at the cerebrum; between the crura is *the interpeduncular space*, containing posterior perforated spot, corpora albicantia, and tuber cinereum. The grey matter of the interior, from its dark color, is called *locus niger*.

Posterior perforated spot: is placed between the two crura; it is perforated by vessels passing to the optic thalamus; forms part of floor of 3d ventricle.

Corpo´ra albican´tia: two small white bodies, each about size of a pea, formed by doubling upon themselves of the anterior crura of the fornix, composed internally of gray substance, externally of white matter.

Tu´ber cine´reum: a grey body placed behind the *optic commissure*, forming part of floor of 3d ventricle, with which its canal communicates; from its under surface a tubular process, the *infundibulum*, extends, which joins it to the *pituitary body*, a reddish vascular mass, weighing 5 to 10 grains, lying in the sella turcica.

The optic commissure is the point of junction of the two optic nerves.

The anterior perforated spot: triangular shape, of grayish color, situated at the inner end of the fissure of Sylvius, perforated by branches of the middle artery.

Lam´ina cine´rea: a layer of gray matter passing from the end of the corpus callosum, above the optic commissure, to the tuber cinereum.

Cor´pus callo´sum: this bends anteriorly very abruptly and forms the genu or rostrum.

THE INTERIOR.

Section should be made transversely with a scalpel, on the level of the corpus callosum.

The cen´trum ova´le mi´nus is the central white mass of a hemisphere.

The cen´trum ova´le ma´jus is centrum ovale minus of each side, joined by the corpus callosum.

The white mass is studded with red dots, *puncta vasculosa*, caused by the severing of bloodvessels in the brain mass; if the brain be inflamed or congested, these will be more numerous and darker colored than in health.

The cor´pus callo´sum is the commissure of the cerebrum, the fibres passing from one hemisphere to the other, forming the roof of the lateral ventricle in each hemisphere. Along the upper surface in the middle line is a raphè, on either side of which are the longitudinal

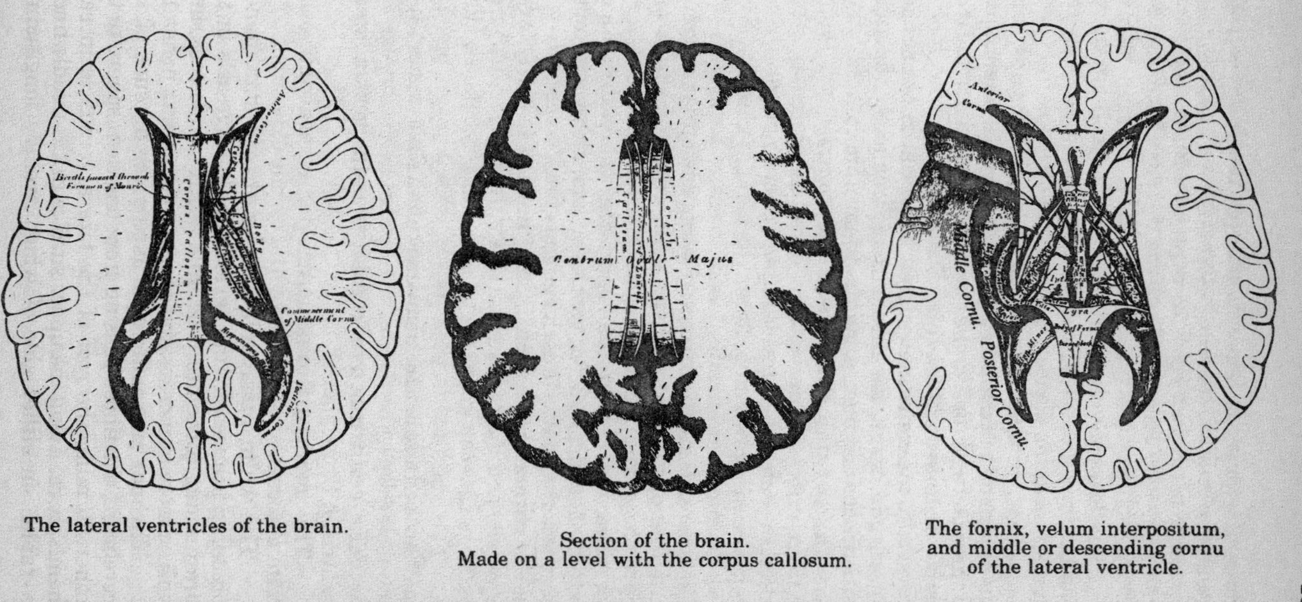

The lateral ventricles of the brain.

Section of the brain.
Made on a level with the corpus callosum.

The fornix, velum interpositum,
and middle or descending cornu
of the lateral ventricle.

fibres, the *nerves of Lanci´si*, external to which are some transverse marks, the *line´æ transvers´æ*.

Make an incision through the corpus callosum, on either side of median line, when the large irregular-shaped cavities, *lateral ventricles*, will be exposed.

The lateral ven´tricles (serous cavities) are two in number, one in each hemisphere, being separated by the *septum lucidum*. Each consists of a body or *central cavity* and *three cornua*, an *anterior*, turning forwards and outwards in the posterior lobe, containing a longitudinal eminence, the *Hippocampus minor*, and a descending one, to be described afterwards. BOUNDARIES.—The *roof* is formed by the corpus callosum, the *floor* from before back by corpus striatum, tænia semicircularis, thalamus opticus, choroid plexus, corpus fimbriatum, fornix.

The cor´pus stria´tum: is the *superior ganglion of the cerebrum*; it is pyriform in shape, with the larger end directed forward. Receives its name from its striped appearance shown on section.

Tænia semicircula´ris: a narrow band of white fibres connecting the corpus striatum and optic thalamus.

The op´tic thal´amus: a white oblong mass resting upon the crus cerebri. BOUNDARIES.—*Externally*. Corpus striatum and tænia semicircularis. *Internally*. Forms lateral boundary of 3d ventricle, and along the upper border is the peduncle of the pineal gland. *Superiorly*. It is partly covered by the fornix, and in front is the anterior tubercle. *Inferiorly*. It projects into the descending cornu and presents the internal and external geniculate bodies.

The cho´roid plex´us, a vascular, fringe-like membrane, formed by a process of the pia mater. It is connected with the one on the opposite side through the *foramen of Monro*, and ascends to middle horn of lateral ventricle.

The cor´pus fimbria´tum is the thin ribbon-like edge of fornix.

The for´nix, triangular in shape, broadest in front, is placed in the middle line beneath the corpus callosum. It divides anteriorly into two crura, which have been seen to form the corpo´ra albica´ntia; posteriorly it joins the hippocampus major. Is made up of white fibrous matter.

The 5th ventricle is situated between the layers of the *septum lucidum*; it is lined by a serous membrane which in the fœtus communicates with the 3d ventricle.

The (middle) descending cornu (the largest)

passes backwards, outwards, and downwards, and curving round the crus cerebri, goes forwards and inwards; the floor is formed by the following:

The hippocam´pus ma´jor: the continuation of the fornix is of white substance of curved, elongated form extending the length of the floor of the middle horn of the lateral ventricle; it has an enlarged anterior extremity, *pes hippocam´pi.*

Tæ´nia hippocam´pi: the continuation of the tænia semicircularis, under which is the *fas´cia denta´ta.*

The cho´roid ple´xus: continuous with that of the lateral ventricle.

The pes accesso´rius or *eminen´tia collatera´lis*: a projection between the hippocampus major and the minor just at the beginning of the descending cornu.

The transverse fissure is opposite the interval between the cerebrum and cerebellum, and, through the pia mater, passes to the interior of the brain.

Sep´tum Lu´cidum forms the internal boundary of the lateral ventricle. Is a thin transparent membrane of triangular shape, and consists of two laminæ, between which is the 5th ventricle.

The ve´lum interpos´itum is a triangular process of pia mater which passes into the brain by the transverse fissure. In the centre of it are the two *venæ Galeni*, and on each side the *choroid plexus*. Is a part of upper boundary of 3d ventricle.

The 3d ventricle is the narrow, oblong fissure between the optic thalami, extending to the base of the brain.

BOUNDARIES.—*Roof*, formed by fornix and velum interpositum. *Floor*, by structures at base of brain within the circle of Willis. *Anteriorly*, is the anterior commissure, connecting the corpora striata. *Posteriorly*, the posterior commissure, connecting the optic thalami. The middle commissure, also connecting the optic thalami, passes across the ventricle.

COMMISSURES: its cavity is crossed by three commissures, viz: *the Anterior*, a rounded white cord in front of the ant. crura of the fornix; *the Middle*, is soft and composed of gray matter, and connects the optic thalami; *the Posterior*, a flattened white band, connecting optic thalami posteriorly.

OPENINGS: the 3d Ventricle has 4 openings, viz: the *two apertures* of the foramina of Monro, in front; *a third* into 4th ventricle by aqueduct of Sylvius; *a fourth*, deep pit in front leading down to the infundibulum. Gray matter covers most of the surface of this ventricle.

Fora´men of Mun´ro is a Y shaped passage from the lateral ventricles down to the 3d ventricle.

The pin´eal gland is a conical, reddish-gray, vascular body placed between and upon the nates. Its base is connected with the optic thalami by two anterior peduncles, and to the posterior commissure by small inferior peduncles. Is 4 *lines* long and 2 to 3 wide, and contains a transparent viscid fluid.

The corpo´ra quadrigem´ina are four small bodies composed of white matter outwardly, gray within, which are placed in pairs behind the 3d ventricle, *the anterior* pair (the larger) being called the *nates* and the *posterior* the *testes*. There are two bands passing from the cerebellum to the testes, sometimes called *brachia*, or *proces´sus a cerebel´lo ad tes´tes*, and between these is the *valve of Vieussens*. The corpora quadrigemina receive from below the *fillet of the olivary body*.

Valve of Vieus´sens, a thin translucent membrane of medullary substance stretched between the processus a cerebello ad testes, forming part of roof of 4th ventricle. It gives origin to the 4th nerve on either side.

Corpo´ra genic´ulata are two small, flattened, oblong masses on the under side and back part of each optic thalamus at the outer side of the corpora quadrigemina. One lies to the outside of the optic thalamus; the other to the inside of the same body.

THE CEREBEL´LUM is contained in the occipital fossa, and is separated from the cerebrum by the tentorium. The surface is divided into laminæ, which are separated by sulci.

Weighs, in the male, on the average, 5 ozs. and 4 drs.; a little lighter in the female. The proportion between it and the cerebrum is, in the male, as 1 to 8 4/7; in the female, as 1 to 8¼; in the infant, as 1 to 20.

Form is oblong, flattened from above downwards, its greatest diameter being from side to side.

Size: 2 to 2½ inches long; 3¼ to 4 inches wide; 2 inches thick at the centre and at the circumference only 6 lines in thickness.

Substance: consists mainly of gray matter (darker than that of the cerebrum) on the outside, and white matter within.

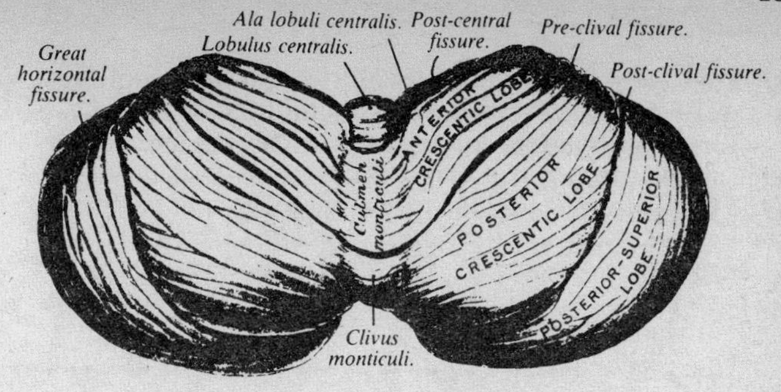

Upper surface of the cerebellum.

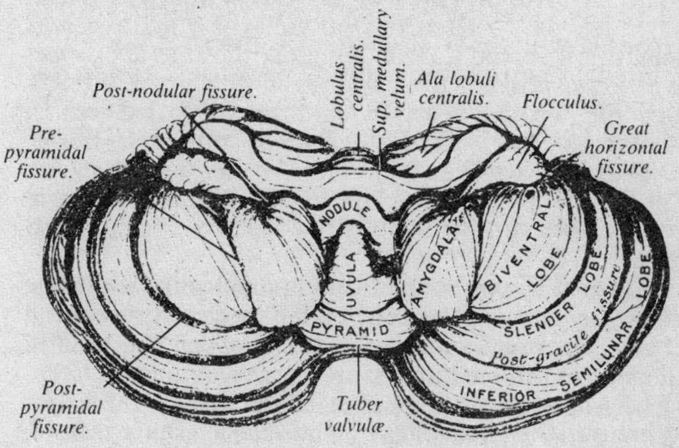

Under surface of the cerebellum.

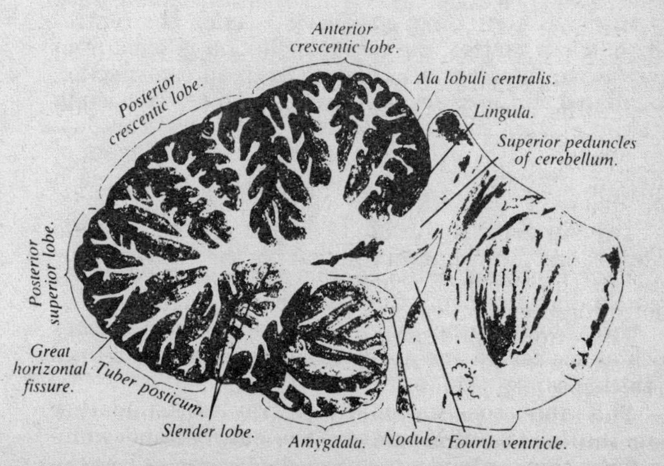

Sagittal section of the cerebellum,
near the point of junction of the worm with the hemisphere.

The cerebellum is not convoluted like cerebrum, but it is quite deeply marked with circulo-transverse striæ.

The upper surface presents in the median line a ridge called the *superior vermiform process*, and each half of the cerebellum is divided into an anterior and posterior lobe by the fissures, the *incisura cerebelli anterior* and the *incisura cerebelli posterior. The anterior lobe* reaches from the posterior part of the vermiform process forwards. The *posterior* lobe is the remaining part.

The cerebellum is connected to the cerebrum and cord by three peduncles:

The inferior surface of the cerebellum is divided into two lobes by a fissure, "the valley" (which receives the medulla), on the floor of which is the *inferior vermiform process*, which is divided as follows: most anteriorly is the *uvula*, with a projection forwards of it called *nodule*, and behind is the *pyramid* and a few transverse fibres. *Post. medullary velum* of white matter lies, one on either side of the nodule. *Nidus hirundinis* (swallow's nest) is a deep fossa between the nodule and post. med. velum and nodule.

Lobes: the under surface is divided into *five* lobes, viz: 1. *Flocculus*, a prominent tuft of gray matter behind and below the middle peduncle. 2. The *amygdala* (tonsil), on the side of the median fissure, by the uvula. 3. *Digastric* lobe, situated to the outside of the tonsil. 4. *Slender lobe*, behind the digastric, by the side of the pyramid. 5. *The Inferior Posterior lobe*, joining the *commissura brevis* in the valley.

Internal structure: on vertical section the central body of white matter, the *corpus dentatum* is seen, from which arise the 10 or 12 *Laminæ* (seen on cross-section are foliated and so called *arbor vitæ*), each one consisting of white matter covered with gray substance; this last also appears to be in two imperfectly defined layers of different consistence. From the anterior part of each hemisphere arise the three following peduncles:

The superior peduncle is the *processus a cerebello ad testes*, and forms lateral boundary of the 4th ventricle; between the two peduncles is a layer of white fibres connecting them, called the valve of Vieussens.

The middle peduncle (largest of the three) forms the transverse fibres of the pons varolii, and connects the 2 cerebellar hemispheres.

The inferior peduncle connects the cerebellum and the medulla and forms part of the restiform bodies.

The 4th ventricle (or cerebellar ventricle) is the

space between the posterior surface of the medulla oblongata and pons, in front, which forms its floor, and the cerebellum behind (see figure page 128). It is a lozenge-shaped cavity, being broadest at the middle.

BOUNDARIES.—*Floor* as above. *Roof* by valve of Vieussens and inferior vermiform process. Laterally by the superior peduncles. *Below*. Restiform body. The cavity communicates with the 3d ventricle by the aqueduct of Sylvius.

The ventricle is closed below by a reflection of pia mater, which joins the *choroid plexus* of the 4th ventricle. In the floor is a median groove continuous with the central canal of the cord, and on each side of the groove is a small eminence, the *fasculus teres*. The lower part is bounded by the ends of the posterior pyramids and is termed the *calamus scriptorius*.

The *locus cæru'leus* is a small bluish-gray eminence opposite the crus cerebelli. The lower part of the floor of this ventricle is crossed transversely by various lines, the *lineæ transversæ*.

The lining membrane of this ventricle is continued up into that of the 3d ventricle through the aqueduct of Sylvius.

The *choroid plexuses* of this ventricle are two in number—slight vascular fringes on either side passing to the outer margins of the restiform bodies.

THE ORGANS OF DIGESTION.

THE TONGUE.

The tongue, consisting of two symmetrical halves, occupies the floor of the mouth; posteriorly it is connected with the hyoid bone, the epiglottis, the soft palate, and the pharynx; inferiorly it is attached to the lower jaw by the genio-glossi muscles; is thicker behind than in front, and sometimes contains a small fibro-cartilage.

The mucous membrane: on the under surface is smooth, forming a median fold, the *frænum linguæ*; on the sides it is continuous with the mucous membrane of the mouth. On the dorsum there is a *raphé* along the middle line, which ends posteriorly in the *foramen cæcum*. Posteriorly the epiglottis is connected to the tongue by three glossi-epiglottic folds. The anterior two-thirds of the dorsum of the tongue is covered with papillæ; they are of three kinds:

The *circumvallate* or *papillæ maximæ* (seven to ten),

are of large size, and form a row on each side at the back of the tongue, meeting in the middle line, thus, Λ

The *fungiform papillæ*; found principally at the apex and on the sides, of large size, and of deep red color.

The *filiform* or *papillæ minimæ* are numerous, and are arranged in rows parallel to the circumvallate, but towards the tip of the tongue their direction becomes more transverse.

Taste buds: supposed to be the organs of taste, are flask-shaped bodies found buried in the epithelium around the circumvallate papillæ.

Mucous glands, are found chiefly below the membrane on the posterior dorsum at its third; their ducts either open on the surface, or into depressions about the large papillæ.

Lymphoid tissue is mostly at the back of the tongue, though collected into numerous masses known as *follicles*.

Epithelium: this is of the scaly variety, though thinner than in skin, and runs down into the large papillæ.

Muscles: see page 26.

Arteries: branches from the lingual, facial and pharyngeal, (see page 33).

Nerves: 3 in number in each half—the gustatory, a branch from the 5th; the lingual, branch of glossipharyngeal; and the hypoglossal (see pages 45 and 26). The first two are for common sensation and taste; the last for mobility.

THE PAL'ATE.

The palate forms the roof of the mouth, and consists of two parts, *hard* and *soft*.

The hard palate consists of the palatal processes of the superior maxilla together with mucous membrane and periosteum lining them.

Along the middle line is a ridge or *raphé*, terminating anteriorly in a papilla which receives filaments of the naso-palatine and anterior palatine nerves; on either side the mucous membrane is corrugated and pale, while behind it is smoother and darker; is covered with squamous epithelium, which is continuous with that covering the soft palate.

The soft palate, consisting of muscles, aponeurosis, vessels, nerves, etc., enclosed in a layer of mucous membrane, is attached in front to the posterior margin of the hard palate, and the sides blend with the pharynx; its anterior surface is concave, the posterior convex. From

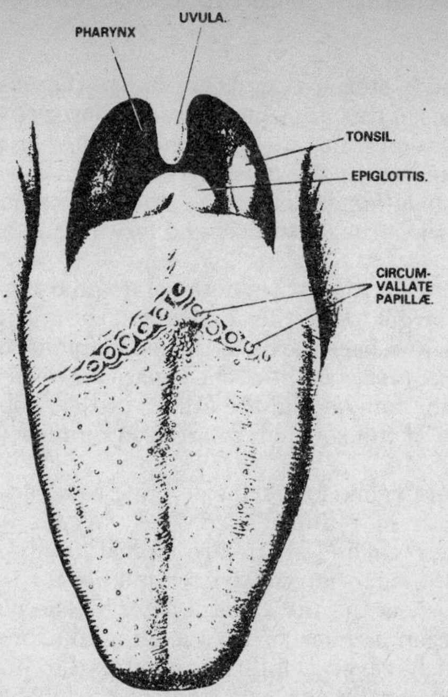

Upper surface of the tongue.

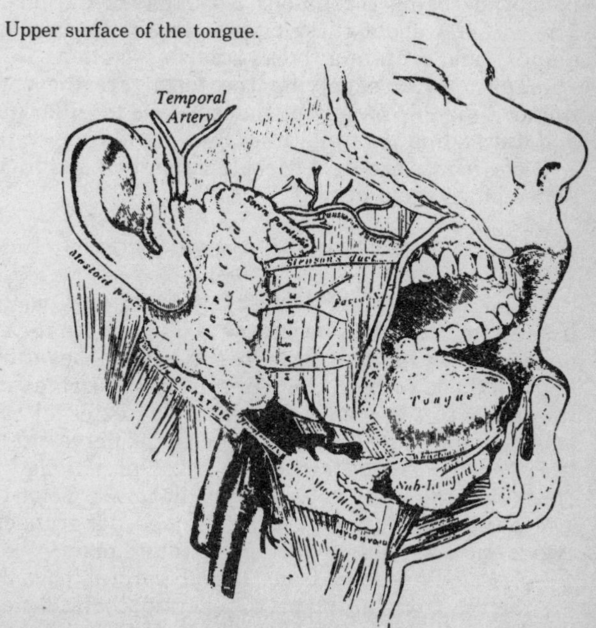

The salivary glands.

the middle of the posterior edge there hangs the *uvula*, and from the basis of this arch, on each side there are two folds of mucous membrane, etc., the *pillars*, anterior and posterior, between which the *tonsil* lies.

The anterior pillar is formed of the plato-glossus muscle, runs downwards, outwards and forwards to base of tongue.

The posterior pillar is formed of the plato-pharyngeus muscle, and is larger than the anterior; it runs downwards, outwards and backwards to the side of pharynx.

Isthmus of the fauces is the arched space between the soft palate and the tongue and the pillars just described.

The muscles of the soft palate are 5 in number (see page 28).

Tonsils: there is one for each side lying between the palate pillars; are of rounded form and variable size, though from their almond shape are named *amygdalæ*. They lie upon the superior constrictor muscle for their bed (beyond which is the int. carotid artery). The inner surface has a dozen or more orifices leading to as many crypts into which several follicles empty, the lining membrane being continuous with that of the pharynx. The *capsules* about these crypts are similar to "Peyer's glands," and contain a thick, grayish secretion.

The arteries supplying the tonsil are the dorsalis linguæ (lingual), ascending palatine and tonsillar (facial) and descending palatine (from int. maxillary).

The nerves are from Meckle's ganglion and from the glosso-pharyngeal nerve.

THE THREE SAL´IVARY GLANDS.

The paro´tid (see page 141) is the largest (weighing from ½ to 2 ozs.), and lies below and in front of the external ear, being limited above by the zygoma, below by the angle of the lower jaw, and a line drawn horizontally from this point to the mastoid process. The external carotid artery is imbedded in it, and the facial nerve crosses it transversely. The *duct* (Steno's) is 2½ in. in length, and opens into the mouth opposite the upper 2d molar tooth. The *Socia parotidis* is a separate lobe lying under the zygomatic arch, whose duct opens into Steno's.

Arteries: branches from the ext. carotid (page 30).

Nerves: from carotid plexus of sympathetic, facial and brs. from auriculo temporal and gt. auricular (page 49).

The submax´illary, weighing about 2 drachms, is placed under the lower jaw, lying upon the mylo-hyoid, stylo-hyoid, and hyoglossus muscles, and separated from the parotid by the stylo-maxillary ligament. The facial artery is imbedded in a groove on the posterior surface. The *duct* (Wharton's) 2 inches long, opens at the summit of a papilla by the side of the frænum linguæ.

Arteries: brs. from facial and lingual (page 30).

Nerves: brs. from submaxillary ganglion, sympathetic and mylo-hyoid br. of inf. dental.

The sublin´gual, the smallest, weighing about ½ drachm, is placed under the mucous membrane at the floor of the mouth. It is almond shaped, and its *ducts* (duct Raviani) (18 to 20) open separately on the floor of the mouth; generally one or two joined together (Bartholini's duct) go to join Wharton's duct.

Arteries: brs. from sublingual and submental.

Nerves: brs. from gustatory (see page 44).

Structure: the structure of all these glands is of the compound racemose order, joined together by the dense areolar tissue ducts and vessels.

Secretion: alkaline in reaction, watery, and contains, especially, ptyalin. It acts upon starch, changing it into dextrine and grape sugar.

Mucous glands also abound in the mouth, and are of the ordinary compound racemose type.

THE PHAR´YNX.

The pharynx is situated behind the nose, the mouth, and the larynx, and extends from the bases of the skull to the cricoid cartilage in front, and the 5th cervical vertebra behind. It is about 4½ inches in length and broader transversely than from before backwards, being broadest at line of hyoid bone, and narrowest at the œsophageal juncture.

Openings, 7: the *posterior nares* (2), placed in the upper part of the anterior wall. *Eustachian tubes* (2), open one on each side at the upper part. The *mouth*, situated just below the posterior nares. The *laryngeal* and *œsophageal*.

Structure: 3 coats; 1. *Mucous* continuous with that of nares, mouth and larynx. Squamous epithelium covers it to level of the floor of the nares, where the columnar ciliated variety begins and finishes covering its surface.

Racemose mucous glands are found throughout the extent, most numerous in its upper portion. There are also numerous crypts, or recesses, similar to those of the tonsils in their lymphoid character; this is especially so at the upper portion between the two Eustachian tubes.

2. *The fibrous coat*, between the mucous and muscular, called *pharyngeal aponeurosis*; it is thickest above, gradually diminishing as you descend to the œsophagus.

3. *Muscular*. See page 26.

THE ŒSOPH´AGUS.

The œsophagus extends from pharynx to stomach, and is 9 inches long; is slightly curved from before backwards and to the left side; it begins at the lower border of cricoid cart. and passes through the posterior mediastinum and the diaphragm to the cardiac orifice of the stomach opposite the 9th dorsal vertebra.

Relations in the neck.—*In front.* The trachea, thyroid gland, and thoracic duct. *Behind.* Vertebral column and longus colli. *Laterally.* Common carotid artery, the thyroid gland, recurrent laryngeal nerves.

In the thorax.—*In front.* Trachea, arch of aorta, left carotid and left subclavian arteries, left bronchus, pericardium, left pneumogastric. *Behind.* Vertebræ, longus colli, intercostal vessels, aorta, right pneumogastric. *Laterally.* Pleuræ, venæ azygos major on the right, and descending aorta.

Structure: has 3 coats; 1. *Mucous*, a thick, reddish coat above, paler below, and disposed more or less into small folds, with the surface studded with minute papillæ; the whole covered with squamous epithelium. 2. *Cellular* coat, which loosely connects the mucous with the muscular. 3. *The Muscular coat*, made up of two planes of fibres, the internal or Circular, and external or Longitudinal. Above, these are of striped voluntary fibres; below of involuntary fibres. The Circular fibres are continuous with the inf. constrictor. The Longitudinal fibres inosculate with each other and with the inf. constrictor of the pharynx.

Glands: throughout the tube are numerous small racemose glands, being more numerous near the cardiac end of the tube; they have quite long excretory ducts.

THE STOM´ACH.

This is the principal organ of digestion, and the most dilated portion of the alimentary tract. The LEFT EXTREMITY extending 2 or 3 inches to the left of the

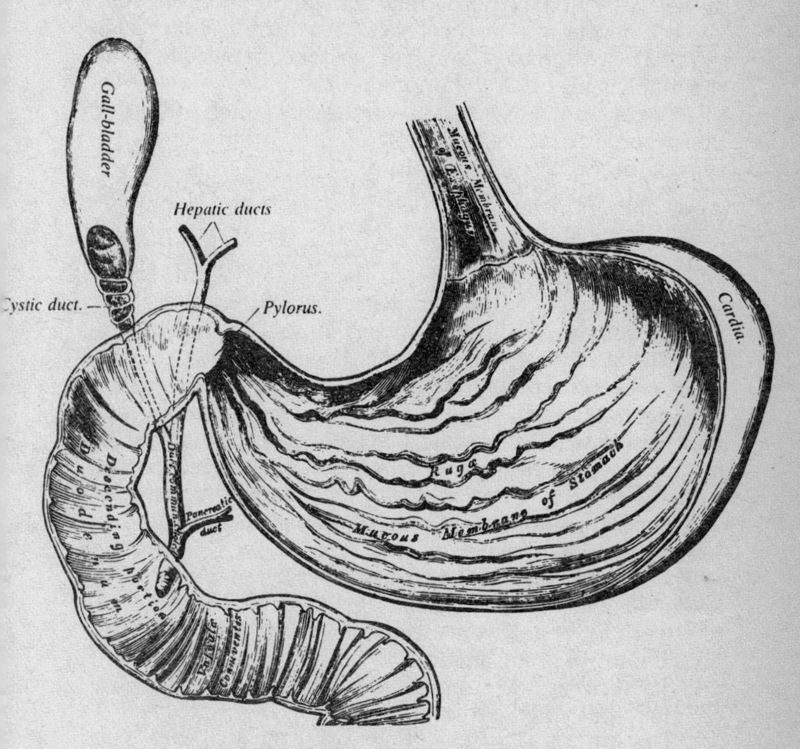

The mucous membrane of the stomach and duodenum with the bile-ducts.

œsophagus, is also called the *greater* or *cardiac* or *splenic end*. The RIGHT EXTREMITY, much the smaller, lying in contact with abdominal wall, and the under surface of the liver, corresponding with the 8th rib-cartilage, is called the *lesser* or *pyloric* end. The stomach *is held in position* by the lesser omentum (which extends from the transverse liver fissure to the lesser curvature) and by the *gastophrenic ligament*, a fold of peritoneum passing from the diaphragm on to the œsophagus.

Form: conical, with base or fundus to left side; the upper border concave, and called *lesser* curvature: lower, convex, named *greater* curvature.

Position: occupies left hypochondriac and epigastric regions.

Orifices: *Cardiac*, at the left end communicating with the œsophagus; *pyloric*, at the right extremity passing into the duodenum, guarded by pyloric *valve* or fold.

Dimensions: 10 to 12 inches long; 4 to 5 inches in diameter at widest part. *Weight*, 4½ ounces.

Connections: *Left* or *cardiac end*: fixed by œsophagus to diaphragm, lying beneath the ribs and connected with the spleen by the gastro-splenic omentum. *Right* or *pyloric end*: reaches gall-bladder, touching under part of left lobe of liver. *Anterior surface* is in contact with, from left to right, diaphragm, abdominal parietes (epigastric region), under surface of left lobe of liver. *Posterior surface* is connected with pancreas, crura of diaphragm, aorta, vena cava, solar plexus. *Superior border*: attached to liver by small omentum. *Inferior border*: gives attachment to great omentum.

Structure: The stomach has 4 coats, viz: a mucous, cellular, muscular and serous.

THE MUCOUS COAT, is thick, its surface velvety, and it is thrown, when the organ is not disturbed, into numerous *rugæ*, as shown in the cut. It is of a reddish tinge in youth, a straw or ash-gray tinge in old age. It is thinnest at the cardiac portion. Microscopically this coat is seen studded with alveoli, containing at the bottom the *gastric follicles* or the *mucous* and *peptic glands*. The alveoli are of hexagonal shape and vary from 1/100 to 1/350 of an inch in diameter. The *mucous glands* have quite long ducts; are lined with a delicate basement membrane covered with flattened epithelium. The *peptic glands* have shorter ducts, and are lined with columnar epithelium. Between this epithelium and the basement membrane are numerous spheroidal granular cells, called *Peptic Cells.*

CELLULAR COAT, sometimes called *sub-mucous,* is a loose, thin layer of areolar tissue connecting the mucous with the muscular coat; it furnishes a support to the bloodvessels as they ramify to the mucous layer.

THE MUSCULAR COAT has 3 sets of fibres; *The Longitudinal* (most external) running lengthwise of the stomach, and most prominent at the greater and lesser curvatures. *The Circular* fibres run *around* the stomach beneath the longitudinal ones. At the pylorus they are most abundant, and form a sort of a ring which projects into the cavity; this "ring," covered with mucous membrane, is the so-called "pyloric valve." *The Oblique fibres* are mainly found at the cardiac portion, partially surrounding this part.

THE SEROUS COAT is derived from the peritoneum and covers the entire organ, except at the lesser and greater curvatures, at the points of attachment of the 2 omenta; the layers here separate into a triangular space for the passage of the bloodvessels and nerves.

Arteries: (see page 87). *Coronary* and *superior pyloric* run along lesser curvature. *Right* and *left gastro-epiploic* along inferior or greater curvature. *Vasa brevia* from the splenic to the fundus.

Nerves: (see pages 45, 46). *Right pneumogastric* to posterior surface. *Left pneumogastric* to anterior surface.

THE SMALL INTESTINE.

The Small Intestine is a convoluted tube about 20 feet long, in which the *chyme* is mixed with the pancreatic juice, bile and other secretions of the intestinal canal, and where the nutrient portion of food, *chyle,* is formed.

It is divided into three portions: the *Duode´num, Jeju´num* and *Il´eum.*

The following will serve to distinguish the three parts of the small intestine:

Duodenum.	*Jejunum.*	*Ileum.*
The largest part.	*More vascular than ileum*	*Villi, small.*
Thickest coats.	*Valvulæ conniventes.*	*Valvulæ conniventes not present, or only slightly.*
Brunner's glands.	*Villi, well marked.*	*Peyer's patches.*
Valvulæ conniventes.		
No mesentery.		

THE DUODE´NUM, so called because the equal in length to the breadth of 12 fingers. It is the shortest, widest and most fixed part of the small intestine. Has no mesentery, and only partially covered, in front, with peritoneum.

Length: 8 to 10 inches.

Shape: Horse shoe, with the convexity to the right side, the concavity enclosing the head of the pancreas.

Position: Occupies right hypochrondriac and epigastric regions.

Divided into three parts: ascending, descending and transverse.

Ascending part: freely movable; 2 inches long; directed from pylorus to gall-bladder. *In front.* Liver, gall-bladder. *Behind.* Bile duct, vena porta, hepatic artery.

Descending part: firmly fixed; 3 inches long, passes from gall-bladder down to 3d lumbar vertebra, in front of right kidney. Ducts of liver and pancreas enter this part. *In front.* Hepatic flexure of colon. *Behind.* Right kidney. *Inner side.* Head of pancreas, common bile duct.

Transverse part: longest and narrowest part, is about 5 inches long; passes across spine, ascending from 3d to 2d lumbar vertebra, ends in jejunum on left side of spinal column, lying between layers of transverse meso-colon. *In front.* Superior mesenteric vessels and plexuses of nerves. *Behind.* Aorta, vena cava, crura of diaphragm. *Above.* Pancreas.

Arteries: pyloric and pancreatico-duodenal of hepatic, inferior pancreatico-duodenal of superior mesenteric (page 87).

Nerves from solar plexus.

Veins terminate in the splenic and superior mesenteric.

THE JEJU´NUM, so-called from *jejunus*, meaning "empty," because usually found empty after death; it occupies 2/5 of the rest of the small intestine, or 8 feet; commencing on the left side of the 2d lumbar vertebra, it terminates in the ileum; it is wider, coats thicker, more vascular and of a deeper color than the ileum.

THE IL´EUM consists of the remaining 3/5 of the small intestine, or about 11 feet, and terminates in the right iliac fossa by opening into the cæcum.

It occupies chiefly the umbilical, hypogastric and iliac regions. It is narrower, and its coats thinner and lighter than the jejunum.

Structure: the small intestine has 4 coats; the *Mucous, Cellular, Muscular* and *Serous.*

THE MUCOUS COAT is thick, highly vascular at upper portion, less so below; on the free surface are columnar epithelial cells, granular, and with oval nucleus. Beneath this is layer of retiform tissue, in which are the ramifications of the bloodvessels and nerves, and numerous lymph corpuscles. Still further beneath is a layer of unstriped muscular fibre, *the muscula′res muco′sæ.*

The Val′vulæ Conniven′tes are permanent reduplications or foldings of the mucous and submucous tissues; they extend transversely around the intestinal cylinder for from ½ to ⅔ of its circumference; they alternate in small and large folds, commencing about 2 inches from the pylorus; are more prominent in the upper portion of the intestine, varying from 2 inches in length and ⅓ of an inch in depth to slight transverse wrinkles. Their office is to retard the downward passage of food, and so favor digestion and absorption.

The Vil′li are minute, vascular processes projecting from this coat, and give it a velvety appearance. Most frequent in the upper portion of the intestine and are of various shapes, cylindrical, conical, triangular, etc. The essential structure is the *lacteal vessel* (single or multiple) in the center, terminating in a blind extremity at the summit; *the muscular fibres*, from the musculares mucosæ, surround the lacteals; *the bloodvessels* form a cone of inosculating branches about the lacteal tufts and muscle fibres; *the basement membrane*, formed of a stratum of endothelial cells, covers the whole and over the whole surface of this membrane is a layer of columnar epithelium.

The Simple Follicles (crypts of Lieberkühn) are found pretty generally over the mucous surface. They are small tubular depressions in the mucous membrane, with thin walls lined with columnar epithelium; outwardly they are encrusted with fine capillaries. Their contents vary, and their purpose is unknown.

The Duode′nal Glands (Brunner's), limited to the Duodenum and upper portion of the jejunum, and in structure resembling the pancreas, are small, flattened, granular bodies, with minute excretory ducts.

The Solitary Glands are most numerous at lower portion of ileum, are found throughout this tract; are small (½ to 1 line in diameter) round, whitish bodies, surrounded by openings of the simple follicles, their free surfaces being covered with villi. They are supposed to be lymphoid follicles, being packed with lymph corpuscles.

Peyer's Glands are aggregated patches of these

solitary glands that vary in length from ½ to 4 inches, and from 20 to 30 in number in this tract, being largest and most numerous in the ileum. They run lengthwise with the intestine, and are placed to the opposite side of the mesenteric attachment; are covered with mucous membrane, which is highly vascular about them; are largest during digestion.

CELLULAR COAT: this connects the mucous with the muscular coat; it is loose areolar tissue wherein the vessels and nerves ramify.

THE MUSCULAR COAT consists of two layers of fibres; the *circular*, or internal layer, is thick and uniform, but the fibres do not entirely surround the intestinal cylinder; the *longitudinal* fibres are thinly scattered over the intestinal surface, and are more external.

SEROUS COAT is derived from the peritoneum, which almost surrounds the upper duodenal portion, but only partially so the lower portion; the jejunum and ileum are surrounded by this membrane, except at the mesenteric border, where there is a free passage left for the bloodvessels and nerves.

Arteries and veins: see pages 89 and 95.

THE LARGE INTESTINE.

THE LARGE INTESTINE, or Colon: *Extent.* From the ileum to the anus, or about 5 feet.

Characteristics. Larger sized and more fixed than the small intestine, and sacculated. It commences in right iliac fossa (see cuts, page 88) in a dilated part (cæcum) ascends through right lumbar and hypochrondriac region to liver, then transversely across to left hypochondriac region, and descends to left iliac fossa, where it becomes convoluted (*sigmoid flexure*) then enters the pelvis where it descends to the anus (*rectum*).

The cæ´cum (*cæcus*, blind) is a dilated pouch (measuring 2½ inches in diameter) in which the large intestine commences, situated in the right iliac fossa and well bound down by peritoneum; at the lower end and back part is the *appen´dix vermifor´mis*, a blind tubular projection, from 3 to 6 inches in length.

The *Ileo-cæcal valve* (*Val´vula Bauhini*) is formed by the ileum passing through the wall of the cæcum. The upper fold is horizontal and called the ileo-colic. The lower is vertical and termed the ileo-cæcal. The ridge on either side is called the *frænum*. Each valvular segment is a reduplication of the mucous membrane and circular muscle-fibres of the intestine, and just below is the opening into the vermiform appendix.

The Co´lon is divided into *ascending, transverse, descending* and *sigmoid flexure*.

THE ASCENDING portion (smaller than cæcum) extends from the cæcum to the under surface of the liver, to the right of the gall-bladder, where it turns to the left, forming the *hepatic flexure*. The peritoneum covers the anterior and lateral surfaces.

Relations: *In front*. The convolutions of the ileum. *Behind*. Quadratus lumborum, right kidney.

THE TRANSVERSE portion, the longest part of the large intestine, passes transversely from right to left from the gall-bladder to the spleen. It forms an arch, convex anteriorly, *the transverse arch of colon*. It is surrounded by peritoneum, which is attached to the spine by the meso-colon. It is the most movable part of the colon.

Relations: *Above*. Liver, gall-bladder, stomach, lower end of spleen. *Below*. small intestines. *Anteriorly*. Anterior layers of great omentum, parietes. *Posteriorly*. Transverse meso-colon.

THE DESCENDING portion passes vertically downwards from the spleen to the left iliac fossa, ending in the sigmoid flexure. The peritoneum invests its anterior and lateral surfaces. It is smaller and more deeply placed than the ascending portion.

Relations: *Behind*. Left crus, left kidney, quadratus lumborum.

THE SIG´MOID FLEXURE, the narrowest part of the colon, is placed in the left iliac fossa; it commences at the margin of the crista ilii, curves like an **S**, and ends in the rectum, opposite the left sacro-iliac articulation. It is retained in place by the *Sigmoid meso-colon*, and has the small intestines in front.

The rec´tum, the terminal part of large intestine, extends from the sigmoid flexure to the anus. It is not sacculated, like the rest of the large intestine, and varies from 6 to 8 inches in length. It commences opposite to the left sacro-iliac junction, passes in a gentle curve obliquely down to the right to the middle of the sacrum, then descends in a curve to the coccyx, the convexity of this curve looking backwards; from this point it curves backwards, for a short distance, to the anus, the convexity of this curve looking forward. The rectum (from *rectus*, meaning straight) is anything but a straight tube. It is cylindrical, non-sacculated, and capable of wide dilation. It is divided into *three* parts:

Part I extends from the left sacro-iliac articulation to the middle of the 3d piece of the sacrum. *Relations.* Completely surrounded by peritoneum and attached to the sacrum by meso-rectum. *Behind.* Pyriformis, sacral plexus, branches of left internal iliac artery. *In front.* Posterior surface of the bladder (male), posterior surface of uterus (female).

Part II extends from the ending of the 1st part to the tip of the coccyx. *Relations.* It has peritoneum on the upper part of anterior surface only. *In front.* Triangular part at base of bladder, vesiculæ seminales, vasa deferentia, under surface of prostate (male), posterior wall of vagina (female). *Laterally.* coccygæus.

Part III extends from the tip of the coccyx to anus.

Relation: has no peritoneum.

In front: fore part of prostate, membranous part of the urethra, bulb of corpus spongiosum; in the female, vagina.

Laterally and behind: levatores ani.

Structure: the large intestine has 4 coats; *mucous, cellular, muscular* and *serous.*

THE MUCOUS MEMBRANE in the Cæcum and Colon is smooth, lacking villi, raised into numerous folds, and of a pale or grayish color; in the Rectum it is more vascular, of darker color, and at the lower part is thrown into numerous longitudinal folds, simulating pouches somewhat, though are effaced when the organ is distended. Besides, there are 3 or 4 prominent, permanent folds of semi-lunar shape, arranged in a valve-like manner, their office being to assist in holding supported the rectal contents. The description of the formation of the mucous membrane of the large intestine is exactly similar to that of the small intestine.

Simple follicles are longer and more numerous than those in the small intestine; are tubular prolongations downward of the mucous membrane; have minute rounded orifices at the surface of the membrane.

The Solitary glands are most abundant in the cæcum and vermiform (worm-like) appendix, but are scattered over the whole mucous surface. Are similar to those found in the small intestine.

THE CELLULAR COAT connects the mucous with the muscular coat.

THE MUSCULAR COAT consists of two layers of fibres: the first, or *circular* fibres, lying just beneath the mucous coat; and the second, or *longitudinal fibres, lying more externally.*

The circular fibres are thickly placed in the rectum, forming the *Internal Sphincter*; are more thinly placed in the colon and cæcum.

The longitudinal fibres in the cæcum and colon are collected into three flat bands, about ½ inch in width, surrounding, in part, the tube; being shorter than the other structures they contract the tube into numerous *sacculi.* In the sigmoid flexure they become more scattered, and about the rectum spread out in a uniform layer of some thickness.

The arteries and veins: see pages 88 and 94.

The nerves are from the several plexuses of the *Sympathetic* system which surround the mesenteric arteries and rectum. They penetrate to the muscular layer, between the circular and longitudinal fibres, they inosculate with others, and ganglia, forming *Auerbach's plexus*; from this one a secondary plexus is formed (*Meissner's plexus*) by branches perforating the circular fibres, and spreading out beneath the mucous surface and inosculating freely and with other ganglia.

THE LIVER.

The Liver is a large glandular organ, whose main function is the secretion of bile.

Situation: right hypochondriac and epigastric regions.

Average weight: three to four pounds.

Average size: 10 to 12 inches in its transverse diameter; 6 to 7 inches antero-posteriorly; 3 inches thick at back part of the right lobe, which is its thickest part.

Upper surface: convex, smooth, covered by peritoneum, directed upwards and forwards; above is the diaphragm, below abdominal parietes. It is divided into two unequal parts by a fold of peritoneum, called the *suspensory* or *broad ligament.*

Under surface: concave, and is connected with the stomach, duodenum, hepatic flexure, right kidney, and supra-renal body; divided by a longitudinal fissure into a right and left lobe.

Posterior border: connected to diaphragm by the coronary ligament; is broad and round; is in relation with the aorta, inferior vena cava and diaphragmatic crura.

Anterior border: sharp, thin and free, and marked by a notch opposite attachment of suspensory ligament. In women and children this border is usually *below* the ribs; in men, above the ribs.

The right extremity of this organ is thick and rounded; the left, flattened and thin.

It is well to remember that the Liver changes with the position of the body; with the state of the stomach; with the inflation of the lungs, etc.

The ligaments are five in number; four are composed of double layers of peritoneum and are:

The suspensory, fal´ciform, or *broad* ligament is sickle-shaped, with the base forward. It is attached above to the diaphragm, extending on to the sheath of rectus as far as the umbilicus, and below from the notch in front to the posterior edge of the liver; it consists of closely united double-fold of the peritoneum, and the anterior edge closes the round ligament.

The lateral ligaments, right and left, extend from the sides of the diaphragm to the posterior border of the liver; are of triangular shape, the left being the larger.

The cor´onary ligament is continuous with the lateral ligaments and attaches the posterior margin of the liver to the diaphragm. Between the folds of this ligament is a large oval space divided into parts by a notch which lodges the inferior vena cava into which the hepatic veins open.

The round ligament (*ligamentum teres*) is the obliterated umbilical vein. It ascends from the umbilicus, in the longitudinal ligament, to the anterior border of the liver, and from there on along the longitudinal fissure to the inferior vena cava.

The fissures are 5 in number, on the under surface, dividing the liver into **5** lobes, and are arranged somewhat in the shape of the letter **A**, the apex being at the liver's posterior margin.

The longitudinal fissure divides the body into right and left lobes; it commences at the notch on the anterior border and ends at the posterior edge; its anterior half is called the *umbilical* fissure, being deeper than the posterior part, and lodges the umbilical vein, in the fœtus, or round ligament in the adult. It is frequently "bridged over" by liver substance, called the *pons hepat´icus.*

The fissure of the duc´tus veno´sus is the posterior half of the longitudinal fissure, and contains a fibrous cord, the *ductus venosus* of the fœtus.

The transverse or *portal fissure* is placed at right

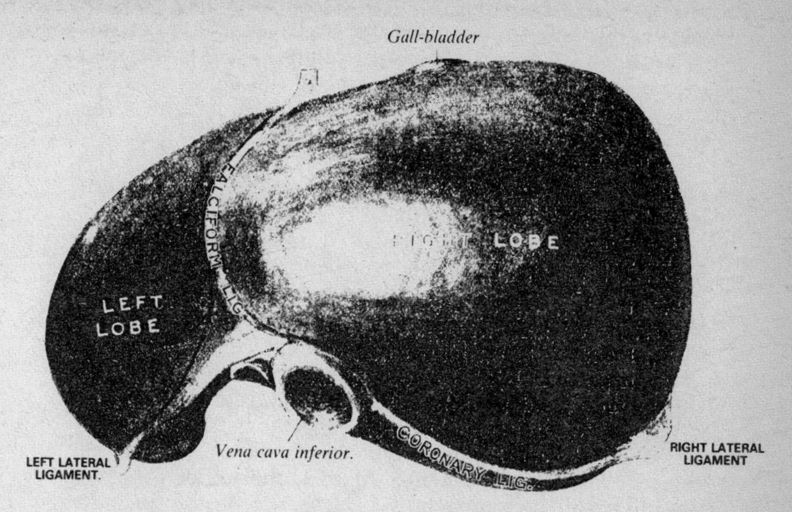

The liver. Upper surface.

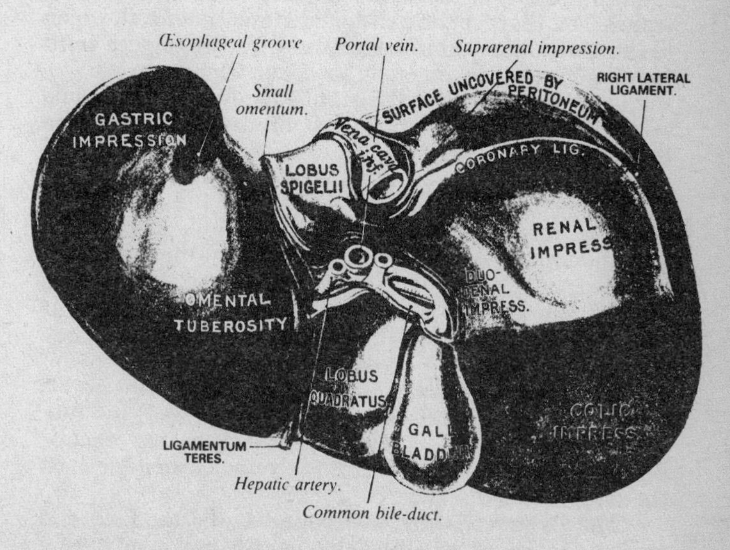

Posterior and under surfaces of the liver.

angles to the longitudinal, and lodges, from before backwards, the hepatic duct, artery, and portal vein. It is a short, deep fissure, two inches in length, and is confined to the right lobe. Was considered by the old anatomists the gateway (porta) to the liver, hence the origin of the name to the portal vein.

The fissure for the gall-bladder is parallel to the longitudinal fissure, on the under surface of the right lobe; is a shallow, oblong fossa.

The fissure for the ve´na ca´va is placed obliquely at the posterior margin of the liver behind the gall-bladder; is a short, deep fissure; is separated by the *lobus quadratus* from the transverse fissure, and by the *lobus Spigelii* from the longitudinal fissure.

The lobes are also 5 in number, corresponding with the number of ligaments and the number of fissures.

The left is smaller than the right lobe; is more flattened and is in the left hypochondriac and epigastric regions; its upper surface is convex, the under concave, resting upon the stomach.

The right lobe is much larger than the left (as 6 to 1), thicker, is of quadrilateral shape, and occupies the right hypochondrium; is separated from the left by the longitudinal fissure; on its under surface are 3 fissures: the transverse, that for the gall-bladder and that for the vena cava; it also has 2 shallow depressions: the one in front for the hepatic flexion of the colon (*impressio colica*), the other in behind for the right kidney and its capsule (*impressio renalis*). It contains on its under side the following lobes:

Lob´ulus quadra´tus, bounded behind by transverse fissure; in front, by the free margin; on the left, by the umbilical fissure; on the right, by the gall-bladder fissure.

Lob´ulus Spige´lii, at the back part of the right lobe; bounded, in front, by the transverse fissure; at the back, its own free margin; on the left by the ductus venosus fissure; on the right by the vena cava fissure.

Lob´ulus cauda´tus (tailed lobe) runs from the under surface of the right lobe to the lobulus Spigelli, separating the transverse fissure from the vena cava fissure.

Vessels: these are also 5 in number, viz: hepatic artery; the portal vein; the hepatic vein; the hepatic duct; the lymphatics.

The Hepatic Artery (see page 94), *Portal Vein* and *Hepatic Duct*, accompanied by nerves and lymphatics, ascends to the transverse fissure, between the layers of gastro-hepatic omentum; the artery lies to the right, the duct to the left, and the vein behind the two. The *capsule of Glisson* is a layer of areolar tissue enveloping them, and continuing on into the *portal canals*, and liver substance with them.

The hepatic veins return the blood from the liver substance, collecting, in the deep fossa, into three large and several smaller branches to empty into the inferior vena cava.

The lymphatics are of two sets; one, the internal, originates around the capillaries of the lobules and passes outwards; the external set are beneath the peritoneal surface; both sets are connected to the various glands of similar matter in the thoracic and gastric regions.

The nerves are mainly from the sympathetic system with a hepatic branch from the pneumogastrics (right and left), and from the right phrenic.

Histological structure: the liver has a *serous* and *fibrous coat.* The former is absent from the posterior border and from the portal fissure, where the latter, which elsewhere is thin, is most developed. A strong sheath of areolar tissue (Glisson's capsule) surrounds the vessels of the organ as they ramify in it, and, at the transverse fissure, becomes continuous with its fibrous coat.

The liver substance proper consists of lobules about the size of a millet seed, which are closely packed polyhedral masses more or less distinct, arranged round the sides of the branches (sub-lobular) of the hepatic veins, and connected to them by minute veins which begin in the centre of the lobules (intra-lobular veins).

Each lobule consists of a mass of compressed spheroidal or polyhedral nucleated and nucleolated cells from 1/1030th to 1/830th of an inch in diameter, often containing oil globules. Surrounding the lobules is a variable amount of a fine connective tissue, in which is contained a minute branch (intra-lobular) of the portal vein, a branch of the hepatic artery, and of the hepatic duct, together with minute lymphatic vessels covering them.

Interior arterial blood supply: the hepatic artery entering through the transverse fissure and portal canals gives off *vaginal* branches; these ramify in Glisson's capsule and furnish nutrient branches to the large vessels, ducts and membrane; they also give branches (*capsular*) which terminate in stellate plexuses on the surface in the fibrous coat; the artery finally terminates in the *inter-lobular* plexuses on the outer surface of the lobules.

The internal venous supply is by the portal vein; this is also in Glisson's capsule, and, subdividing into smaller branches, finally forms into the *inter-lobular plexuses* between the different lobules; in their course they receive the *vaginal* and *capsular* (see ¶ above) veins. All blood, either by the hepatic artery or portal vein, at last reaches these inter-lobular plexuses; from here it is carried *into* the lobule's centre by small capillaries, which are enmeshed by liver cells; from here it is collected into a vein, running from apex to base through the centre of each lobule, called the *intra-lobular vein*; this empties into the *sub-lobular vein*, at the base of each lobule; these sub-lobular veins at last unite into larger trunks, forming the *hepatic veins*; these converge to 3 large trunks and empty into the vena cava inferior.

The ducts: these commence by minute passages between the cells (*inter-cellular biliary passages*), which radiate towards the lobule's circumference, there forming (after piercing its walls) between the lobules the *inter-lobular plexus*; ducts arise from these plexuses that pass into the portal canals, enclosed in Glisson's capsule, accompanied by the hepatic artery and portal vein; they then unite into two main trunks which leave the liver at the transverse fissure as the hepatic duct.

THE GALL-BLADDER

Is a conical or pear-shaped bag placed in a fossa on the under surface of the right lobe of the liver. It is held in position by peritoneum; it is 4 inches long, 1 inch in width at the widest portion, and will hold 8 to 10 drachms. It has a *fundus*, or broad extremity, placed at the right, and a *body* and *neck* (page 155).

Relations: the body is in relation in front with the liver, the 1st part of duodenum, the pylorus, hepatic flexure of colon. The fundus is in contact with the parietes opposite the 10th costal cartilage.

The biliary ducts are 3 in number, namely:

The hepat´ic duct issues from the liver at the transverse fissure, and joins the cystic to form the common bile duct; it is formed by the union of a duct from the right and from the left liver lobes; is about 1½ inches long.

The cys´tic duct passes from the neck of the gall-bladder to join the preceding; is the smallest of the three ducts, being about 1 inch in length, and lies in the gastro-hepatic omentum. Its lining membrane is thrown into 5 to 12 consecutive folds extending obliquely around the tube.

The duc'tus commu'nis choled'ochus or common bile duct, the largest of the ducts, is the result of the union of the hepatic and cystic ducts. It is about 3 inches long, and the size of a goose-quill. It descends behind the 1st part of the duodenum, in front of the vena portæ, and to the right of the hepatic artery, and passing between the pancreas and 2d part of the duodenum, enters the small intestine obliquely with the pancreatic duct, a little below the middle of the descending part of the duodenum. This duct, with the pancreatic duct, empties from a common orifice, at the summit of a papilla, into the intestine.

Structure: there are 2 coats to the gall-bladder and its ducts, viz: the fibrous and the mucous.

The mucous or *internal coat* is continuous from the duodenum up the ducts and about the gall-bladder; its epithelium is of the columnar variety; it has numerous lobulated mucous glands.

The fibrous or *external coat* is of a strong areolar tissue with a few interspersed muscular fibres.

Artery: the cystic artery, a branch from the hepatic, gives this organ its blood supply.

THE PANCREAS

Is a compound racemose gland (see page 94) analogous to the salivary glands in structure. It is transversely oblong, flattened and with the right end, or head, bent downwards, covered by the duodenum. The left end is tapering and straight. It lies horizontally across the epigastric space at the back of the stomach, reaching into both hypochondriac organs.

Size: *Length*: 6 to 8 inches. *Breadth*: 1½ inches. *Thickness*: ½ inch.

Relations: *In front*: ascending transverse mesocolon. *Behind*: Aorta, vena cava, crura of diaphragm, splenic vein, commencement of vena portæ, left kidney. *Upper border, from right to left*: 1st part of duodenum and hepatic artery, cœliac axis, splenic vessels. *Lower border, from right to left*: 3d part of duodenum, superior mesenteric vessels, inferior mesenteric vessels. *Left end or tail*: touches spleen, above left kidney. *Right end or head*: Embraced by duodenum, partly separated, behind by bile duct, and in front by pancreatico-duodenal arteries.

Duct: (canal of Wirsung) extends transversely from left to right, opens into 2d part of the duodenum. Begins as two small ducts at the tail, which coalesce near the middle of the gland; it increases in size, from additions of numerous small ducts, till it reaches the duodenum, where it is as large as a goose-quill. Sometimes this duct, and the common bile duct, open separately into the small intestine, but usually not.

The lesser pancreatic duct is the one from the head of the pancreas, when this portion is separated from the main gland.

The walls are thin and consist of a fibrous (external) coat, and a mucous (internal) coat.

Structure: it is not surrounded by a capsule, as most glands are, but loose areolar tissue dips down into it, forming lobes; each lobule is formed by the ramifications of the duct surrounded by *acini*. The short ducts are lined by short columnar epithelium and cells.

The fluid secreted resembles saliva and it digests starches like saliva. It also emulsifies fats, and changes albumenoids into peptones.

Arteries: splenic, pancreatico-duodenal of hepatic, superior mesenteric (page 87).

Veins: open into splenic and superior mesenteric (page 94).

Nerves: splenic plexus.

THE SPLEEN

Is of an oblong, flattened form, situated in the left hypochondriac region. It is covered by peritoneum and connected with the stomach by the gastro-splenic omentum. It is usually classified with the other ductless glands—the thyroid, thymus and supra-renal capsules. It is of very brittle consistency, soft, vascular, and of a bluish-red color. Its *external surface* is smooth and convex. Its *internal surface* is slightly concave, and is divided by a vertical fissure—*the hilum*. The *upper end* is thick and rounded; the *lower end* pointed. *It is held in position* by the gastro-splenic omentum, and by the suspensory ligament to the diaphragm.

Size and weight: these vary greatly though it usually measures in adults, 5 inches in length; 1 or 1½ inches in thickness; 3 or 4 in width; *weighs* about 7 ounces. At birth it is as 1 to 350: in adult life as 1 to 320 or 400; in old age, 1 to 700. Is increased during digestion and fevers, especially intermittent, when it may weigh 18 to 20 pounds.

Relations: *Externally:* diaphragm, which separates it from the 9th, 10th and 11th left ribs. *Internally*: cardiac end of stomach, tail of pancreas, left crus, left supra-renal body. *Above*: connected by a suspensory ligament to the diaphragm. *Below*: splenic flexure. *Posterior margin*: left kidney.

Coverings: 2; serous and fibro-cellular.

The serous or *external coat* is thin, smooth, and derived from the peritoneum; it is intimately attached to the internal coat, and surrounds almost the entire organ, being reflected at the *hilum* upon the stomach and diaphragm at the upper end.

The fibro-elastic or *internal coat* invests nearly the whole spleen, and at the hilum is reflected inwards as sheaths for the vessels entering its substance; from these sheaths there are given off numerous *trabeculæ* that unite to form the areolar framework of the organ; in these areolar spaces is contained the *splenic pulp*. These trabeculæ, the vessel sheaths and the covering of the organ are made up of yellow, elastic fibrous tissue, hence the possibility of its fluctuation in size.

Spleen pulp: this is a soft, dark, reddish-brown mass, something like clotted blood, which, under the microscope, proves to be branching cells and intercellular substance. The cells are connective tissue corpuscles. Blood corpuscles, in various stages of disintegration, are found freely intermixed in these intercellular meshes; indeed, Prof. C.H. Stowell regards this organ as the graveyard for the red blood corpuscle.

Malpighian bodies: these appear to be cylindrical masses of adenoid tissue, which are found throughout this organ. These are so intimately connected with the arterioles that they are regarded by some as the altered coats of them. They are from 1/25 to 1/100 of an inch in diameter, and resemble the adenoid tissue of lymphatic glands.

The blood supply: this is from the remarkably large and tortuous splenic artery (see page 87), which divides into several branches at the hilum, receiving sheaths from the internal covering of the spleen. These branches, piercing the substance, divide into *arterioles* finally, and these give origin to the Malpighian bodies, in their finer ramifications.

The splenic vein: the rootlets of the minute veins about the Malpighian bodies gather the changed blood up, these unite to form larger vessels, remarkable for their numerous anastomoses (the arteries are lacking these anastomoses), and while not specially accompanying the arterial branches, emerge from the hilum in from 4 to 6 radicles, which finally unite to form the splenic vein (see page 94), the largest branch of the portal.

The lymphatics originate from the arterial sheaths and the trabeculæ; these unite and pass through the hilum to empty into the thoracic duct.

Nerves: branches of right and left semi-lunar ganglia and right pneumogastric nerve.

THE THORAX.

THORAX.

It is of conical shape, the base downwards, formed of an osseous and bony frame-work. It is flattened from before backwards.

Boundaries: *In front*: the sternum, the 6 costal cartilages, ribs and intercostal muscles. *At the sides*: by the ribs and intercostal muscles. *At the back*: the spine, ribs and muscles.

The superior opening, in front, is bounded by the manubrium; at each side by 1st rib: behind by the 1st dorsal vertebra and 1st ribs.

The inferior opening, or base, is bounded, in front, by the ensiform cartilage; on each side by the last rib and part of diaphragm; at the back by the last dorsal vertebra and part of diaphragm.

Contents: the heart and pericardium; the lungs and pleuræ.

Through the superior opening pass, in order from before backwards, the sterno-hyoid and sterno-thyroid muscles; the remains of the thyroid gland; the trachea; œsophagus; thoracic duct; the longus colli muscles. To the sides of this opening, the innominate artery; the left carotid; the left subclavian; the internal mammary and superior intercostal arteries; the right and left innominate and inferior thyroid veins; the pneumogastric, sympathetic, and cardiac nerves and the left recurrent laryngeal nerve. The apex of each lung, with its pleura, also projects through this opening.

The mediastinum is a subdivision of this space; see chapter on Triangles and Spaces at end of the book.

THE PERICAR'DIUM.

The pericardium is a conical fibro-serous membrane, placed behind the sternum, containing the heart and the commencement of the great vessels. The *apex* points upwards and surrounds the vessels coming from the heart for two inches; the *base* is fixed to the central tendon of the diaphragm.

In front: thymus gland, overlapped by left lung.

Behind: bronchi, œsophagus, descending aorta.

Laterally: pleura, phrenic vessels, phrenic nerve.

The serous layer of the pericardium surrounds the heart and is continued *on to the inner surface* of the pericardium. It encloses the pulmonary artery and aorta as a single tube, but only partially covers the venæ cavæ inferior and superior, and the 4 pulmonary veins. It is a smooth, glistening membrane, covered with squamous endothelium on the inside, and secretes a thin fluid, to facilitate the heart's movement. This fluid is called the *pericardial fluid*, and is present, normally, in quantities just to prevent friction.

The fibrous layer is a strong, dense membrane surrounding the heart, and has prolongations upwards upon the aorta, pulmonary artery and veins, and superior vena cava. (The inferior vena cava receives no covering, as it passes through the central tendon of the diaphragm). It is attached below to the diaphragm (central tendon), and on the left side to its muscular fibres.

Arteries: these are branches derived from internal mammary, bronchial, œsophageal and phrenic arteries.

(COR) THE HEART.

Position: The heart, a hollow muscular organ, is placed obliquely in the chest, the base being directed upwards, backwards, and to the right, corresponding to the space between the 5th and 8th ribs. The apex points downwards, forwards, and to the left, and corresponds to a point one inch to the inner side, and two inches below the *left* mamilla.

The upper border of the heart corresponds to a line drawn across the sternum on a line with the upper border of the 3d costal cartilage.

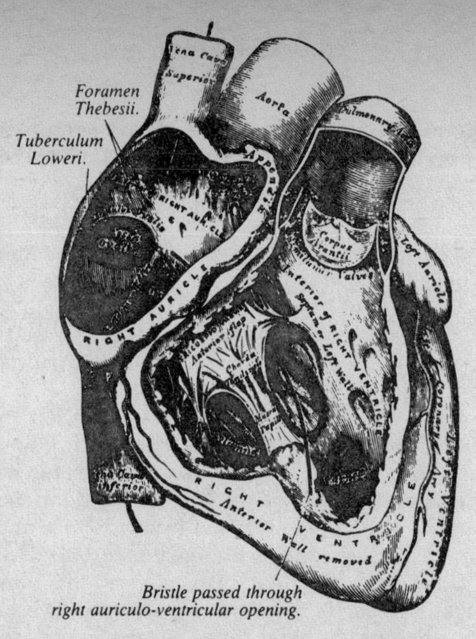

The right auricle and ventricle laid open, the anterior walls of both being removed.

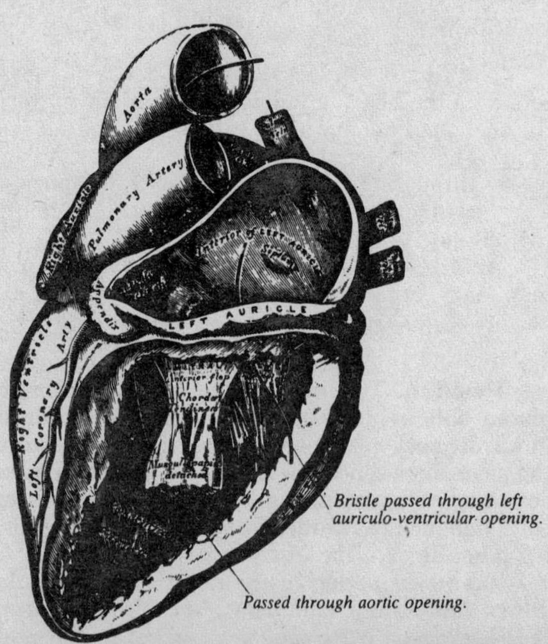

The left auricle and ventricle laid open, the posterior walls of both being removed.

The lower border, to a line drawn across the lower end of the gladiolus from the right costo-xiphoid articulation to the apex.

Its anterior surface is convex, rounded, and is directed upwards and forwards, and is formed chiefly by the right ventricle, aided by part of the left.

Its posterior surface is flattened, and is against the diaphragm, and is mainly formed by left ventricle.

The left border is short, thick and round. *The right* is long, thin and sharp.

Size: it measures, in adults 5 inches in length, 2½ inches in thickness.

Weight: in the *adult male*, from 10 to 12 ounces; in the *female*, from 8 to 10 ounces. *The proportionate* weight is 1 to 169 for males and 1 to 149 for females.

Divisions: the heart is divided longitudinally by a septum into two halves, right and left, each of which is subdivided transversely into two cavities, by the *auriculo-ventricular groove.* These four divisions are indicated on the heart's surface by grooves. The upper are called the *auricles* and the lower the *ventricles.*

The circulation: the right auricle receives venous blood from the vena cava and coronary sinus; thence is passes into the right ventricle, whence it is conveyed to the lungs by the pulmonary artery, (the only *artery* in the adult carrying *venous* blood; in the fœtus the umbilical artery also carries it). After being oxygenised the blood passes into the left auricle by the pulmonary veins (the only *veins* in the adult carrying *arterial* blood; in the fœtus the umbilical, hepatic veins and the inf. vena cava carry arterial blood; thence it is conveyed into the left ventricle, and from there to the aorta, whence it passes through the body.

THE RIGHT AUR´ICLE is larger than the left (holding about 2 ounces), though its walls are thinner, measuring but a line in thickness. It consists of a principal cavity (*si´nus*) and an *appen´dix auric´ulæ.*

The principal cavity or **sinus** is of an irregular quadrangular form placed between the two venæ cavæ, connected below with right ventricle, and internally with left auricle; has very thin walls.

The appen´dix auric´ulæ (dog's ear) is a small muscular pouch overlapping the root of the pulmonary artery and aorta.

Openings: *Supe´rior ve´na ca´va*, in the upper and

front part, directed downwards and forwards; returns the blood from the upper half of the body.

Inferior vena cava opens at the lowest part of the auricle; between the two is the *tuberculum Loweri.* It is larger than the superior, and its opening is directed upwards and inwards. It returns the blood from the lower half of the body.

Corona´ry sinus opens between the inferior cava and auriculo-ventricular opening. It returns the blood from the neart's substance. Before entering the auricle this sinus is dilated to size of little finger, and it is protected at the orifice by a semi-circular fold, the "coronary valve;" where the coronary vein enters it there is a valve of two unequal segments. Has some muscular fibres in its walls.

Foram´ina Thebe´sii, are the mouths of small veins (*ve´næ cor´dis min´imæ*). They return the blood from the muscle substance of the heart.

The auric´ulo ventric´ular opening: this is the large oval aperture, one inch in diameter, connecting the auricle with the ventricle. It corresponds with centre of sternum between the 4th costal cartilages. It is oval from side to side, and is surrounded by a fibrous ring covered with the heart's lining membrane. It is somewhat larger than the left auriculo-ventricular opening, and is guarded by the *Tricuspid valve.*

Valves: *The Eustachian valve* is a semi-lunar fold between the anterior margin of the inferior vena cava and the auriculo-ventricular orifice. Its convexity is attached to the vena cava, and its concavity is free, the left terminating-cornua is attached to anterior edge of the *an´nulus oval´is.* In the fœtus this valve is of large size and directs the blood through the *fora´men ova´le* into the left auricle. Occasionally in the adult it is wanting.

The coronary valve protects the opening of the coronary sinus, and is a semi-circular fold of the lining membrane. Is occasionally double.

The mus´culi pectina´ti (comb tooth) are prominent muscular columns running over the surface of the appendix auriculæ.

REMNANTS OF FŒTAL STRUCTURE: *Fos´sa ova´lis* is the remains of the foramen ovale on the septum auricularum.

The an´nulus ova´lis is an elevated margin of the fossa.

THE RIGHT VEN´TRICLE consists of a cavity and a funnel-shaped cavity leading to the pulmonary artery.

Is of triangular form, and it forms the largest part of the front of the heart. Its walls are thinner (as 1 to 2) than those of the left ventricle. It will hold about three fluid-ounces.

On the wall are projections, **colum´næ car´neæ**, of which there are three varieties: *the first* are merely prominent ridges; *the second* are attached at both their ends merely; *the third* are the *musculi-papilla´res*, which project forwards, and to which are attached the **chor´dæ ten´dine´æ**, or cords attached to the auriculo-ventricular valve. This cavity has 2 openings; that of the *auriculo-ventricular*, described on p.166, and that of the *pulmonary* artery; also 2 valves; the *tricuspid* and *semi-lunar*.

The tricuspid valve (behind middle of sternum on level of 3d left rib), which guards the right auriculo-ventricular opening, consists of three flaps, formed by a reduplication of the endocardium, together with some muscular fibres. The bases of the flaps are attached to a tendinous ring about the orifice, while to their free ends are attached chordæ tendinæ. One segment corresponds to front of the ventricle, another (the largest) is placed towards the left of the opening, and the third to the back. The valve prevents regurgitation of blood into the auricle during the heart's contraction.

Semi-lunar valve: the opening of the *pulmonary* artery is circular in form, at the summit of the funnel-shaped cavity, and is guarded by the pulmonary semi-lunar valve. There are three folds of the lining membrane which guard the orifice of the pulmonary artery, which opens at the left of the auriculo-ventricular opening; 2 of the folds are placed anteriorly, the other posteriorly. The free margin of each has in its middle a small nodule or *cor´pus aran´tii*, and between each valve and the beginning of the pulmonary artery is a dilatation called the pulmonary sinus, or *sinus of Valsalva.* The point corresponding externally to these valves, is the junction of the third *left* rib cartilage with the sternum. These valves serve to prevent the regurgitation of the blood during the ventricle's contraction, up the artery; the 3 *corpora arantii* close the centre of the pulmonary artery when the valves are shut.

THE LEFT AU´RICLE, smaller than the right, consists of a principal cavity or *sinus*, and an *appendix auriculæ*; the latter looks forwards and to the right side, projecting over the commencement of the pulmonary artery. *The sinus* is of cuboidal form, and concealed, in front, by the pulmonary artery; behind it receives the 4 pulmonary veins.

Openings: *the pulmonary veins* (4) open into the cavity, two on either side. They have no valves. (Sometimes these veins terminate by a common opening).

The auriculo-ventricular opening is smaller than that on the right side.

The mus'culi pectina'ti: these are fewer and smaller than on the right side, and are confined to the appendix.

THE LEFT VEN'TRICLE is longer and more conical than the right, with its wall nearly twice as thick, being thickest at the broadest part of the ventricle. It forms a little of the anterior and much of the posterior surface of the heart.

The auric'ulo-ventric'ular opening corresponds to the 3d *left* intercostal space, and is smaller than the right.

The mitral valve closes this opening, being attached to its circumference. It consists of 2 flaps of unequal size, the larger being anterior; both are formed from a doubling of the lining membrane with muscle fibres and fibrous tissue. This valve is thicker and stronger than the tricuspid. It has chordæ tendinæ attached, similarly to the tricuspid, to the apex of each flap. This valve lies an inch to the left of the sternum in the 3d intercostal space.

The aor'tic opening (a small circular aperature) is placed in front, and to the right side of the preceding (being separated from it by one of the mitral valve segments), and its position may be marked externally by a line drawn through the sternum, level with the lower border of the 3d *left* costal cartilage.

The semi-lunar valves guard the aorta, surrounding its orifice. Are similar in placement and structure to those guarding the pulmonary artery, though are thicker and stronger, and their corpora arantii are larger. The Valsalvaian sinuses are also larger than those of the right side.

The columnæ carneæ are similar to those on the right side, though smaller and more numerous; but two have only one attachment, being large and supporting, each, a chorda tendina at their free extremities.

The endocar'dium is the thin, serous membrane lining the whole of the interior of the heart, and is continuous with the lining of the bloodvessels. By its reduplications it forms the valves. It is smooth, transparent, and gives a glistening appearance to the heart-cavities when opened.

Structure: the heart is built up mainly of fibrous rings and muscle fibres.

The fibrous rings are stronger on the left than on the right side of the heart, and surround the various openings, giving attachments to the various vessels entering thereat, and to the valves which may guard such openings.

THE MUSCULAR STRUCTURE admits of two main divisions, the ventricular fibres and the auricular fibres.

The auricular fibres are of two layers, a superficial and deep; the former run in a transverse direction and form a thin layer; the latter are arranged in *annular* and *looped* groups. The *annular* surround the appendices, and are continued upon the walls of the vessels entering or leaving this cavity. The *looped* pass upwards over each auricle, having their anterior and posterior attachments to the auriculo-ventricular opening.

The ventricular fibres are arranged in numerous layers, though not independent ones; Pettigrew gives as high a number as 7. They can be divided into the two groups, *superficial* or *longitudinal fibres*, and *deep* or *circular fibres*. The superficial ones take, frequently, a spiral direction, and at the apex assume a looped condition. The circular ones are deeply placed, and at the base may surround each ventricular cavity separately; though some of them are continued across the furrows, and so surround both ventricles; more of this furrow-crossing is seen posteriorly. They are attached to the fibrous rings at the ventricular base.

The nerves come from the cardiac plexuses, which are formed partly from the spinal and partly from the sympathetic systems. The filaments are freely distributed upon the surface and within the substance, with numerous small ganglionic attachments.

The lymphatics empty into the right lymphatic and thoracic ducts.

The arteries supplying the heart are the anterior and posterior coronary. See page 84.

The veins accompany the arteries and terminate in the right auricle, and are the *great cardiac vein*, the *anterior cardiac vein* and the small or *venæ Thebesii*.

Circulation of the blood: See page 246.

ORGANS OF VOICE AND RESPIRATION.

THE NOSE.

The organ of smell consists of an Anterior Prominent Part and two Nasal Fossæ.

The nose, of triangular form, is the anterior part projecting from the face; it is constructed of bones and cartilages, covered with muscles and skin externally, and with mucous membrane internally. Inferiorly are the two nostrils, separated by the *columna*, around which orifices are arranged stiff hairs, *vibrissæ*, which arrest the entrance of foreign bodies during inspiration.

The *bony framework* occupies the upper portion, and consists of the *nasal bones*, and *nasal processes of the superior maxillary*. (See sections on Osteology).

The cartilages of the nose are five in number:

The *upper lateral cartilages* (2): situated just below the free margins of the nasal bones. Each cartilage is triangular in shape, flattened, thicker anteriorly than where it joins its fellow and the cartilage of the septum; posteriorly it is in connection with the nasal process of the superior maxilla, and inferiorly it joins the lower lateral cartilage.

The *lower lateral cartilages* (2): are thin, flexible plates peculiarly curved to form the nostrils; posteriorly each cartilage is connected to the nasal process of the superior maxilla by fibrous membrane, in which are two or three sesamoid cartilages; above, it joins the upper cartilage and the cartilage of the septum.

The *cartilage of the septum*, of triangular shape, is thicker at the edges than at the centre; its connections are: anteriorly with the nasal bones, the two upper lateral cartilages and the lower lateral cartilages; posteriorly with the perpendicular plate of the ethmoid; inferiorly with the vomer and the palatal process of the superior maxillæ.

All these cartilages are loosely connected together, and to the bones, by a tough fibrous membrane, or perichondrium, so that there is considerable freedom of motion.

The nasal muscles: see page 22.

The arteries are the *lateralis nasi*, from the facial; the *nasal artery of the septum*, supplying the alæ and septum, a branch from the superior coronary; the *nasal*

SEEN FROM BELOW

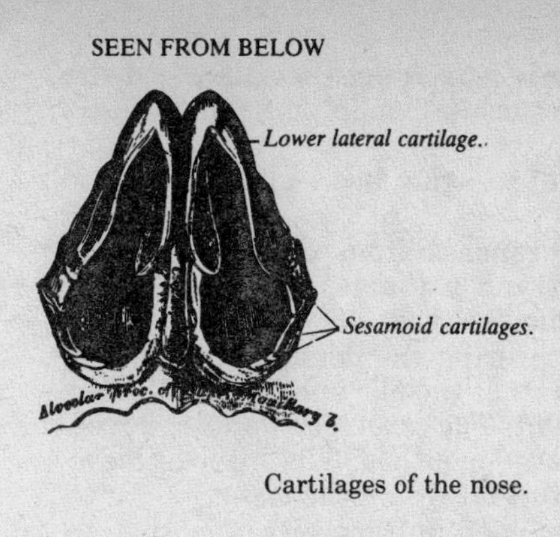

SIDE VIEW

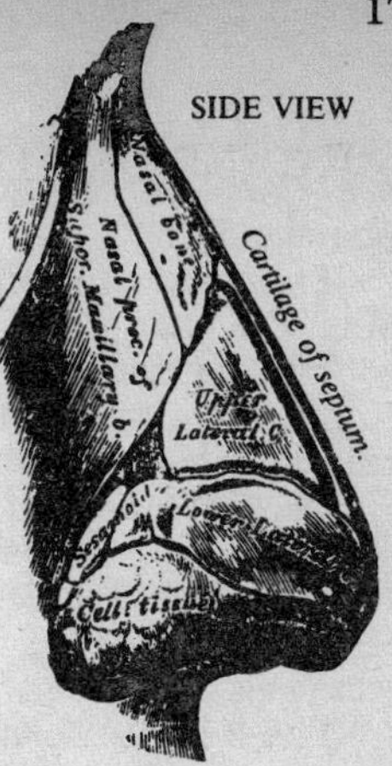

Cartilages of the nose.

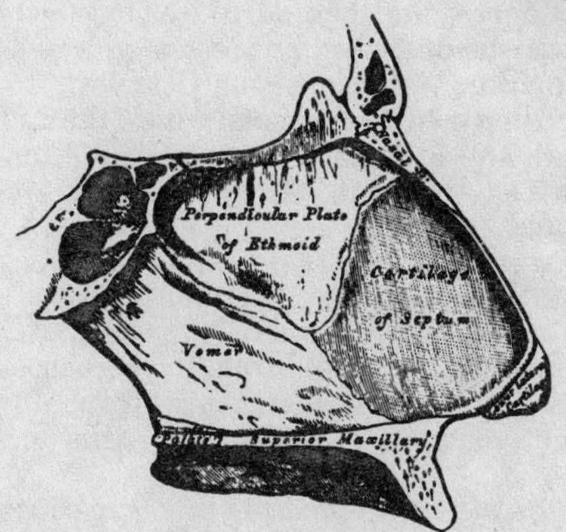

Bones and cartilages of septum of nose. Right side.

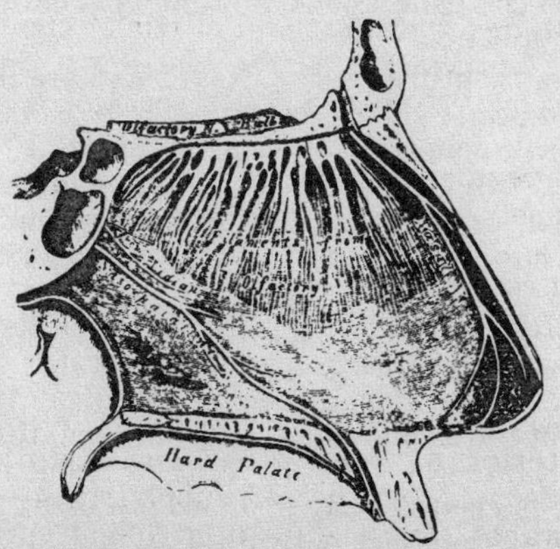

Nerves of septum of nose. Right side.

branch of the ophthalmic (supplying the sides) and the *nasal branch* of the infra-orbital (supplying the dorsum). See page 32.

The veins terminate in the facial and ophthalmic veins. See page 41.

The nerves are branches from the facial, infra-orbital, infra-trochlear, and a filament from the nasal branch of the ophthalmic. See page 48.

The integument covering the dorsum and sides is thin and loosely adherent; at the alæ and tip it is thicker and more firmly adherent. *The sebaceous follicles*, covering the nose-tip, are large-mouthed. The *mucous membrane* is continuous with that lining the fossæ.

The 2 nasal fossæ: open in front, on either side, by the anterior nares, and behind by the posterior nares. Each fossa may be described as possessing a roof, a floor, an inner and outer wall.

The *roof* is formed by the nasal bones, the nasal spine of the frontal, the cribriform plate of the ethmoid, the under surface of the body of the sphenoid, sphenoidal turbinate bones.

The *floor* consists of the palatal processes of the superior maxilla and of the palate bones.

The *inner wall* is constructed by the crest of the nasal bones, the nasal spines of the frontal, the perpendicular plate of the ethmoid, the vomer, the rostrum of the sphenoid, and the crests of the superior maxilla and palate bones.

The *outer wall*: nasal process of the superior maxilla, the lachrymal bones, the ethmoid, inner surface of the superior maxillæ, the inferior turbinate bones, the vertical plate of the palate bone, and the internal pterygoid plate of the sphenoid.

Mucous membrane: the nasal fossæ are lined by a mucous membrane called the *pituitary* or *Schneiderian*, which is continuous with that of the pharynx, with the conjunctiva, with the lining of the tympanum and mastoid cells, the frontal, ethmoidal and sphenoidal sinuses, and the antrum. It is thickest and most vascular over the turbinated bones.

The epithelium: near the orifices, it is of the common pavement variety; in the respiratory tract it is columnar and ciliated; in the olfactory region (the upper tract) it is columnar, but without cilia, and between these cells, lying loosely are the *olfactory cells* of Schultze, which consist of a nucleated body and two processes, one running inwardly, the other outwardly.

Pigmentation of the membrane is seen in the olfactory region.

Branched mucous glands are freely scattered over the surface.

The superior meatus, the *middle meatus* and the *inferior meatus* are more or less encroached upon by the thickness of this mucous membrane, studded with these mucous cells.

The arteries are the ant. and post. ethmoid, from the ophthalmic; the spheno-palatine and alveolar from the int. maxillary.

The veins follow loosely the reverse course of the arteries.

The nerves: these are the olfactory (the special sense nerve), being distributed over the upper 3d of the septum, and then over the superior and middle spongy bones; then the nasal branch of the ophthalmic; the Vidian; filaments from the ant. detal branch of the sup. maxillary; the naso-palatine and the ant. palatine to the middle and lower spongy bones.

THE LARYNX.

The Larynx (the organ of voice) is composed of cartilages connected together with ligaments and moved by muscles, the whole being lined with mucous membrane. It is situated at the upper portion of the air passage between the trachea and the base of the tongue. It is narrow and cylindrical below, but broader above, presenting here a somewhat triangular shape.

The cartilages of the larynx are 9 in number:

The thy′roid (the largest) consists of two wings united in front, forming the projection known as the *o′mum Ada′mi.* Each wing is quadrilateral in shape, the posterior border being rounded and prolonged into a *superior* and *inferior cornu*; the latter articulates with the cricoid cartilage, but the former is free. On the external surface there is an oblique ridge, giving attachment to sterno-thyroid, thyro-hyoid and part of inf. constrictor muscles. The posterior border receives insertion of stylo-pharyngeus and palato-pharyngeus muscles. The inner surface of each ala is smooth, and at the point of junction with the other are attached the epiglottis, the true and false vocal cords, the thyro-arytenoid and thyro-epiglottidean muscles. (See page 28.) It *articulates* with the following cartilage by the crico-thyroid membrane and muscle.

The cri´coid cartilage (resembling a signet-ring) is shallow in front, but deep behind; between this and the thyroid cartilage in front, is the crico-thyroid membrane. On the upper border of the posterior part are *two articular surfaces* for the arytænoid cartilages; whilst on each side are two facets for articulation with the inferior cornu of the thyroid cartilage. Is connected with the trachea by a fibrous membrane.

The muscular attachments are as follows: anteriorly, the crico-thyroid and inf. constrictor; posteriorly, the œsophagus and crico-arytænoideus post. The inner surface is smooth and lined with mucous membrane.

The ary´tænoid or *pitcher-shaped cartilages* (2) are pyramidal in shape; the bases articulate with the cricoid cartilage, the true vocal cords being attached to their anterior angles. The apex looks inwards and backwards, and on it is the *cornic´ulum laryn´gis*. The muscular attachments are the thyro-arytenoid to the ant. surface; the post. and lateral crico-arytenoid muscles to the base; the aryteno-epiglottidean fold to the apex; the arytenoid to the post. surface. The internal surface is covered with mucous membrane.

The cornic´ula laryn´gis, (2), or cartilages of Santorini, are two small cartilages of conical shape attached to the apices of the arytænoid cartilages.

The cune´iform cartilages, (2), or cartilages of Wrisberg, are two small cartilages often found in the arytæ´no-epiglot´tide´an folds.

The epiglot´tis, a thin lamella of yellowish color that covers the superior opening of the larynx; it is shaped like a leaf, the apex being attached to the angle of union of the alæ of the thyroid cartilages; it is connected also to the hyoid bone by the hyo-epiglottic ligament; its free extremity is broad and rounded; its lingual surface is curved forwards, and covered by mucous membrane. The posterior (pharyngeal) surface is concave laterally, convex from top to bottom, and the covering membrane is studded with small mucous glands.

Structure: the cuneiform and cornicula cartilages, with the epiglottis, are of yellow fibro-cartilage with little tendency to ossification; the others become more or less ossified in old age, and resemble the costal cartilages.

The ligaments of the larynx are of 2 classes the *extrinsic*, connecting the thyroid cartilage to the hyoid bone; the *intrinsic*, connecting the several cartilages together.

The thyroid-hyoid membrane (a broad, fibro elastic membrane), passing from the upper border of the thyroid

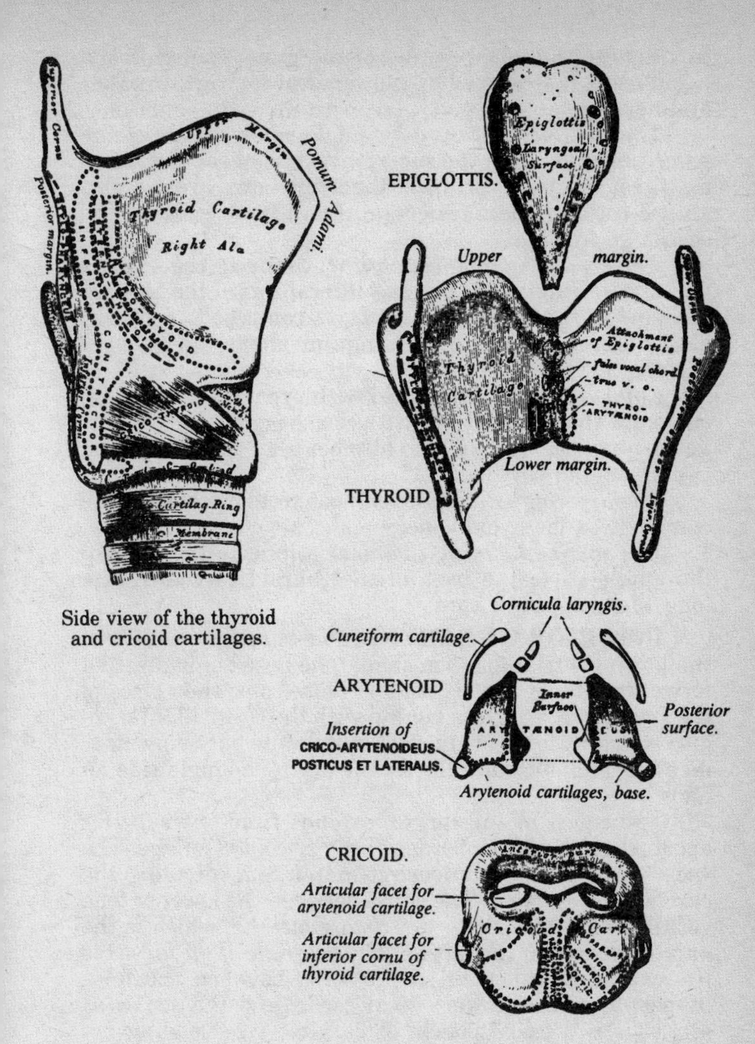

Side view of the thyroid and cricoid cartilages.

The cartilages of the larynx. Posterior view.

cartilage to the upper border of the inner surface of the hyoid bone. It is pierced by the superior laryngeal vessels and nerve.

The two lateral thy´ro-hy´oid ligaments, part of the preceding, pass from the superior cornua of the thyroid to the tip of the great cornua of the hyoid bone; are rounded elastic cords, sometimes containing a bony nodule—the *cartila´go triti´cea.*

The cri´co-thy´roid membrane: connects the thyroid and cricoid cartilages, passing laterally into the inferior margins of the true vocal cords; is composed of yellow, elastic fibres, and is of a triangular shape, its anterior surface being convex.

Cap´sular ligaments: lined with synovial membrane, surround the articulations between the *cricoid* and inferior cornu of the thyroid, and also between the *cricoid* and two *arytænoid* cartilages.

The hyo-epiglot´tic ligament: connects the apex of the epiglottis to the hyoid bone.

The thyro-epiglot´tic ligament: connects the apex of the epiglottis to the back of the thyroid cartilage; is a long, slender elastic cord.

INTERIOR OF LARYNX: *The superior aperture* of the larynx is triangular in shape, the base being directed forwards. The epiglottis bounds it in front; the apices of the arytenoid cartilages behind with the cornicular laryngis; laterally folds of mucous membrane inclosing ligamentous and muscular fibres–the ary´teno-epiglottide´an folds.

The cavity of the larynx extends from the *superior* aperture to the lower border of the cricoid cartilage. The vocal cords form an imperfect diaphragm, dividing the cavity into two parts. The chink between the lower or true vocal cords is the *glottis* or *ri´ma glotti´dis*, which is the narrowest part of the larynx. The *ventricle of the larynx* is the oval depression, on either side, between the false (upper) and true (lower) vocal cords; and the *sac´culus laryn´gis* is a cæcal pouch, of variable size, leading upwards on the outer side of the super-vocal cord; its office is to furnish a lubricating secretion to the vocal cords, which is forced out upon them by the contraction (compression) of the inferior arytæ´no-epiglottide´us muscle—(the *compressor sacculi laryngis* of Hilton).

The *ri´ma glotti´dis* is the narrowest part of the cavity, and is at the lower level of the arytenoid cartilages; its *length* (in the male) is little short of an inch; its *width* (when dilated) from ⅓ to ½ of an inch. (*In the female* these measurements are 2 or 3 lines less). *Its form* varies; when in repose and quiet breathing, it is dilated and somewhat triangular, with the base backwards; in forcible expiration it is smaller than in inspiration; when sound is produced, it is narrowed, the edges of the vocal cords being brought parallel and closer together, being closest when the note is high-pitched.

The superior or *false vocal cords* are two folds of mucous membrane enclosing the superior thyro-arytænoid ligament. They consist of a thin band of elastic tissue (*sup. thyro-arytenoid ligament*), the front attachment being at the angle of the thyroid cartilage, below the epiglottis, and the back attachment being at the anterior surface of the arytenoid cartilage.

The inferior or *true vocal cords* are attached in front to the receding angle of the thyroid cartilage, and behind to the anterior angle at the base of the arytænoid cartilage. They are two strong, yellow, fibrous bands (*inferior thyro-arytænoid ligaments*).

The muscles of the larynx have been described on page 28.

The mucous membrane is similar to that in the mouth, pharynx, trachea and bronchi; by its reduplication it forms the most part of the false (superior) vocal cords; it is thin over the true (inferior) vocal cords. It is covered with columnar, ciliated epithelium below the false vocal cords; above, the cilia are limited to the front of the larynx, the rest of the surface being covered with squamous epithelium.

The glands are muciperous, found in large quantities in the sacculus, and along the posterior margin of the aryteno-epiglottidean fold.

The arteries of the larynx are the laryngeal branches of the superior and inferior thyroid. See page 59.

The veins empty into the superior, middle and inferior thyroid veins.

The lymphatics terminate in the deep cervical glands.

The nerves are the superior laryngeal, inferior or recurrent laryngeal, and branches of the sympathetic. The first supplies the mucous membrane and crico-thyroid muscles; the inf. laryngeal the balance of the structures.

THE TRACHEA.

Or air tube, extends from the lower border of the larynx (5th cerv. vert.) to opposite the 3d dorsal vertebra, there dividing into two bronchi, one for each lung. It is a cartilaginous and membranous cylinder, flattened posteriorly, and is 4½ inches in length, and from ¾ to 1 inch in diameter—being a little smaller in the female.

Relations in the neck: *In front.* Isthmus of thyroid, inferior thyroid veins, sterno-hyoid muscles, sterno-thyroid muscles, cervical fascia, anas. of ant. jugular veins. *Laterally.* Common carotid artery, lateral lobes of thyroid, inferior thyroid artery, infer. laryngeal nerve.

Relations in the thorax: *In front.* 1st piece of sternum, thymus gland, arch of aorta, innominate arteries, left carotid arteries, deep cardiac plexus. *Laterally.* Pneumogastric nerve. *Posteriorly.* Œsophagus.

The cartilages, from sixteen to twenty in number, forming imperfect rings, the anterior ⅔ or convex part being cartilage, the posterior of fibrous membrane. They measure ½ line in thickness and 2 lines in depth; are flattened outwardly, but are convex inwardly. They are all inclosed in a fibrous elastic membrane.

The peculiar ones are the First, which is broader, and is sometimes divided at one end. The Last one, which is thick and broad in the middle, and with a hooked process, curving downwards and inwards between the two bronchi.

Sometimes two or more of the cartilages unite together; they are elastic, and seldom ossify.

The right bron´chus, about one inch long, is shorter and more horizontal in direction than the left. It has 6 to 8 rings. The right pulmonary artery is below; then in front of it, and the vena azygos arches over it from behind.

The left bron´chus is nearly two inches long, and enters the lung about an inch lower than the former. It is smaller, more oblique, and contains from 9 to 12 rings. It crosses the œsophagus, thoracic duct and descending aorta in front, and passes beneath the aortic arch; has the left pulmonary artery at first above, then in front of it.

Transverse section: if this is made a little distance above point of bifurcation, on looking down the tube it will be noticed that the right bronchus opens, or starts, almost directly in the axis-cylinder of the trachea; this being so, it follows that a foreign body drawn into the trachea would fall, or be drawn, into the *right* bronchus instead of the left; this tendency to the right bronchus is also increased by the larger size of the right over the left one.

The mucous membrane contains a large quantity of lymphoid tissue, and has several layers of epithelial cells, the uppermost being the ciliated columnar variety; is continuous with the membrane above and with that of the lungs.

Glands: these are in great abundance in the posterior wall, and are small, ovid bodies, with an excretory duct opening at mucous surface. The secretion is supposed to lubricate the tracheal surface.

Vessels and nerves: *the arteries* are the inferior thyroids. *The veins* open into the thyroid plexus. *The nerves* are branches from the pneumogastric and sympathetic system.

THE PLEURÆ AND LUNGS.

The pleu´ra is a delicate, serous membrane covering the lung-substance as far as the "roots," and is then reflected upon the thoracic walls, making a sac of itself. The part upon the lungs is named the *pleu´ra pulmona´lis*, and that upon the chest-wall the *pleu´ra costa´lis*. *The right* pleural sac is shorter, reaches higher in the neck, and wider than the left.

VESSELS AND NERVES: *the arteries* are branches from the intercostal, internal mammary, musculo-phrenic, thymic, pericardiac and bronchial. *The veins* correspond with the arterial supply; *the nerves* come from the phrenic and sympathetic. *The lymphatics* are numerous.

The lungs, two in number, occupy the thorax, and are separated from each other by the heart and the mediastinum. They are conical in shape and are covered with the pleuræ.

The *apex* projects under the clavicle, into the root of the neck, a distance of 1 or 1½ inches.

The *base* is broad, concave and rests upon the diaphragm, and following the attachment of the midriff, is placed lower posteriorly than anteriorly. The *anterior margin* is thin and sharp of the left lung and presents a notch for the apex of the heart. The *outer surface* of each lung is convex. The *inner* is concave and about its middle presents a slit (*hi´lum pulmo´nis*) where the root of the lung is attached.

Lobes: each lung is divided, by a long, deep fissure, into two main lobes. In the *right* lung a shorter fissure subdivides its upper lobe into two smaller ones, thus making three for it.

The right lung is the larger and shorter of the two, and has three lobes: it is also broader than the left.

The left lung is smaller, narrower and longer than the right, and is divided into two lobes.

The roof of each lung lies a little above the middle and nearer the posterior than the anterior border of the inner surface, and connects it with the trachea and heart. It is formed by the bronchial tube, the pulmonary artery and veins, the bronchial arteries and veins, pulmonary nerve plexus, lymphatics, bronchial glands and areolar tissue, all closed in by a pleural reflection.

The root of the *right* lung lies behind the ascending aorta superior vena cava, and below vena azygos; that of the *left* lies beneath the aortic arch and before the descending aorta.

Topography: *right lung*: anterior border corresponds with median line of chest; extends down from junction of 1st and 2d pieces of sternum to 6th costal cartilage. *Left* is shorter, only extending down to 4th costal cartilage.

The weight of both lungs together averages 42 ounces for adult males, the right weighing 2 ounces more than the left. They are heavier in the male than the female, the bodily proportion being as 1 to 37 in the male, and 1 to 43 in the female. The Sp. Gr. varies from .345 to .746 for the lung tissue, water being 1,000.

Size: when fully inflated, for the average male, there is a capacity for 282 cubic inches of air; "residual" air is about 57 cubic inches.

Color, at birth, pinkish; adult life, a mottled dark slate; old age, mottled bluish-black. The posterior border is usually the darker, and the lungs of males are darker than those of females. The surface has dark polyhedral markings, indicating the lobules.

The substance is of a light, spongy, porous nature, floating in water; crepitates, and is highly elastic.

Structure: the lungs have an external or serous coat, a subserous and the parenchyma.

The serous, or external coat, is a thin investing membrane, transparent, derived from the pleura.

The subserous is a layer of elastic tissue fibres extending into the substance between the lobules.

The parenchyma is of lobules, closely connected together and vary in form and size, the more external being pyramidal and the larger. Each lobule is made up of the termination air-cells of the bronchus and the investing vessels and nerves.

The bron´chus divides and subdivides dichotomously, until the terminal alveoli, or air-cells, are reached. The lining mucous membrane is covered with columnar ciliated epithelium; the muscle fibres are of the unstriped

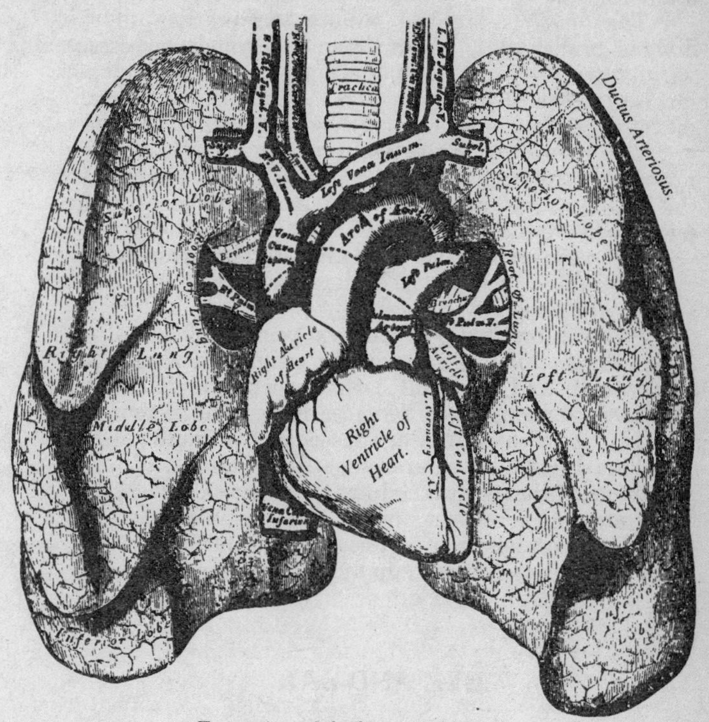

Front view of the heart and lungs.

variety, and in annular layers. The cartilages of the bronchial tubes are of thin laminæ of various forms and sizes, and are found even in tubes of ¼ of a line in diameter.

The air-cells are the minute terminations of the divided bronchioles, and are small polyhedral recesses varying from 1/75 to 1/120 of an inch in diameter, being largest at the surface. They are lined with a delicate mucous membrane covered with squamous epithelium.

The bronchial arteries, from the aorta, supplying nutrition to the lungs. They end in minute vessels upon the walls of the smallest tubes, and terminate in the pulmonary veins.

The veins: the superficial and deep bronchial veins, on *the right side*, terminate in the vena azygos; on the left side, in the superior intercostal vein.

Pulmonary artery conveys venous blood to the lungs, for æration; divides into a minute network upon the air-cells and inter-cellular passages. *The pulmonary capillaries* form plexuses beneath the mucous membrane, on the air-cell walls; they are exceedingly minute vessels.

The lymphatics are superficial and deep, and terminate at the seat of the lung, in the bronchial glands.

Nerves: branches from the sympathetic and pneumogastric form the anterior and posterior plexuses, and filaments therefrom supply the lung structures.

EYE AND EAR.

THE EYE.

The eyeball is contained in the orbit; its shape is spherical with the segment of a smaller sphere (cornea) placed anteriorly. The antero-posterior diameter is about one inch; the transverse, shorter by a line. The optic nerves enter the balls on the nasal side. The eyeball is composed of several tunics and media.

The tunics are 3, viz: 1. Sclerotic and Cornea; 2. Choroid, Iris and Ciliary Processes; 3. Retina.

The refracting media, or humors, are 3, viz: 1. Aqueous. 2. Lens and Capsule. 3. Vitreous.

The sclerotic (meaning hard) is the thick, tough membrane surrounding the eyeball; it is thicker posteriorly than anteriorly; is white externally, and receives the insertion of the muscles which act on the globe;

internally it is of a dark brown color, with grooves for the ciliary nerves, and connected with the external surface of the choroid by the *lam´ina fus´ca.* The optic nerve passes through this membrane behind and to the inner side; this spot is the *lam´ina cribro´sa,* as it is transversed by fibrous septa; an opening in the centre of the lamina, the *po´rous op´ticus,* transmits the central artery of the retina; anteriorly the sclerotic is continuous with the cornea, overlapping it.

Structure: white fibrous tissue, with elastic fibres and fusiform nucleated cells. The capillaries are very small, and it is almost nerveless.

The cornea is the anterior transparent part of the outer coat of the eyeball, forming 1/6 of the globe; it is convex anteriorly, and has been likened to a watch-glass projecting from its case; the anterior surface is consequently smaller than the posterior. The cornea is constructed of five layers, which are arranged from without in as follows:

(1) The conjunctiva, (2) anterior elastic lamina (3) cornea proper, (4) posterior elastic lamina, (5) posterior epithelial layer.

Epithelial lining of aqueous chamber is a layer of transparent nuclear cells.

Arteries and nerves: is non-vascular, the capillaries terminating in loops. The nerves are numerous (20 or 30), and are branches from the ciliary.

SECOND COAT: *choroid, iris, ciliary processes, ciliary muscle and ligament.*

The cho´roid, vascular, or pigment coat extends as far forwards as the cornea, terminating at the ciliary ligament by the ciliary processes; invests 5/6 of the globe. Behind, the optic nerve pierces it; the inner surface is dark brown and covered with the pigment cells of the retina. This coat is resolvable into 3 layers: the *external* consists of curved branches of the ciliary arteries, and the veins (*venæ vorticosæ*) collecting into 4 or 5 groups. *The middle,* consisting of a fine capillary plexus of the short ciliary vessels (*tunica Ruyschiana*). The *internal coat* is the pigmentary layer—hexagonal nucleated cells loaded with pigment granules. (In albinos these cells contain no pigment).

The cil´iary processes (60 to 80) are formed by a folding inwards of the choroid; they are arranged in a circle round the edge of the lens behind the iris; the larger ones are one line in length. Their structure is similar to the choroid.

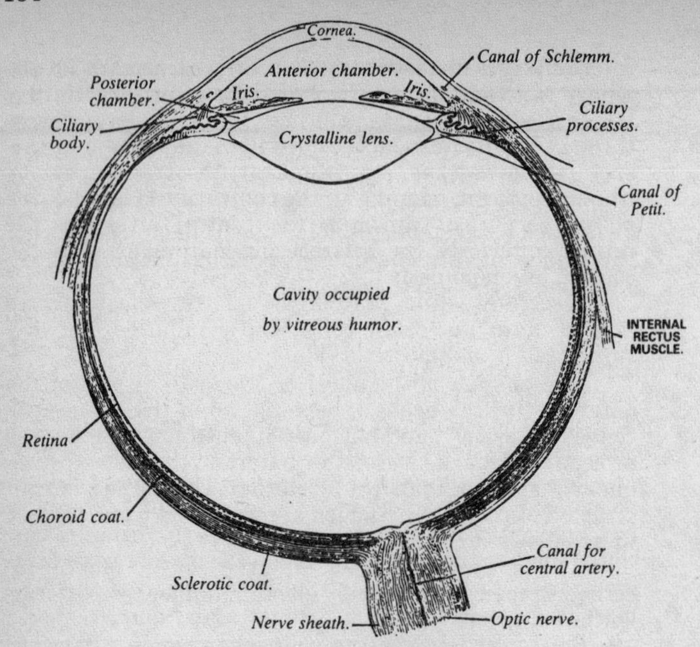

A horizontal section of the eyeball.

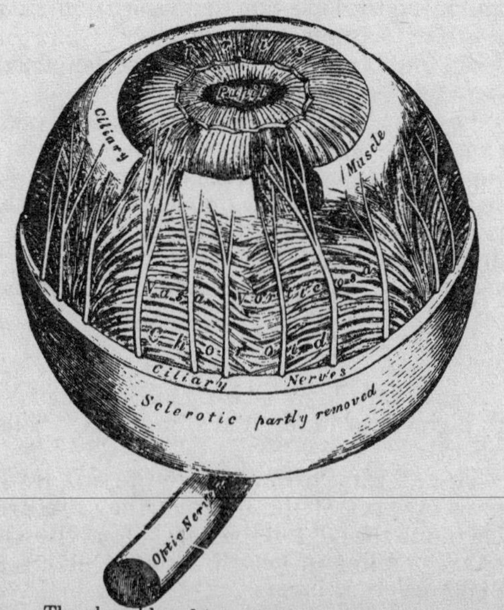

The choroid and iris. (Enlarged.)

The iris (rainbow) is the thin, colored membrane suspended in the aqueous humor in front of the lens; in the centre is an aperture, the *pupil*. By the circumference it is connected to the choroid, and anterior to this to the ciliary ligament, by which it is connected to the cornea and sclerotic. While variously colored anteriorly, its posterior surface is of a deep purple color (uvea) like a ripe grape. *In its structure* are found delicate fibrous tissue bundles, muscular fibres and tissue cells.

The muscle fibres are involuntary and consist of *circular* and *radiating* groups. The former contract the pupil (sphincters); the latter expand it (dilators).

The pupil (*membra′na pupilla′ris*) in the fœtus is closed by a delicate membrane; this disappears at the 8th month, absorption beginning at the centre.

The arteries of the iris are from the long and anterior ciliary and vessels of ciliary processes. (Page 189.)

The nerves are derived from ciliary branches of the lenticular ganglion, and the long ciliary, a branch from the nasal branch of the ophthalmic division of the 5th. They form a plexus around the attached margin of the iris.

The ciliary muscle surrounds the circumference of the iris; is of unstriped fibres, grayish, a circular semi-transparent band ⅛ of an inch broad; thickest in front; it arises from the junction of the sclerotic and cornea, and is inserted into the choroid opposite the ciliary processes. It consists of two sets of fibres, *radiating* and *circular*. This muscle is the chief agent in "accommodation," as by its contraction it is supposed to compress the lens, thus increasing its convexity. The origin of this muscle used to be described as the ciliary ligament.

The ret′ina contains the terminations of the optic nerve; it lies within the choroid coat (thickest behind), and the vitreous humor lies inside; it extends forwards as far as the ciliary muscle, where it ends with a saw-edged border, the *or′a serra′ta*. The outer surface is covered with pigment cells. On the inner surface in the axis of the eye is an elevated yellow spot, *mac′ula lu′tea*, and in the middle of this a depression, *fo′vea centra′lis*. About one-tenth of an inch to the inner side is the *po′rus op′ticus*, transmitting the central artery of the retina. It is semi-transparent, and of pinkish color in health.

STRUCTURE: this is exceedingly complex, being made up of 10 different layers.

1. *Membra′na lim′itans inter′na* is derived from the supporting framework of the retina, and is in contact with the hyaloid membrane of the vitreous humor; is the most internal layer.

2. *Fibrous layer*, continuation of nerve-fibres of the optic nerve; is thickest at optic nerve entrance.

3. *Vesicular layer* is made up of large, flask-shaped ganglionic cells; is in a single layer, except at macula lutea, where there are several layers.

4. *Inner molecular layer* is made up of a granular looking substance and a dense fibrillar reticulum.

5. *Inner nuclear layer*, made up of 3 kinds of nuclear bodies: (*a*) oval nuclei or bi-polar (branched) nerve-cells surrounded with protoplasm; (*b*) unbranched nerve-cells; (*c*) cells connected with Müller's fibres.

6. *Outer molecular layer*, thinner than the inner, though made up of much the same structures, with the addition of branched, stellate cells, which Schultze considers as ganglion cells.

7. *Outer nuclear layer*, somewhat similar to the inner, but has a distinct division of its layers into *rods* and *cones*. The *rod-granules*, quite numerous, have a peculiar striped appearance, and from either end a fine process. The *cone-granules*, less numerous than the rod, contain a large pyriform nucleus; have no stripings, and are placed close to the following layer.

8. *Membra′na lim′itans exter′na*, like the *interna*, is derived from Müller's fibres.

9. *Layer of rods and cones* (Jacob's membrane): the *rods* are solid, of uniform size, and stand perpendicular to the surface; each rod consists of two portions, outer and inner, of equal length, cemented together; these portions differ in refraction and in taking coloring of re-agents, the inner becoming more easily stained; the outer portion, showing striæ, is made up of super-imposed discs; the inner portion partakes more of a granular nature.

The cones are flask-shaped, the pointed end towards the choroid, and are made up of two portions, similar in markings and structure, to the rods.

10. *The pigmentary layer* (formerly regarded as part of choroid) is the most external layer, and consists of hexagonal epithelial cells.

All the above ten layers are connected together by a framework of connective tissue fibres (Müller's fibres). As they pass through the several layers numerous roughnesses show, as if processes were broken off.

Exceptions: at the *macula lutea* the nerve fibres are wanting, but there are several layers of cells (vesicular), and there are no rods in Jacob's membrane, and only long, curved cone fibres in the outer nuclear layer. The staining of this spot imbues all the layers except Jacob's membrane, and it does not seem to consist of pigment granules; it is of a rich yellow color.

At the *fovea centralis* the only parts are the cones of Jacob's membrane, the outer nuclear layer, and a very thin inner granular layer.

At the *ora serrata* the layers terminate abruptly, and the fibres of Müller assume the appearance of columnar epithelial cells.

Arte´ria centra´lis ret´inæ, with its vein, pierces the nerve and enters the eye at the *po´rus op´ticus*: the artery divides into 4 or 5 branches, running forwards between hyaloid and nervous layer, pierces the latter and, dichotomously, gives origin to a minute capillary plexus in the inner nuclear layer.

Aqueous humor fills both the ant. and post. eye chambers; is of alkaline reaction; weighs about 5 grains, and is water (chiefly) and a little chloride of sodium.

The anterior chamber is bounded anteriorly by the cornea, and posteriorly by the iris.

The posterior chamber, much smaller than the anterior, is bounded anteriorly by the iris, and posteriorly by the capsule of the lens, the suspensory ligament, and the ciliary processes.

The vitreous body forming 4/5 of the globe, occupies the concavity of the retina, and is enclosed in a *hyaloid* membrane. In front it is closely adherent to the lens and its capsule. It is a transparent, jelly-like mass of albuminous fluid, resembling closely pure water in its composition. In the centre of the vitreous, running from before backwards, is the empty *canal of Stilling*.

The hy´aloid membrane encloses all the vitreous except the anterior surface, hollowed out for reception of the lens; it is reflected to the lens margin, forming *the suspensory ligament of the lens*. It is a delicate membrane, structureless, except where it forms the suspensory ligament; it there has a few elastic fibres.

Nourishment: in the fœtus a small artery passes through the vitreous to the lens; in the adult no vessel penetrates it, hence it must be nourished by the retinal and ciliary vessels.

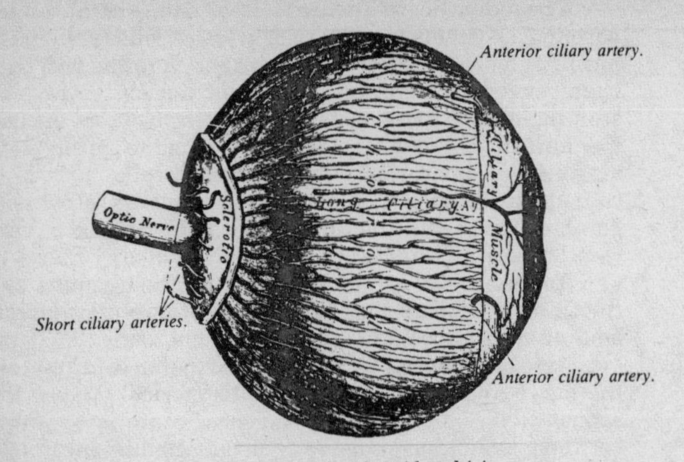

The arteries of the choroid and iris.
The sclerotic has been mostly removed. (Enlarged.)

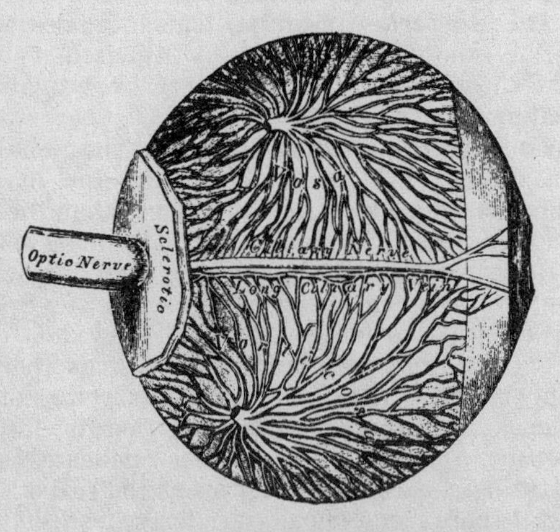

The veins of the choroid. (Enlarged.)

The capsule of the lens is a thin, brittle, transparent, elastic membrane closely surrounding the lens, forming, by its anterior surface, the back of the posterior chamber. It is maintained in its position by the suspensory ligament of the lens, thicker in front than behind. The ant. surface of lens is connected with the capsule by a single layer of transparent, many-sided cells; no epithelium on post. surface. The *liquor Morgagni* is found only *post mortem*, and is from the breaking down of these cells.

The lens is a transparent body, convex anteriorly and more so posteriorly; it is 1/3 of an inch in diameter and 1/4 of an inch in thickness. It consists of concentric layers, pealing off as does an onion, after being boiled or hardened in alcohol. (It is also demonstrated by these reagents that the lens consists of 3 spherical, triangular segments.)

The laminæ are made up of hexagonal prisms laid parallel, their breadth being 1/5000 of an inch; the fibres of the outer layers are nucleated.

Changes in age: in the *fœtus*, it is pinkish and spherical and soft. In the *adult*, it is transparent, colorless, harder, and with the posterior surface having greatest convexity. In *old age*, is slightly opaque, flattened on both sides, denser, and of a straw tint; it is enclosed in the capsule. The *canal of Petit* surrounds its circumference.

The suspensory ligament is placed between the anterior surface of the vitreous body and the ciliary processes; is a transparent, thin structure, assisting in holding the lens. Its outer surface has numerous stained folds, or plaitings, arranged around the lens, which receive between them the folds of the ciliary processes.

The canal of Petit, about 1 line wide, is bounded in front by the suspensory ligament; behind, by the vitreous humor.

Arteries: *short ciliary*, to the choroid and ciliary processes; pierce the sclerotic about the optic nerve entrance.

Long ciliary (2) run forward between sclerotic and choroid to the ciliary muscle, and form a vascular circle about the iris, giving off therefrom numerous smaller branches to muscular structures.

Anterior ciliary (5 or 6) are branches from the ophthalmic piercing the eyeball at just back of the cornea, and go to the ciliary processes and the vascular circle about the iris, *arte'ria centra'lis ret'ina*. See page 187.

Veins, (usually 4), are formed from the choroid

surface and, piercing the sclerotic midway between the cornea and optic nerve, empty into the ophthalmic vein.

Nerves: *optic* (page 43), with its decussating fibres, the nerve of sight; *long ciliary*, from the nasal branch of the ophthalmic; *short ciliary*, from the ciliary ganglion.

APPENDAGES OF THE EYE.

The eyebrows (*supercil´ia*) are two arched eminences over each orbit, consisting of thickened integuments and muscles (page 21) surmounted by hairs.

The eyelids (*palpebræ*) are two movable folds, an upper and a lower, the upper one being more movable, which, by their closure, protect the eye from injury. When the eyelids are open the angles of junction of the upper and lower lids are called *external* and *internal can´thi*. In the inner canthus the lids are separated by the *la´cus lachryma´lis*, which is occupied by the *carun´cula lachryma´lis*, and opposite the commencement of this, on each lid, is the *lachrymal papilla*, which is pierced by the *punc´tum lachryma´le*, the commencement of the *lachrymal canal*. When the eyelids are opened an oval fissure (*fissu´ra palpebra´um*) is left. (For the eyelid muscles see page 22).

Structure from without inwards: skin, areolar tissue, orbicularis muscle, tarsal cartilage, fibrous membrane, Meibo´mian glands, conjuncti´va; the upper lid has also the aponeurosis of the leva´tor palpe´bræ.

The tarsal cartilages: the *superior one* is the larger, being 1/3 of an inch broad at the centre, and is of semi-lunar shape. The *inferior* is thinner and of elliptical shape. Both help to support the lids, and the ciliary margins are the thicker. The outer angle of each is attached by a ligament (ext. palpebral) to the malar bone.

The tarsal ligament is a layer of fibrous membrane giving support to the lids, and retaining the cartilages in position.

The Meibo´mian glands, on the inner surface of the eyelids, between the cartilages and conjunctiva, are about 30 in number in the upper lid, and longer; are shorter and some less in the lower lid. They each consist of a single, straight, cæcal tube, with numerous secondary follicles opening into them. The cartilages are grooved to contain them. They run in parallel rows with the short axis of the lid. Their secretion lubricates the lids.

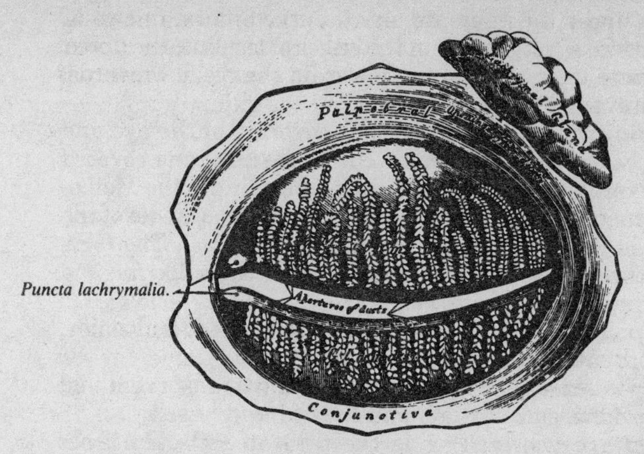

The Meibomian glands, etc., seen from the inner surface of the eyelids.

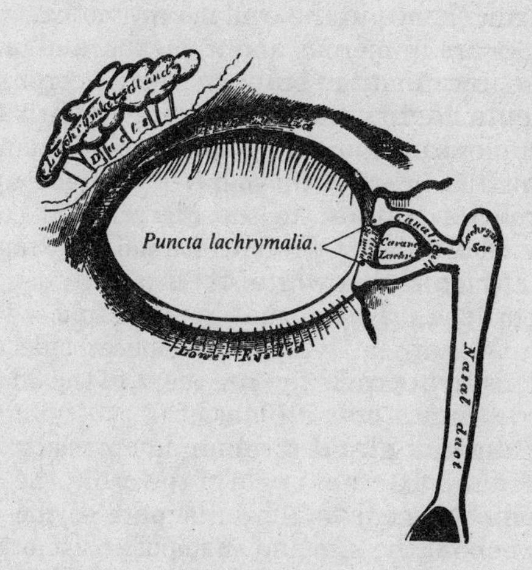

The lachrymal apparatus. Right side.

The eyelashes (*cil´ia*) are more numerous and larger on the upper lid; they are short, curved hairs, in two or three rows, springing from the edges of the lids; the upper lid's curve upwards and the lower lid's curve downwards, thus preventing interlocking.

Conjunctiva is the mucous covering of the eye and lids. *The palpebral portion* is thick, vascular and covered with papillæ, and continuous with that lining the Meibomian ducts and lachrymal canal. At the inner angle of the eye it is folded on itself (*pli´ca semi-luna´ris*). The folds upon the ball are called, respectively, the *superior* and *inferior* palpebral folds.

Upon the sclerotic it is loosely connected, is thinner, transparent, and slightly vascular.

Upon the cornea it is extremely thin, transparent and closely adherent, with no papillæ or bloodvessels.

For´nix conjuncti´væ is the point of reflection from the lid to the ball, and hereat are numerous mucous convoluted glands; these are more numerous in the upper lid. Henle's *trachoma glands* are found usually near the inner canthus.

Lymphatics arise from a delicate zone about the cornea, and run thence to the ocular conjunctiva.

The nerves are numerous and in plexus, and (according to Krause) terminate in bulbs or "tactile corpuscles."

Carun´cula lachryma´lis, a small, reddish elevation at each inner canthus, filling up the small triangular space thereat (the *la´cus lachryma´lis*). Is made up of a cluster of follicles (similar to the Meibomian) covered with mucous membrane. A few slender hairs spring from the surface. It furnishes a white secretion.

Pli´ca semi-luna´ris is a small mucous fold, with the concavity to the cornea, lying to the outer side of the caruncula. This structure is the rudiment to the 3d eyelid in birds—the *membra´na nicti´tans.*

The lach´rymal gland occupies a depression in the frontal bone in the external angle of the orbit; the anterior margin is connected to the back part of the upper eyelid. It is of about the size and shape of an almond, and its under (concave) surface rests on the eyeball upon the superior and external rectus muscles. The fore part of the gland is separated from the main body by a small depression, and this lobe is described as the *palpebral portion of the gland.*

Structure: is similar to the salivary glands.

Ducts are 6 or 7 in number, which open by minute orifices arranged in a row on the upper (and outer) half of the conjunctiva, near its juncture with the ball.

The lach´rymal canals commence as minute orifices, the *puncta lachrymalia*, which are the openings of the canaliculi on slightly elevated papillæ (*papil´la lachryma´lis*), which join to pass inwardly and enter the *lachrymal sac.*

The superior canal is the smaller and longer of the two, ascending at first, then tending downwards to the sac.

The inferior, at first descends, then ascends to the sac. They are both made up of elastic and somewhat dense materials.

The lach´rymal sac is placed in a groove formed by the lachrymal bone and the nasal process of the superior maxilla; it is the dilated upper end of the nasal duct. Is oval, the upper extremity being somewhat bulbous and closed in, and covered by the tensor tarsi muscle. It has a fibrous elastic coat, and is lined with mucous membrane, joining that in the nose and the conjunctiva.

The nasal duct leads from the lachrymal sac to the inferior meatus of the nose, where it opens by a valve formed of the mucous membrane. Is about ¾ of an inch in length, somewhat expanded at both ends, and has a direction downwards, backwards and slightly outwards. Is coverd with ciliated epithelium (as in the nose) except in the canaliculi, where it is of the scaly variety.

THE EAR.

The ear is divided, for the purpose of description, into three parts, external, middle, and internal.

EXTERNAL EAR consists of the auricle or *pinna*, the expanded portion for collecting sound vibrations, and the *external auditory meatus* for conducting the same to the ear-drum.

The pin´na is a plate of yellow cartilage of ovoid form, covered with integument and attached to the commencement of the meatus; it has numerous ridges and depressions, as follows: the external rim is the *helix*, and anterior and parallel to it is another ridge, the *anti-helix*, which is bifurcated above to enclose the *fossa of the anti-helix*; between the helix and anti-helix is the *fossa of the helix.* Anterior to the anti-helix is a depression, the *concha*; projecting backwards over the meatus is the

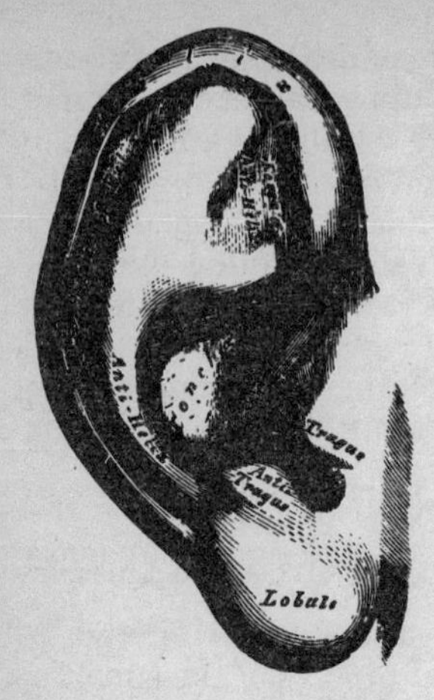

The pinna, or auricle.
Outer surfaces.

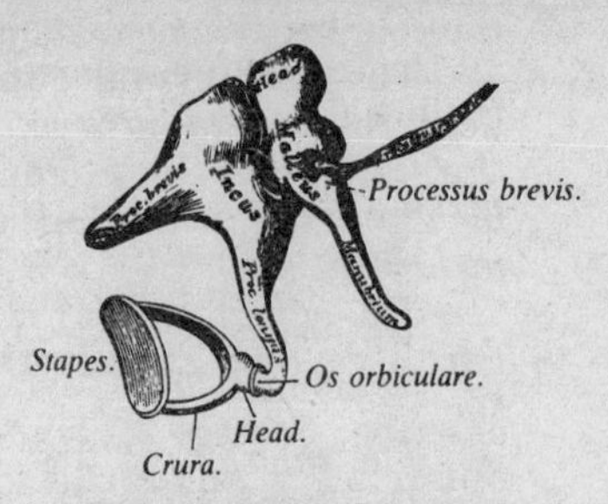

The small bones of the ear,
seen from the outside.
(Enlarged.)

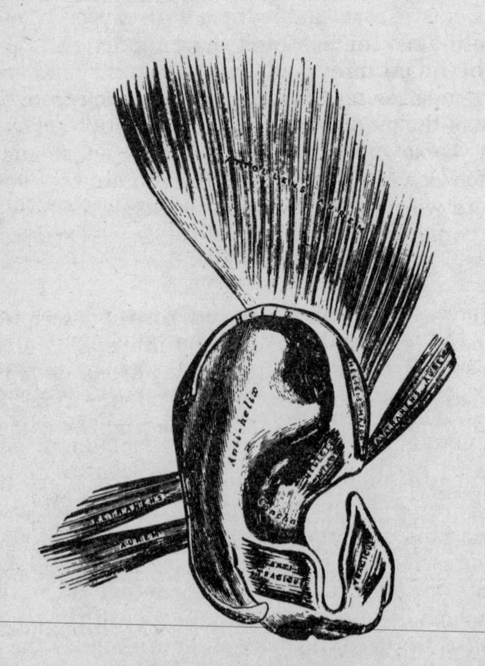

The muscles of the pinna.

tragus (goat's beard), and opposite to the latter is the *anti-tragus*; the lowest part of the pinna is called the *lobule*.

The cartilage is in one piece, though it is not found in all the external ear parts; is of the variety known as the "yellow fibro-cartilage."

Ligaments: these are in two sets: 1st. Those connecting the various cartilages together, two in number. 2d. Those connecting the pinna to the head, also two in number—the *anterior* and the *posterior*.

Muscles: these are also in **2** sets similar to those of the ligaments. Those connecting the ear to the head are described on page 21, while the pinna muscles proper are 6 in number, and are as follows:

Hel´icis ma´jor, from process of helix—into anterior border of helix.

Hel´icis mi´nor, helix—into concha.

Trag´icus, outer part of tragus—same.

Anti-trag´icus, outer part of anti-tragus—processus caudatus of helix.

Transver´sus auric´ulæ, on cranial surface of pinna—same.

Obliq´uus au´ris, back part of concha—posterior convexity of pinna.

Arteries: posterior auricular, a branch from external carotid; anterior auricular, from temporal; also a branch from occipital artery.

Veins accompany above arteries.

Nerves: auricularis magnus, from the cervical plexus; posterior auricular, from facial; auricular branch, from pneumo-gastric; auriculo-temporal, from inferior maxillary.

The external auditory mea´tus reaches from the bottom of the concha to the membrana tympani, and is 1¼ inches in length. It is arched slightly upwards, and is directed forwards and inwards; at the entrance the greatest diameter is vertical, but transversely at the tympanum, and it is smallest at the middle; it is formed in part by cartilage, and in part by bone. The outer or cartilaginous part is continuous with the pinna, and is about half an inch long; the inner or osseous part is ¼ inch longer than the preceding, and at its inner end there is a groove round the sides and floor for the insertion of the membrani tympani.

The skin lining the meatus is very thin, and adheres very closely to the tube proper.

In the outer part of the meatus are hairs and ceruminous glands; the latter secrete the ear-wax.

The arteries are branches from post. auricular, internal maxillary and temporal.

The nerves are branches from the auriculo-temporal branch of the inferior maxillary.

THE MIDDLE EAR or TYM´PANUM: the tympanum is contained in the temporal bone, petrous portion; is bounded in front by carotid canal; behind by mastoid cells; below by jugular fossa; internally by labyrinth; externally by meatus auditorius. It communicates with the pharynx by the Eustachian tube, and is traversed by a chain of bones, which connect the membrana tympani with the internal ear; it is filled with air.

The cavity of the tympanum (3 by 5 lines in extent, being longest from before backwards) is bounded *externally* by the meatus and membrana tympani, *internally* by the external surface of the internal ear, and it communicates posteriorly with the mastoid cells.

The roof is formed by a thin plate of bone separating the tympanum and the cranium.

The floor is formed by the roof of the jugular fossa.

THE OUTER WALL is formed by the membrana tympani and the bone around it; the *following fissures* are seen:

The Glasserian fissure: through which the processus gracilis of the malleus, and the laxator tympani, pass.

I´ter chor´da poste´rius: leading to a canal, which opens into the aquæductus Fallopii.

The i´ter chor´da ante´rius: leading to the canal of Huguier; the chorda tympani leaves the tympanum here.

THE INNER WALL is vertical and looks outwards, and presents the following:

The fenestra ovalis: a reniform opening leading into the vestibule and occupied by base of stapes.

The ridge of the aquæductus Fallo´pii: placed just above the preceding and curving downwards at posterior wall.

The prom´ontory: a hollow prominence placed below the fenestra ovalis, formed by the projecting cochlea.

The fenes´tra rotun´da lies at the bottom of a funnel-shaped depression, and leads to the cochlea; is below and behind the fenestra ovalis, and closed by a 3-layer membrane (*membra´na tympa´ni secunda´ria*); internal layer is mucous (from tympanum); middle layer, fibrous; external layer, serous (from cochlea).

The pyramid is placed just behind the fenestra ovalis; it contains the stapedius, the tendon of which projects through the apex; its cavity is prolonged minutely to the aquæductus Fallopii for passage of nerve to stapedius.

THE POSTERIOR WALL, wider above than below, presents many irregular apertures, which are the **openings of the mastoid cells**; there is one large, irregular aperture, others are smaller, and all are lined with mucous membrane continuous with the tympanum.

THE ANTERIOR WALL, wider above than below, and corresponding with carotid canal, from which it is separated by a thin plate of bone, shows the following:

The **canal for the ten´sor tym´pani**, the **Eustach´ian tube**, and the **proces´sus cochlearifor´mis**; the latter is a process of bone separating the two canals; the canal for the tensor is the smaller of the two, and is rounded; it transmits the tensor tympani muscle and tendon.

The Eustach´ian tube leads into the pharynx; is partly cartilaginous and partly osseous; the internal or cartilaginous part is trumpet-shaped, and terminates in an oval opening. Is from 1½ to 2 inches in length, the osseous portion being ½ inch. The mucous membrane lining is continuous from pharynx to tympanum, and is covered with ciliated epithelium.

The membra´na tympan´i is the membrane which divides the external and middle ears; it contains, between its layers, the handle of the malleus, which is attached near the center, and which makes the membrane concave externally. Is thin, semi-transparent, oval, and directed obliquely downwards and inwards.

Structure: is of three layers: the internal or mucous, from the tympanum; the middle, fibrous and some elastic tissue; the external, or cuticular, from the meatus auditorius.

The os´sicles of the tympanum: the tympanum is traversed by a chain of small bones, 3 in number: the Malleus, Incus and Stapes.

The mal´leus or hammer, attached by a thin process, *the handle* (*manu´brium*) to membrana tympani; has *a head* separated from the handle by *the neck*, on which latter are *two processes*, one short for the tensor tympani, the other, *processus gracilis*, extending down the Glasserian fissure for the laxator tympani.

The in´cus or anvil is like a bicuspid tooth, with the part answering to the crown articulating with the malleus; has *two processes*; the *short one* is attached to the margin of the mastoid opening, and the *long one*, terminating in the os orbiculare, articulates with the stapes.

The sta´pes closely resembles a stirrup; the *head* articulates with the incus; the *neck* receives the stapedius, and the *base* is fixed to the margins of the fenestra ovalis.

The ligaments: (1) the *Suspensory of the Malleus*, a round bundle from tympanum to the head. (2) *Post. of Incus*, short, thick band to posterior wall of tympanum. (3) *Annular of Stapes*, from base to margin of fenestra ovalis. (4) *Suspensory of Incus*, from roof of tympanum to incus, upper part.

Muscles: (see page 21). *Action*: the Tensor draws the membrane inwards and so increases tension. The laxator draws it outwards and so loosens tension. The Stapedius draws head of stapes backwards and causes the base to rotate, and (probably) compresses contents of vestibule.

Mucous membrane of the tympanum is slightly vascular, thin, and continuous with that of pharynx through that in the Eustachian tube. It covers the ossicles, muscles and nerves and forms the internal layer of the membrana tympani. It is thickest and reddest in the cartilaginous (lower) portion of the Eustachian tube.

Arteries of the tympanum are 5 in number; the two larger are: the branch from the int. maxillary (to the membrana) and the stylo-hyoid branch from the post. auricular (to the tympanum and mastoid cells). The other branches are the petrosal, from the middle meningeal; a branch up the tube from the ascending pharyngeal; a branch from the int. carotid that perforates the thin anterior wall of the tympanum.

The veins terminate in the middle meningeal and pharyngeal, then to the int. jugular.

Nerves are of 3 classes: (*a*) those to the muscles; (*b*) those to the lining membrane; (*c*) those to other nerves. The *first* are given on page 21; the *second* are branches from the tympanic plexus; the *third* are branches of the glosso-pharyngeal (Jacobson's), with the sympathetic, etc. This nerve enters the tympanum by an aperture in the floor, close to the inner wall, and after supplying adjacent structures divides into 3 branches of communication: one to the carotid plexus; one to the petrosal nerve in the hiatus Fallopii; one through the petrous portion of the temporal bone to join the otic ganglion.

The chor´da tym´pani quits the facial near stylomastoid foramen, enters the tympanum and arches forwards across it to the Glasserian fissure, being covered in its course with mucous membrane.

THE INTERNAL EAR, or LABYRINTH, is divided into *osseous* and *membranous* portions, the former enclosed within the latter. Within the membranous labyrinth is a fluid, the *endolymph*, and outside, between the membranous and osseous labyrinths, is a fluid, the *perilymph*.

THE OSSEOUS LABYRINTH consists of the *vestibule*, the *cochlea* and the *semicircular* canals.

The ves´tibule is the common central part of the labyrinth, of ovoidal shape, about 1/5 of an inch in size. On its outer wall is the *fenes´tra ova´lis*, closed by the base of the stapes; on its *inner wall* is a depression, the *fo´vea hemispherica*, perforated by several holes (*mac´ula cribo´sa*), for the divisions of the auditory nerve, and behind this is a ridge, the *cris´ta vestib´uli*. Behind the crest is the *aqueduct of the vestibule*, which transmits a small vein. On the roof is a depression, the *fo´vea hemi-ellip´tica*. At the posterior part are the five openings of the *semi-circular canals*, and at the anterior part a large oval opening, the *apertu´ra sca´læ vestib´uli coch´leæ*.

The semi-circular canals are three arched osseous canals opening, by 5 mouths, into the vestibule, forming about two-thirds of a circle; each present at one end a dilated part, *the ampulla*. Two of the canals are vertical and the third is horizontal. Their diameter is about 1/20 of an inch. One is called the *superior*, and has a vertical direction. The *posterior* one also has a vertical direction, and is directed backwards; it is also the longest. The *external* one is horizontal, and is the shortest.

The coch´lea is cone-shaped, and consists of a tapering spiral canal (like a snail shell), with the inner wall formed by its axis or *modiolus*. The canal winds about this axis 2½ times; is about 1½ inches in length and 1/10 of an inch in diameter, and is divided into two scalæ by a partition of bone and membrane, the *lamina spiralis*. The enclosed arched extremity of the cochlea is called the *cupola*, and the first turn of the canal bulging into the tympanum forms the promontory; the cupola presents the smallest portion of the canal. The *lam´ina spira´lis os´sea* ends at the apex of the cochlea in a small *hamulus*, which, when detached, leaves a small opening, the *helicotrema*, by which the two scalæ communicate.

This canal is divided into 3 chambers or *sca´læ*: the *sca´la tympa´ni* is the lower one, it commences at the fenestra rotunda.

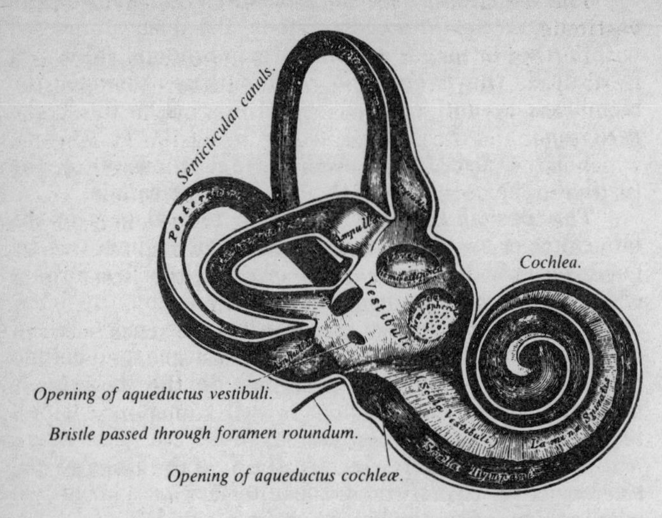

The osseous labyrinth laid open. (Enlarged.)

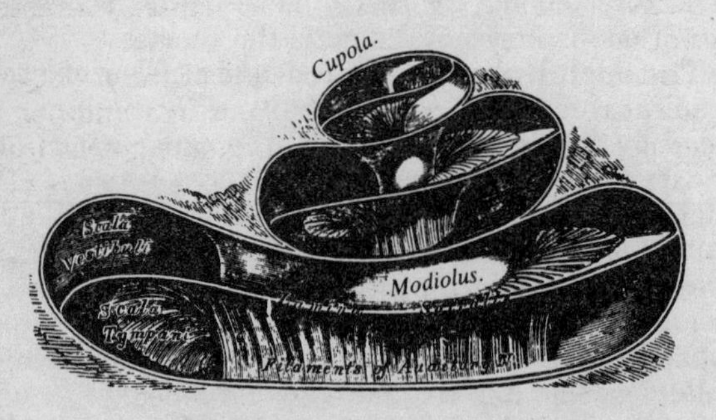

The cochlea laid open. (Enlarged.)

The sca´la vestib´uli commences at the cavity of the vestibule.

The sca´la media: besides these two scala there is a third space, the scala media (closed above and below), being separated from the scala vestibuli by the *membrane of Reissner* and the floor formed by the basilar membrane, which is the part of the lamina spiralis; this swells out at its extremity, forming the *ligamen´tum spira´le.*

The lim´bus lam´inæ spira´lis is the swollen periosteum at the edge of the lamina spiralis, which terminates in a grooved edge, the *sul´cus spira´lis*, the lower lip of which attaches to the basilar membrane.

The rods of Corti are contained in the space between the upper and lower lips of the sulcus spiralis. Waldemeyer estimates 6,000 of these rods for the inner layer and 4,500 for the outer layer; they are supposed to be the terminal apparatus for hearing.

Lining: the entire osseous cavity of the labyrinth is lined with a thin fibro-serous membrane, its surface covered with epithelium and secreting a thin limpid fluid, the *perilymph* (*aqua labyrinthi*), which separates it from the membranous labyrinth. This membrane has no communication with the lining of the tympanum, though it is continuous with that in the other cavities.

THE MEMBRANOUS LABYRINTH is a closed sac within the osseous labyrinth, containing the *endolymph*, and on the wall of which are the ramifications of the auditory nerve. In shape it resembles the osseous labyrinth.

The vestib´ular portion consists of the *u´tricle* and the *sac´cule*, which are distinct from one another; the former is of oblong form, and the larger, communicating with the 5 openings of the labyrinth. The latter is globular, and is quite distinct from the former.

The membranous semi-cir´cular canals are about one-third the size of the osseous ones; they open by five openings into the utricle.

The sca´la me´dia forms the membranous part of the cochlea, and has been already described.

Stays: this membranous portion is held in position by fibrous bands to the osseous framework, these bands carrying the nerves and bloodvessels as well.

Structure: is semi-transparent and consists of 3 coats; the *external*, of flocculent structure, with numerous pigment cells (similar to those in retina); the *middle* resembles the hyaloid membrane; the *inner* is polygonal, nucleated epithelium secreting the endolymph.

Otoliths: 2 small, round bodies of minute carbonate of lime crystals in the wall of the ventricle and saccule opposite the nerve distribution.

Arteries: *int. auditory*, from the basilar; *stylo-mastoid*, from the post. auricular; occasionally, *branches from occipital.*

Veins terminate in sup. petrosal sinus.

Nerves: *the auditory* (*portio mollis* of 7th pair), the special nerve of hearing, divides at bottom of meatus (internal) into 2 branches, the Cochlear and Vestibular.

The Coch'lear divides into numerous filaments at base of modiolus, which ascend along its canals to the rotunda, having numerous ganglionic enlargements and plexuses.

The Vestib'ular divides into superior, middle and inferior branches: the former, the largest, distributes branches to the utricle, ampulla and ext. and sup. semi-circular canals. The Middle gives numerous branches to the saccule. The Inferior, and smallest branch, is distributed to the ampulla and post. semi-circular canal.

THE URINARY ORGANS.

THE KIDNEYS.

The kidneys, the two largest tubular glands of the body, secrete the urine, and are situated in the posterior part of the lumbar region of the abdomen behind the peritoneum, extending from the 11th rib to nearly the crista ilii, the right being placed lower than the left.

The average length of each kidney is four inches; breadth, two inches, and thickness, one inch.

The average weight, for the adult male, is from 4½ to 6 ounces; for the adult female, from 4 to 5 ounces. The *left* is usually 2 drachms heavier than the right one. Their combined weights are, to the whole body, as 1 to 240.

Color: is dark red, and the texture firm and granular, though easily lacerated under pressure.

Relations: the relations of the two kidneys differ somewhat, though each is covered with peritoneum anteriorly.

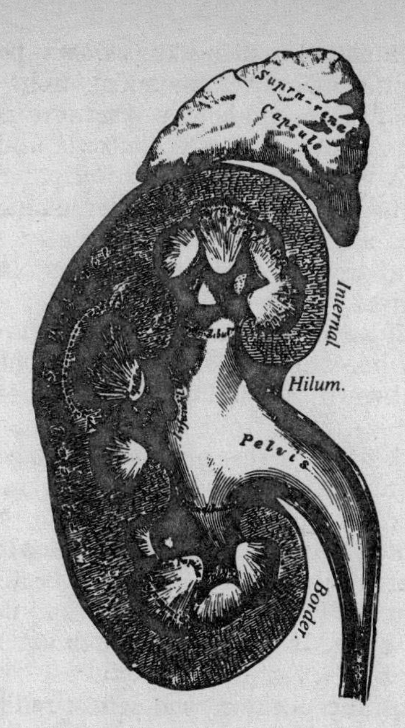

Vertical section of kidney.

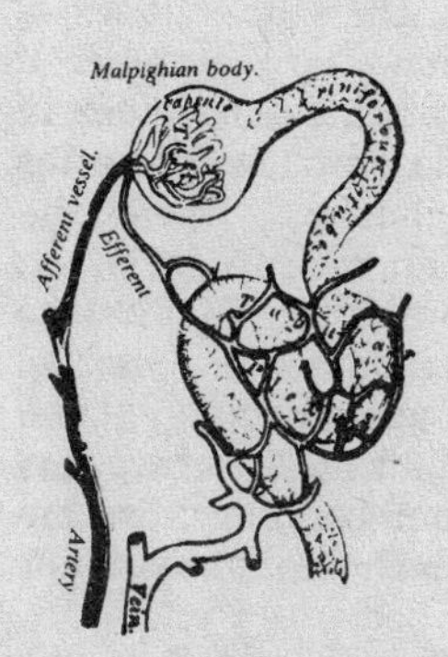

Minute structure of kidney.

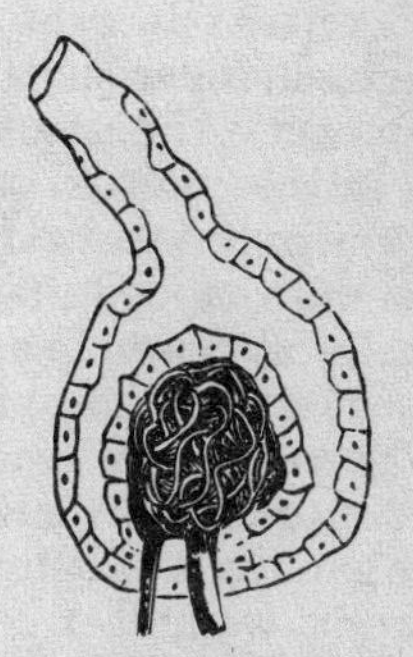

Malpighian body.

RELATIONS OF THE RIGHT KIDNEY: *In front.* Right lobe of liver, 2d part of duodenum, ascending colon. *Behind.* Right crus of diaphragm, quadratus lumborum, psoas.

RELATIONS OF THE LEFT KIDNEY: *In front.* Cardiac end of stomach, lower border of spleen, tail of pancreas, descending colon. *Behind.* Left crus, quadratus lumborum, psoas.

The external border is convex, and is placed outwards and forwards.

The internal border is concave, and at the center is the fissure or *hilum*, where the vessels enter, here lying from before backwards as follows: renal artery, vein, ureter.

Above each kidney is the supra-renal body slightly embracing it.

Below each kidney is the iliac crest.

GENERAL STRUCTURE: 1st. *Outer* or **Cortical portion**: this is closely covered in by the *Capsule* (a fibrous coat reflected inwards at the hilum), and is soft, granular and of a reddish-brown color. It is composed of *tubuli uriniferi*, bloodvessels, lymphatics and *Malpighian bodies.*

2d, or **Medullary portion**, consists of reddish, striated conical masses (*Malpighian pyramids*) from 8 to 18 in number; the base of each is towards the kidney surface, and is surrounded by a cortical arch, and the apex is covered by mucous membrane (projecting into the calices), and is called *papilla* or *mamilla.* Besides these pyramids, arteries and veins, the *looped tubes of Henle*, there enter into its composition a large number of straight uriniferous tubes passing from base to apex of the pyramids, inosculating freely, so that fewer mouths open at the mamilla surface.

Malpigh´ian bodies, 1/104 of an inch in diameter, are small rounded masses of convoluted tubes, of a deep red color, enclosed in a membranous capsule. (*Malpighian capsule*). The *Malpighian tuft* of tubes are the *afferent* and *efferent* renals, derived from the interlobular artery and arteriæ propriæ renales, these anastomosing form an enmeshing venous plexus about the uriniferous tube adjacent.

The Malpighian capsule is lined with flattened epithelial cells without cilia.

Tu´buli urinif´eri commence as a cæcal dilation of the Malpighian capsule, and they terminate in the opening on the summit of the papillæ in the calices. The contracted portion at capsule is called the neck; it then becomes convoluted, and finally, in the medullary portion of the kidney, assumes a spiral condition and finally straight. The urine is secreted in the capsule.

Structure: The tubuli consist of basement-membrane, lined with epithelium, of various characters, according to the portion of the tube examined.

Arteries: the renal divides into 4 or 5 branches at the hilum, these subdivide until the minuter terminal branches of the *arteriæ propriæ venales*, which surround the Malpighian bodies, are reached.

The interlobular arteries are branches from the a propriæ renales, and supply the capsule and Malpighian bodies.

The arterio'læ rectæ are a second set from the propriæ renales; supply the medullary pyramids, and terminate in the venous plexuses thereabouts.

Veins: these arise from 3 sources: 1st. Those beneath the capsule being terminations of the interlobular arteries. These join with the 2d set (those around the tubuli contorti) to form the *venæ interlobulares*; these then pass to the bases of the Malpighian pyramids, and join with the 3d set, the *venæ rectæ*, which set is formed from the terminals of the arteriolæ rectæ. These venæ rectæ pass straight outwards and go to form the proper renal veins. These (*venæ propriæ renales*) run with the arteries, along the sides of the pyramids, receiving the efferents from the Malpighian bodies, to the sinus, there forming *the renal vein*, which passes through the hilum and empties into the vena cava inferior; the left renal being larger than the right one.

Nerves, 15 in number, and small. They have ganglia upon them, and are derived from the solar plexus, semilunar ganglion and lesser and smallest splanchnic.

Lymphatics are in superficial and deep sets, and all terminate in the lumbar glands.

URETER.

Relations of the ureter: *Behind.* Psoas, common or external iliac artery. *In front.* Spermatic vessels, ileum (right side), sigmoid flexure (left).

Each kidney is connected with the bladder by a ureter, which serves to convey urine to the latter viscus; the top of each ureter is expanded and forms the *pelvis of the kidney*, which is divided into three parts called *infundibula*, which are subdivided into *calices*. Into these calices small *papillæ* project, which are the apices of the *pyramids of Ferrein*, which latter form the medullary substance of the kidney.

The *right* ureter lies close to the outer side of the inferior vena cava.

Structure: the ureter has 3 coats, mucous, muscular and fibrous.

The mucous is smooth, with a few longitudinal folds, and is covered with "transitional" epithelium, like that in the bladder. *The muscular coat* is well marked in the tubular and pelvis portions of the ureters. *The fibrous or outer coat* is continuous throughout.

Arteries are branches from the renal, spermatic, internal iliac and inferior vesical.

Nerves are from the inferior mesenteric, spermatic and hypogastric plexuses.

SUPRA-RENAL CAPSULES.

These are ductless glands, resembling in shape a cocked-hat, and which embrace the upper extremity of each kidney. They are of yellowish color, the left being the larger.

Size: 1¼ to 2 inches long, not quite as wide and ¼ inch in thickness. *Weight*, usually from 1 to 2 drachms.

Structure: external, or *cortical*, and internal, or *medullary* substance.

The cortical is made up chiefly of narrow, columnar masses placed perpendicular to the surface; is the chief part of the organ and of deep yellow color.

The medullary part is softer, and of a dark brown color.

Arteries are of quite large size, and are from the aorta, phrenic and renal.

Veins: on the right side, open into inferior vena cava; on the left, into the left renal vein.

Lymphatics terminate in the lumbar glands.

Nerves: quite numerous; are gangliated, and are derived, mainly, from the solar and renal plexuses.

THE BLADDER.

The bladder receives the urine from the kidneys by

the ureters. It is a musculo membranous sac, of *conical* shape in infancy; *triangular* when empty, in the adult, and *ovoid* when distended. It has a summit, body, base and neck.

Size: it is 5 inches in length, and 3 in breadth, when moderately distended, in the adult. *In the female* it is broader transversely than in the male, and has greater capacity. It holds, usually, one pint.

Position: in *infancy* it lies in the abdomen. In the *adult* it lies in the pelvis behind the pubes; *in the male* in front of the rectum; *in the female* it is placed before the uterus and vagina. It may be so distended that the summit will be at the umbilicus or above it.

The summit, or **apex**, is connected to the umbilicus by the urachus, a fibro-muscular cord, and by the **2** obliterated hypogastric arteries; the part posterior to the urachus is covered with peritoneum.

The u´rachus is the obliterated fœtal canal that leads to the *allantois*.

The body is uncovered anteriorly by peritoneum, and in front are the triangular ligament of the urethra, the symphysis pubis and the internal obturator muscles. Posteriorly it is covered by peritoneum, and is in relation with the *rectum* in the male, and *uterus* in the female. Crossing obliquely, one on either side of the bladder, are the obliterated hypogastric arteries which form the limit, laterally, of the peritoneum; the *vas deferens* crosses obliquely the lower part of the lateral surface along the inner side of the ureter in males.

The base or fundus is directed forwards and downwards.

RELATIONS OF THE BASE: *In the male: Below.* Rectum 2nd part. *Behind.* Peritoneum. *In the female: Below.* Cervix uteri.

The cer´vix or neck of the bladder is the part continuous with the urethra. *In the male* it is surrounded by prostate gland, and has an oblique direction. *In the female* its direction is downward and forwards.

Ligaments: there are **2** sets of ligaments of the bladder, *true* and *false*. The *true* are 5 in number, and are formed of the recto-vesical fascia, and the urachus, being *two anterior* and *two posterior* and the *urachus*.

The false ligaments, 5 in number, are formed of peritoneum; there are *two posterior, two anterior* and a *superior*, the latter covering the urachus.

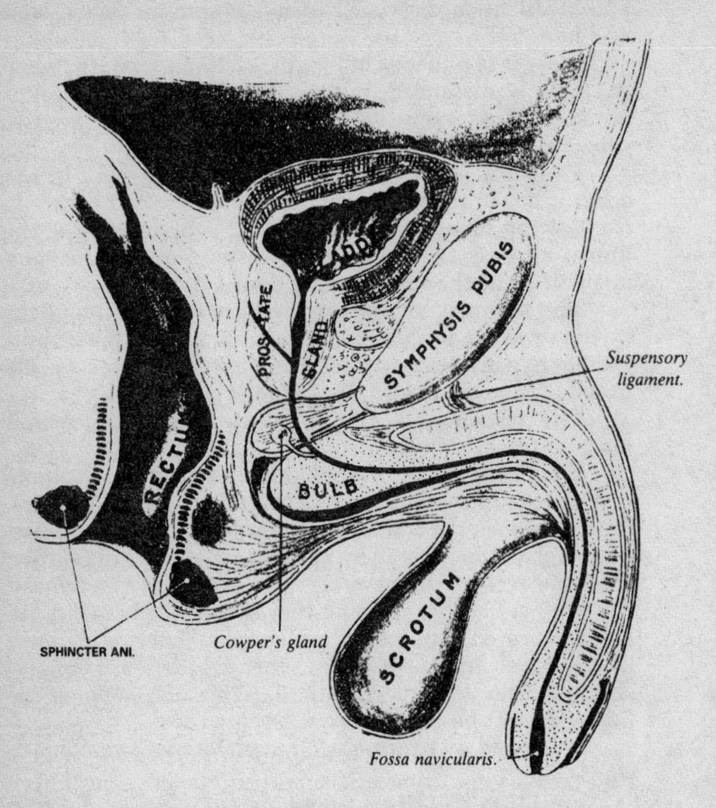

Vertical section of bladder, penis, and urethra.

The tri´gone: upon the inner surface of the base of the bladder, just behind the urethral office, is a triangular smooth surface or trigone, with the apex looking forwards. It is paler, without rugæ, and is bounded, laterally, by two ridges passing to the openings of the ureters, the posterior angles being formed by those openings; at its apex there is an elevation formed by the prostate called the *u´vula ves´icæ*. The ureters are about 2 inches distant from each other, and 1½ inches behind the urethral opening of the bladder.

Structure: the bladder is built up of **4** coats, as follows:

1. *The serous*, derived from the peritonæum, which invests only its posterior surface and part of summit.

2. *The muscular*, composed of two layers of unstriped fibres. The external layer is of *longitudinal fibres*, forming a plexus about the anterior surface arising from the anterior ligaments, and being reflected back over posterior surface, are inserted into prostate gland, or vagina. The *circular fibres* are most thickly distributed around the neck, forming the so-called sphineter of the bladder. The *muscles of the ureters* are two oblique bands supposed to prevent reflux of urine into the ureters during the bladder's contraction.

3. *Cellular coat* is a thin layer of areolar tissue between the mucous and muscular coats.

4. *The mucous coat* is of a light rose color, thin, and is continuous from the meatus of the urethra to the uriniferous tubes. It has a few mucous follicles and numerous small racemose glands. The epithelium covering it is of the "transitional" type, the superficial layer being of flattened polyhedral cells, with two or three nuclei, and beneath is a layer of club-shaped cells, with an oval nucleus.

Arteries: there are the *superior vesical, middle vesical, inferior vesical*, all derived from the anterior trunk of the internal iliac. The *sciatic arteries* also give small branches to the bladder. *In the female* there are also branches from the arterial and vaginal arteries to the bladder.

Veins: from a common plexus about the fundus, body and neck; they terminate in the internal iliac vein.

Lymphatics accompany the bloodvessels.

Nerves: the hypogastric plexus of the sympathetic supplies the upper part, and the 4th sacral nerve supplies the lower part and neck of the bladder.

THE MALE ORGANS OF GENERATION.

THE PROSTATE GLAND

Surrounds the neck of the bladder and the beginning of the urethra. It resembles a horse-chestnut in shape, with the apex directed forwards. *It measures* about 1½ inches across its base, and half that in depth, and weighs 6 drachms. It is held in position by the anterior true ligaments of the bladder, the *pubo-prostatic*, and by the posterior layer of deep perineal fascia, and the anterior portion of the levator ani muscle (*levator prostatæ*). Its smooth under surface is attached to the rectum by dense areolar-fibrous tissue.

The gland consists of three lobes, two lateral and one middle and is perforated from base to apex by the urethra.

The two lateral lobes are of equal size and are separated by a deep notch.

The middle lobe is between the two lateral, and varies in size from a small band (in the young) to a good sized prominence in the aged; it is the enlargement of this lobe that produces obstruction of urine in old men.

The common seminal ducts open into the prostatic portion of the urethra, and are placed between the middle and lateral lobes.

Structure: is of a pale reddish-gray color, friable, though dense, and is encapsulated.

The glandular substance is made up of follicular pouches and canals, the epithelium lining them being of the columnar variety.

The ducts open into the urethra in the floor of the prostatic portion.

Secretion is a milky, acid fluid containing molecular matter, epithelium (squamous and columnar) and granular nuclei. *In the aged*, millet seed concretions, of carbonate of lime and animal matter, may be found.

Arteries are branches from the pudic, vesical and hemorrhoidal.

Veins form an enmeshing plexus; receive the dorsal vein from the penis, and empty into the internal iliac.

Nerves are from the hypogastric plexus.

COWPER'S GLANDS.

These are two small, round bodies, about the size of a

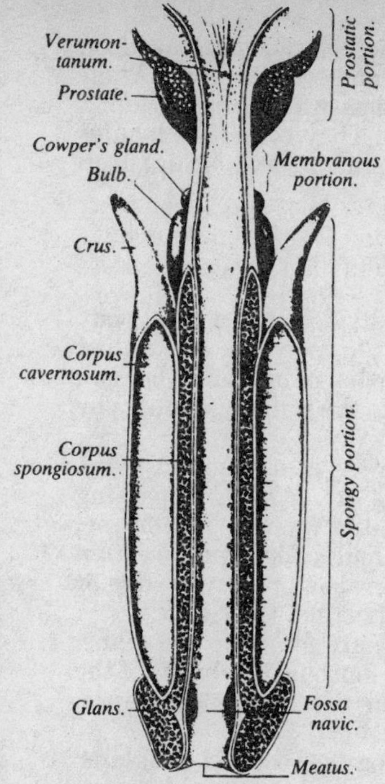

The male urethra, laid open on its anterior (upper) surface.

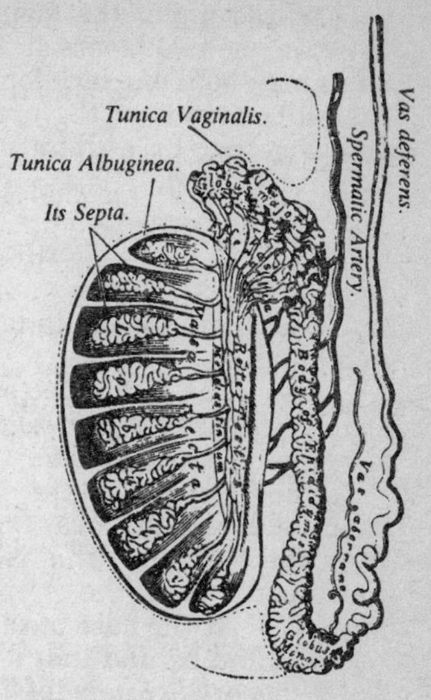

Vertical section of the testicle, to show the arrangement of the ducts.

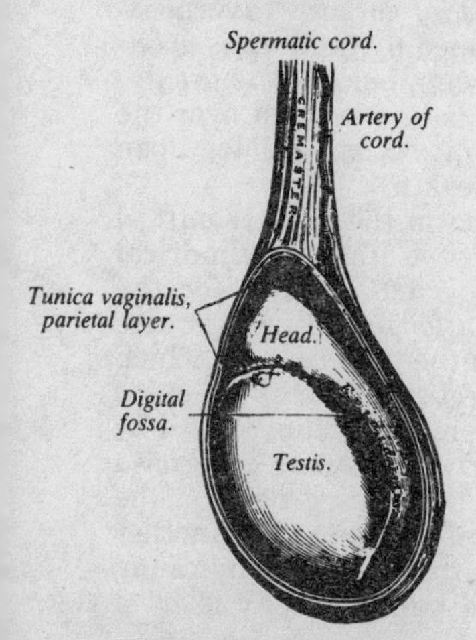

The testis *in situ*, the tunica vaginalis having been laid open.

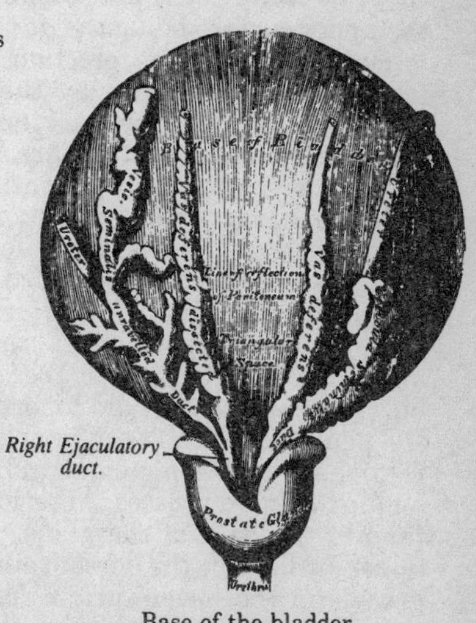

Base of the bladder, with the vasa deferentia and vesiculæ seminales.

pea, placed under the membranous part of the urethra, between the two layers of the deep perineal fascia. Their ducts are about one inch long and pass forwards to open in the bulbous part of the urethra. They consist of lobules held together by a fibrous investment, and diminish in size in the aged.

THE PE´NIS.

The organ of copulation is divided into a root, body and glans.

The root is connected to the pubic rami by two strong processes, *the crura*, and to the symphysis pubis by the *suspensory ligament.*

The glans forms the extremity; at its summit is the opening of the urethra, the *mea´tus urina´rius*; passing from the bottom of this is a fold of mucous membrane, continuous with the prepuce and called the *fre´num præpu´tii.* At the base of the glans is a projecting edge or *coro´na*, and behind that a constriction, the *cervix.*

Sebaceous glands (of Tyson) are found on the glans, though they are not found in the mucous membrane of the foreskin. They secrete a peculiar odoriferous substance, called *smeg´ma.*

The body is the part between the root and the glans; the upper surface being the *dorsum.* In the flaccid state it is cylindrical; during erection it assumes a triangular form, the base (formed by the two corpora cavernosa) being upward, and the apex (formed by the *corpus spongiosum*, enclosing the urethra being below. *The integument* covering the body is continuous with that over the *pubes, scrotum*, and at the glans it becomes folded upon itself, forming the *prepuce* or foreskin.

The corpo´ra caverno´sa form the greater part of the body of the penis; they are two fibrous cylindrical tubes placed side by side, connected together for the anterior ¾th, the *sep´tum pectinifor´me* being between, but are separated behind to form the two crura, by which the penis is attached to the projecting edges of the pubic rami; each crus commences in front of the tuber ischii by a blunt-pointed process. Anteriorly the corpora cavernosa fit into the base of the glans. There is a groove on the upper surface for the dorsal vein of the penis, and another groove on the lower surface for the corpus spongiosum; the corpora are attached to the pubic symphysis by a suspensory ligament.

Structure: a strong, fibrous, elastic *envelope* enclosing a recticular structure and erectile tissue. This envelope throws out numerous bands (*trabec´ulæ*), forming various sized compartments in which the erectile tissue is contained. This trabecular structure fills the interior of the corpora cavernosa, and interspaces which are lined with flattened cells, like the endothelial lining of veins. The septum, investing envelope and trabeculæ are of white fibrous tissue-bands, and enclose a few muscle fibres, and the enclosed arteries and veins.

The sep´tum pectinifor´me (like comb-teeth), is of this material and extends from the dorsum of the penis to the urethra, separating the corpora cavernosa.

The cor´pus spongio´sum, enclosing the urethra, commences at the triangular ligament by an enlargement, the *bulb*, and runs forward in the groove on the under surface of the corpora cavernosa, expanding over their extremities to form the glans. The central portion is cylindrical, but it tapers toward either extremity.

The bulb is covered by the anterior layer of the triangular ligament, and is embraced by the *accelera´tor uri´næ* muscle, and is pierced by the urethra near its upper surface. It varies in size in different subjects. Below the urethra there is a bilobular division of the bulb, which is marked externally by a linear raphé.

Structure: a strong fibrous envelope, with trabeculæ, enclosing erectile tissue. This erectile tissue is made up of various plexuses so numerous and freely intercommunicating that the body has a cellular look on cross section. The veins are smaller in the glans and on the surface of the body, but larger inwardly, and are lined with endothelium; they return the blood from the glans, and empty into the dorsal vein of the penis and in the prostatic plexus.

Arteries are branches from internal pudic. The *cavernosa arteries* are from the dorsal artery of penis; the *spongiosum arteries* are branches from arteries to the bulb. *The helicine arteries* are the twisting arteries in the cavernosa structure, and are most abundant in the back part of the organ, and are not found in the glans. The exact termination of the arteries in the venous spaces of the cavernosa and spongiosum is a matter of dispute.

The lymphatics are of a superficial and deep set, the former terminating in the inguinal glands, and the latter in the deep pelvic lymphatics, beneath the pubic arch.

Nerves are branches from the internal pudic and hypogastric plexus. On the glans and bulb they have Pacinian bodies connected with them.

THE URE´THRA extends from the neck of the bladder to the end of the penis, and has a *length of from eight to nine inches*. It is divided into *three parts*, according to the structures through which it passes.

(1) **The prostatic portion**, the widest and most dilatable, passes through the prostate gland from base to apex; this part is 1¼ *inches long* and spindle-shaped; transverse section shows it to be of a horse shoe shape, and the canal remains closed, except during passage of urine. On the floor is a narrow longitudinal ridge, the *verumonta´num* or *ca´put gallinagi´nis*, and on each side of this promontory is a depression, the *prostatic sinus*, into which the prostatic ducts open by numerous mouths (the middle lobe opens back of this promontory). Towards the anterior part of the verumontanum is a depression, the *si´nus pocula´ris*, upon the elevated edges of which the ejaculatory ducts open; it is a *cul-de-sac* ¼ of an inch long, running upwards and backwards. (By Weber it has been termed, from a fancied resemblance, the *u´terus masculi´nus*).

(2) **The membranous portion** extends from the apex of the prostate to the bulb, and is ¾ inch long above, but only ½ inch long below from the bulb, projecting below it; it is the narrowest part of the urethra, except the orifice, and it is contained principally between the layers of the triangular ligament, and is surrounded by the *compressor urethræ* muscle. Its concave, upper surface is 1 inch below pubic arch, separated from it by the dorsal vessels, nerves, etc. Two layers of the perineal fascia are prolonged around this portion of the canal.

(3) **The spongy** (longest) **portion** is contained in the corpus spongiosum, and occupies the rest of the canal, being six inches in length and ¼ of an inch in diameter; the portion contained in the bulb is somewhat dilated, and the ducts of Cowper's glands open on the floor; the canal enlarges again just below the *mea´tus urina´rius*, which is named the *fos´sa navicula´ris*. The floor is sprinkled with *lacu´næ*, being openings of the *glands of Littre*; one large one in the fossa navicularis is called the *lacu´na mag´na*.

The bulbous portion is the name sometimes given to the dilated spongy portion within the bulb.

Mea´tus urina´rius is a vertical slit of 3 lines, with a small labium on either side. Is the most constricted portion of the urethra.

Structure: three coats, viz: mucous, muscular and erectile.

The mucous coat is continuous with that in the bladder, ureters, kidneys, and with the integument over the glans, and it lines the various ducts. It is arranged in longitudinal folds, when the organ is flaccid. Its epithelium is columnar, except at the meatus, where it is laminated.

The muscular coat is of an external or *longitudinal* layer, and an internal or *circular* layer, and is most abundant at the prostate, being reflected upon the bladder.

The erectile coat surrounds the canal from the corpus spongiosum to the bladder's neck; is in a thin layer.

THE SCRO´TUM

Contains the testicles and spermatic cords. It is divided in the middle line by a *raphe*; the left half is longer than the right, as the left testicle hangs down lower, the left spermatic cord being longer than its fellow. It consists of two layers.

The integument, thin, brownish, wrinkled, studded with sebaceous follicles, and bearing some crisp hairs.

The dar´tos, a contractile layer of loose, reddish tissue, surrounding the whole, and sending a septum (*sep´tum scro´ti*) inwards to divide the cavity into 2 parts, as separate receptacles for the testes. It is very vascular, slowly contractile under cold or mechanical stimuli, but not contractile under electricity.

The intercolum´nar fas´cia is a thin membrane derived, during the descent of the testes, from the margins of the pillars of the external abdominal ring. Is loosely adherent to the dartos.

The cremaster´ic fas´cia, scattered bundles of muscular (*cremaster muscle*) fibres from the lower borders of the internal oblique taken down by the testis.

The tu´nica vagina´lis: described under "testes."

Arteries: these are *superficial* and *external pudic*, from the femoral; *superficial perinæal*, from the internal pudic; the *cremasteric*, from the epigastric.

Veins: these follow the arteries.

Nerves: the ilio-inguinal, *from the lumbar plexus*; 2 superficial perinæal branches, *from the internal pudic*; genital branch, *from the genito-crural*.

Lymphatics terminate in the inguinal glands.

THE SPERMAT´IC CORD.

The spermatic cord consists of the *vas def´erens*, with its vessels and nerves, *spermatic vessels* and nerves, the *cremasteric artery*, the *genital branch* of the genito-crural nerve, lymphatics, together with some areolar tissue; it extends from the internal abdominal ring to the back of the testis. The left cord is longer than the right one.

The vas def´erens is placed at the back of the cord, and may be recognized by its hard and cord-like feeling.

The arteries: *spermatic* (from the aorta), to testicle mainly; *vas deferens* artery, a long, slender vessel (from the superior vesical); *cremasteric* (from the epigastric).

Veins: *spermatic* leave back of testis, receiving those from epididymis; form a plexus (*pampin´iform plexus*), which is a large portion of the cord. They pass up in front of the vas deferens, and, in a single trunk, empty (on the *right* side) into the inferior vena cava, and (on the *left* side) into the left renal vein.

Nerves: the spermatic plexus of the sympathetic.

Lymphatics are of large size and terminate in the lumbar glands.

THE TES´TES.

These are two glandular organs suspended in the scrotum by the spermatic cords, which are attached to their posterior borders; are oval, compressed laterally, and the upper extremity looks forwards and outwards; are invested, except posterior portion, with the *tu´nica vagina´lis*. Each testis consists of two parts: the *body*, which is anterior, and the *epididymis*, which is posterior; from the lower end of this latter the duct or vas deferens is attached.

Size and weight: 1½ to 2 inches long; 1 inch broad; 1¼ inch thick. *Weight*, from 6 to 8 drachms, the *left* being the larger.

The epidid´ymis is a long, narrow body lying posteriorly and to the inside, and consists of 3 parts, viz: *body* or central portion, *head* or *glo´bus ma´jor*, and tail or *globus minor*, with which the *vas def´erens* is continuous. Attached to the upper end are one or more small pedunculated cysts (*hydatids of Morgagni*) supposed to be the remains of the Müllerian ducts.

Coverings of the testis are 3:

1. *Serous* or *tu´nica vagina´lis*: is derived from the peritoneum (during descent of testis), and consists of 2 parts, the visceral and parietal; the former covers the outer surface of testis and epididymis; more extensive than the visceral portion; reaches below the testis and up upon the front of the cord. It is covered with squamous

epithelium. The interval between the layers constitutes "the cavity of the tunica vaginalis."

2. *Tu´nica albugin´ea*, a dense fibrous, bluish-white membrane; covers the body of the testicle, sending in a vertical septum the *mediasti´num* or *cor´pus Highmoria´-num*; this latter gives off secondary processes or *sep´ta*, which serve to separate and support the lobules of the testicle as well as the vessels and testicular ducts.

3. *Tu´nica vasculo´sa* consists of the bloodvessels connected together with areolar tissue. This *pia mater* of the testis invests the inner surface of the tunica albuginea in its numerous processes.

Structure: the glandular structure is made up of from 250 to 400 lobules, conical in shape, and with their apex to the mediastinum. Each *lobule* is made up of one or more convoluted minute tubes—*tu´buli seminif´eri*. The total number being variously estimated from 300 to 840, and their average length being, when straightened, 2¼ feet, (longest 16 feet); their diameter is from 1/150 to 1/200 of an inch. They are pale, but grow darker with age, and are lined with several layers of epithelial cells (*seminal cells*). Near the lumen of the tube these cells become *spermat´oblasts*, which change into *spermatozo´a*. (In the young these epithelial cells do not assume so generous a growth, but resemble an epithelial layer).

These *tubuli* are inclosed in a delicate capillary plexus, and at the lobular apices become less convoluted, and unite to form 20 or 30 larger ducts (1/50 of an inch in diameter) and proceed in a straight course, as the *va´sa rec´ta*. These *vasa recta* pass upwards along the mediastinum, as the *re´te tes´tis*, and at the upper end unite and terminate in the epididymis, in from 12 to 20 vessels, known as the *va´sa efferen´tia*. Their course is straight, at first, but become exceedingly convoluted, and form conical masses (*co´ni vasculo´si*) in the *glo´bus ma´jor*. Each of these "cones," when unravelled, consists of a tube 6 to 8 inches in length.

These efferent vessels empty into a tube which forms, from its many convolutions, the head (*globus major*) body and *globus minor* of the epididymis. When this tube is straightened, it furnishes a tube over 20 feet in length. (The total length of the seminal tract would then be nearly 50 feet).

The vas´culum ab´errans is a narrow tube varying from 1½ to 14 inches in length; is found connected with lower part of the epididymis or commencement of vas deferens. It has a blind extremity extended up into the cord.

The vas def´erens, the excretory duct of testis, commences at the lower part of the globus minor and ascends along the inner side of the posterior part of the epididymis; thence it follows the spermatic cord through the canal and internal abdominal ring, descends into the pelvis, crosses to the inner side of the external iliac artery, and arches over the back of the bladder, crossing the obliterated hypogastric artery to the inner side of the ureter. At the base of the bladder it runs along the inner side of the *vesic´ulæ semina´les*, here becoming sacculated; narrowing again at the base of the prostate, it unites with the duct of the vesicular seminalis, and forms the *common ejaculatory duct*.

It presents a hard, cord-like sensation to the touch; its walls are thick and of dense structure, its canal small, about ½ of a line. Its length is about 2 feet.

Structure: it has 3 coats: external or cellular; a thick, muscular coat, consisting of three layers, two being longitudinal, and the intervening one being of circular fibres. A third coat is the internal or mucous one, of pale color, and covered with columnar epithelium.

THE VESIC´ULÆ SEMINA´LES.

Are two sacculated pouches, placed at the base of the bladder. They are pyramidal in shape, the posterior part being the wider; anteriorly they converge to enter the prostate near the middle line. They are usually 2½ inches in length by 5 lines in breadth and 3 lines in thickness, though they vary considerably in size, even in the same individual.

At the prostate their ducts join with the vasa deferentia to form a common ejaculatory duct, the vasa lying at their inner side.

The ejaculatory duct is about ¾ of an inch in length, and has a slit-opening into the urethra at the sinus pocularis opening.

Structure; each consist of a tube coiled frequently upon itself, and giving off numerous cæcal branches. When uncoiled it is found to be from 4 to 6 inches long, and of the diameter of a quill.

Coats: the vesiculæ seminales have 3 coats; the *outer* being fibro-cellular, and derived from recto-vesical fascia. A *middle* coat of muscle fibres in two layers (transverse and longitudinal), and an *internal* or mucous coat. The latter is pale and covered with columnar epithelium. (The ejaculatory ducts have 2 thin coats, the fibro-cellular being absent).

Vessels and nerves: *the arteries* are branches from the inferior vesical and the middle hemorrhoidal. The *veins* and *lymphatics* accompany their several arteries. The *nerves* are from the hypogastric plexus.

The se´men is a thick, milky fluid of a strong, peculiar odor, and it consists of solid particles, liquor seminis, seminal granules, epithelium and spermatozoa.

Liquor sem´inis is transparent and colorless and albuminous. It contains squamous and columnar epithelial detritus, oil globules, solid particles and granular matter.

The seminal granules are finely granular corpuscles, with a diameter of only 1/4000 of an inch.

The sper´matozo´a are the essential elements for fecundation, and are minute oval particles with a long caudal filament attached, which filament is always in motion during their life. Under the microscope they much resemble an apple-seed with a white thread attached to one end.

DESCENT OF THE TESTES.

This takes place usually between the 5th and 7th month of intra-uterine life; before that time they remain in the abdominal cavity, in front of and a little below the kidneys, their anterior surface and sides invested with peritoneum.
Attached to their lower end is the

Mesor´chium, a fold of peritoneum that covers each testis, and so supports it.

Gubernac´ulum tes´tis, a conical-shaped cord, of soft transparent structure within, covered with the cremaster muscle. It is connected, by one process, *the broadest*, to Poupart's ligament in the inguinal canal. The *middle process*, the longest, extends down the inguinal canal to the bottom of the scrotum. *The internal process*, for attachment, is attached to the os pubis and sheath of the internal rectus muscle.

The descent: at the middle of the 5th month (so Carling believes) the fibres of the gubernaculum begin to contract, and so draw upon each testis; this process continues until by the 7th month the internal abdominal ring is entered, pushing down a small pouch of peritoneum in front (the *proces´sus vagina´lis*). At the end of the

8th month it has entered the scrotum, still carrying with it its peritoneal pouch, which ultimately becomes its covering, the *tu´nica vaginal´is.* At birth the obliteration of the connecting canal between the scrotum and peritoneum is usually consummated.

In the female, a small cord, similar to the gubernaculum, descends in the inguinal canal, and this ultimately becomes the *round ligament.* A pouch of peritoneum also accompanies it, analogous to the processus vaginalis, and it is afterwards known as the *canal of Nuck.*

FEMALE ORGANS OF GENERATION.

THE VUL´VA.

The external organs of generation in the female are: the mons veneris; labia majora; labia minora; clitoris; meatus urinarius, the orifice of the vagina, and the perinæum. The term **vulva** includes the whole of these.

The mons ven´eris is the eminence in front of the pubes, formed of fat, and at puberty is covered with hair (*tressoria*).

The la´bia majo´ra are two prominent folds extending from the mons to the perinæum. Externally they are covered with hair (at puberty) and integument; internally, with mucous membrane; they are joined together anteriorly and posteriorly, forming commissures; are thicker anteriorly (in front) than behind, and are made up of tissue similar to the dartos of the scrotum, areolar tissue, fat, vessels, nerves and glands; altogether they are analogous to the male scrotum. A small transverse fold is found at the posterior commissure called the **fourchette**; the space between this and the commissure is known as the **fos´sa navicula´ris**.

The la´bia mino´ra or *nym´phæ*, are two folds of mucous membrane, extending for 1½ inches downwards and outwards from the clitoris, finally losing themselves below in the labia majora. They surround the clitoris, the upper folds forming the *præpu´tium clitori´dis*; the inferior ones are attached to the glans, forming the *fræ´num clitori´dis.* They have a thin epithelial covering, a plexus of vessels within; the mucous crypts secrete abundant sebaceous matter.

The clit´oris and erectile organ, corresponding somewhat in structure to the penis, is placed just before

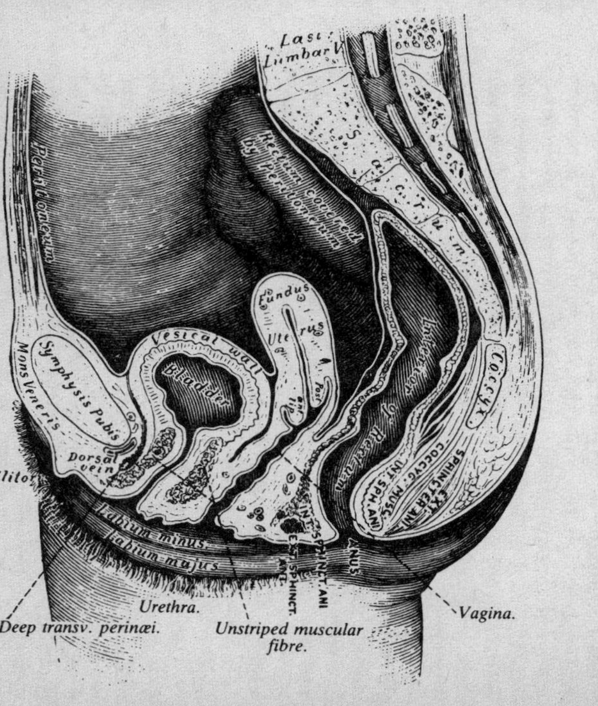

Vertical median section of the female pelvis.

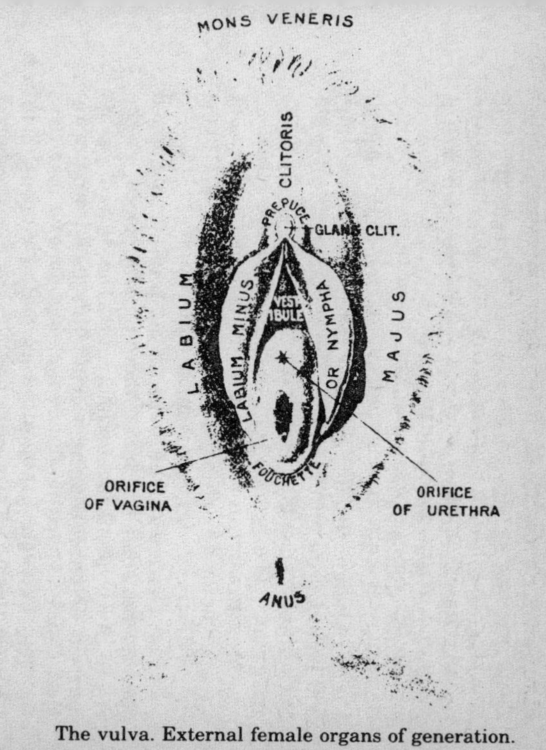

The vulva. External female organs of generation.

the anterior commissure. It consists of 2 corpora cavernosa attached to the pubic rami by 2 crura; the free extremity, or *glans*, is very sensitive, consisting of spongy erectile tissue. Like its analogue, the penis, it has a suspensory ligament and 2 erector muscles, *erecto´res clitori´des*. (See perinæum muscles).

Vestib´ulum: this is a triangular, smooth surface, 1 inch in height (from the clitoris to the vaginal opening, and bounded on either side by the nymphæ. It does not contain sebaceous glands, though has groups of muciparous glands. The *meatus* is at the bottom of this space.

Mea´tus urina´rius, or the urethral orifice, is about one inch below the clitoris. It is marked by a small tubercle or "pursing" of mucous membrane.

The Orifice of the vagi´na is an elliptical-shaped opening between the labia, more or less closed, in the virgin, by a membranous fold called the *hymen*. In the married or child-bearing woman this membrane is replaced by its remains, the *carun´culæ myrtifor´mes*.

The hy´men, a reduplicating fold of mucous membrane at the *introi´tus vagi´næ* of virgins, is usually of a crescentic shape, the concavity looking upwards. Sometimes it remains *imperforated*, thus forming an obstacle to menstruation and copulation. Sometimes it has numerous small openings, and it is then known as *cribriform*; sometimes it is *fimbriated*. Its presence or absence is no *positive* proof of, or against, virginity, as it may remain after copulation, and it is frequently absent in pure women.

The glands of Bartholi´ni, analogous to Cowper's glands, are situated on each side, near the entrance of the vagina, and their ducts (½ inch long) open on the nymphæ, external to the hymen. They are of a reddish-yellow color, and are about the size of a bean; are of the compound racemose order, and secrete a yellowish adhesive fluid, which is poured out abundantly during coitus and labor as a lubricant. Are more developed in young and middle-aged people and become atrophied in old age.

Bul´bi vestib´uli: these are oblong or leech-shaped, about one inch in length, situated beneath the nymphæ, and extend along either side of the vestibuli. They are made up of a plexus of veins enmeshed with a thin, fibrous membrane. They are of an erectile nature, and communicate at their smaller and upper ends with the vessels of the glans clitoridis by a small plexus, the *pars interme´dia*. They are the analogues of the corpora spongiosa of the male.

The perinæ´um is the space between the vaginal orifice and the anus. It is of triangular shape, the apex being at the vulva; the base is about 1½ inches in width, and it is of about the same height. It is composed of a layer of integument, fatty tissue, bloodvessels and nerves, besides the following **13 muscles**, and their aponeuroses, which enter, more or less, into its formation.

EREC´TOR CLITORI´DIS (2, or a pair); arise from the anterior region of the pubic and ischic rami, and are inserted into the clitoris, at the junction of *cru´ra clitori´dis.*

BUL´BO CAVERNO´SUS (2): arise from the perineal body and aponeurosis, superior portions—portion into crus of same side, near insertion of erector muscle; the outer portion winds inwards, under erector muscle to the bulb of the vagina, near its isthmus (under the clitoris). A few fibres pass up over the clitoris, and also up to the pubes.

TRANSVER´SUS PERINÆ´I SUPERFICIA´LIS (2): arise from ramus of ischium, in front of tuberosity, and from anterior aponeurosis of the perineal septum—perineal body and integument in front of anus. (See note below.)

SPHINC´TER A´NI EXTER´NUS: *deep portion*, from tip of coccyx; *superficial portion*, from integument—perineal body, central portion.

PUBO-COCCYGÆ´US (2): from posterior surface of pubes, and aponeurosis—*outer margin*, is inserted into the last two bones of coccyx; *inner margin* of each muscle commingles with its fellow of the opposite side, forming loops that pass between the vagina and rectum, and that unite with the deep sphincter ani.

OBTURA´TO-COCCYGÆ´US (2): from ilio-pubic line of the junction between obturator and recto vesical fascias—sides of last two bones of coccyx. (No rectal connection).

IS´CHIO-COCCYGÆ´US (2): from spine of the ischium and aponeurosis—sides of the bones of the coccyx.

NOTE.—The three last named muscles, which I have named to my classes as THE TRI-FORM MUSCLE, go to form, in the male, what is known as the *levator ani* muscle.

The *pubo-* and *obturato-coccygeal* muscles draw forwards, and assist in closing, the rectum. The *pubo-coccygeal* is the true constrictor of the vagina, *not* the bulbo cavernosus, as usually given; this muscle (the bulbo-cavernous) is the compressor of the vaginal bulb, and in contracting draws the labia together.

In lacerations of the perinæum it is the contraction of the severed *transversus perinæi* muscle that causes the gaping of the vaginal orifice, and so causes deformity.

The ure´thra in the female is only 1½ inches long, and is *embedded in the anterior wall of the vagina*; it perforates the triangular ligament, as in the male. Its *direction* is, from behind the symphisis pubis, obliquely down and forwards, in a slight curve. Its *diameter* is about ¼ of an inch, normally, though is quite readily dilatable to a larger size.

Structure: it has three coats: *the muscular*, continuous with that on the bladder, is in two layers, an outer circular layer and an inner longitudinal layer. The *middle layer* is of spongy, erectile tissue, with a venous plexus, and some unstriped muscular fibres. *The internal*, or *mucous coat*, is pale, and continuous with that of the vulva and that of the bladder. It is in longitudinal folds, the one on the floor being the analogue of the verumontanum. These folds, and the loose attachment of the mucous to the muscular coat allow the formation of a *prolapsus membrana urethræ*, a condition not seen in the male. The formation of tumors of the meatus is also largely due to the loose union of these two coats. It is covered with laminated epithelium, becoming of a spheroidal character at the bladder. *Mucous follicles* are at the meatus.

The bladder (*cys´tis*) has in *front* the pubis; *behind* the uterus and some convolutions of the small intestines, and the vagina; is wider transversely than in the male. See page 207.

THE VAGI´NA.

Is a dilatable membranous canal extending from the vulva to the uterus; the *anterior wall* is about 4 inches, and the *posterior wall* from 5 to 6 inches long. It is curved forwards and downwards, and lies back of the bladder and front of the rectum; it is of a flattened, cylindrical shape, constricted at the first portion, though dilated at the upper end to receive the uterine neck, presenting *cul-de-sacs* both before and behind the neck of the uterus. The *anterior* one is the shallower, the *posterior* the deeper, being about twice that of the anterior. These *cul-de-sacs* are formed by a sort of reduplicated reflexion of the vaginal membranes about the uterine neck.

THE RELATIONS ARE: *anteriorly*. Base of bladder, urethra. *Posteriorly*. Rectum (lower ¾), pouch of Douglas (upper ¼).

The upper portion of the vagina gives attachment, laterally, to the broad ligaments; the lower portion, to the recto-vesical fascia and fibres of the triform muscle. (See note, page 223.)

Structure: the vagina has 3 coats, viz: Internal, or mucous; Muscular; Erectile.

The mucous coat is continuous with that lining the uterus, and with that lining the labia. Along the anterior and posterior walls is a longitudinal raphé, or ridge, called the "vaginal columns." Extending, laterally, outwards from this median raphé are numerous transverse rugæ and furrows of variable depths; these are most marked near the orifice, and in virgins and nulliparæ. *The epithelium* covering the mucous membrane is of the squamous variety. *The submucous tissue* is quite loose, and is, from its plexuses of veins and the muscle fibres present, regarded by Gussenbauer as erectile.

The muscular coat is in 2 layers. *The internal layer* is the stronger, and is built up of longitudinal fibres, which are continuous with the superficial ones of the uterus, the strongest bundles being attached, laterally, to the recto-vesical fascia. *The external layer* of fibres are circular, and have oblique decussating fibres with the longitudinal coat.

The erectile tissue is a layer of loose connective tissue enmeshing the numerous venous plexuses ramifying therein. The circular layer of muscle-fibres sends numerous prolongations into this coat.

THE UTERUS.

The uterus or womb, the organ of gestation, is a pear-shaped body, flattened from before backwards, placed in the pelvis between the bladder and rectum; superiorly it does not reach above the brim of the pelvis. The position corresponds to the pelvic axis. The uterus is covered by peritoneum behind, above, and in front, except where it is attached to the base of the bladder; the peritoneum is reflected from off the sides, forming the *broad ligaments*. The base, or fundus, is directed forwards, and the neck downwards and backwards.

Size: in childhood it is in a complete anteverted position, and is much smaller, relatively, than at puberty. *In the virgin* its internal, long diameter is from 2¼ to 2½ inches. In the multiparæ this is increased slightly, so that 2¾ or 3 inches is no uncommon measurement. *Its breadth*, across the top, laterally, is about 2 inches; its thickness, about 1 inch. Its weight varies, in ordinary conditions, from 1 to 1½ ounces; immediately following parturition, it will be found to weigh many times this, and its size will be proportionately increased. It takes some five or six months for a uterus to regain its normal size following a normal confinement.

For convenience of description, the uterus is divided:

(1) *The fundus*, which is the broad, upper end of the body, projecting into the abdomen between the attachments of the Fallopian tubes.

(2) *The body*, which extends from the fundus to the neck, narrowing as it approaches the latter; at the junction of the fundus and body is an angle to which the Fallopian tube is attached, and a little anteriorly the *round ligaments* are connected, and below and behind the ovarian ligaments.

Its *anterior surface*, flattened, covered on its upper 3/4 by peritoneum, and has its lower 1/4 attached to the bladder; has coils of the small intestine separating its upper portion from the bladder. *Its posterior surface*, convex, covered with peritoneum, is separated from the rectum also by some convolutions of the small intestine. *Its lateral margins* are concave, and furnish attachment to the Fallopian tubes, round ligaments and ovarian ligaments.

(3) The *neck* or *cer´vix u´teri*, the rounded, constricted portion pointing into the vagina, and surrounded by it; it presents a transverse opening, the *os u´teri* the *os u´teri exter´num* or the *os tin´cæ*.

This opening is bounded by *two lips*, the anterior and posterior. *In the virgin* or nulliparæ, this opening or external mouth of the uterus is usually a small, circular depression upon the centre of the vaginal portion of the neck; on the birth of the first child, the mouth becomes lacerated, and, on healing (if not too extensively lacerated), shows a transverse slit. This should not be positively taken as proof of a pregnancy previous to examination, as sometimes (rarely) the transverse opening is seen in virgins.

The size of uterus projecting into the vagina, in virgins, will vary from that of the little finger of the adult male to that of the index finger, and it will be found projecting, usually, about 1/2 to 3/4 of an inch downwards from the vaginal insertion. In the multiparæ this vaginal projection will be somewhat shortened, the anterior and posterior lips will be much more prominently marked, and the organ there presenting at the vaginal vault will be found (in ordinary health) of the diameter of the thumb at the first joint. Extensive laceration gives various sizes and shapes of uterine mouths. In the aged there is no

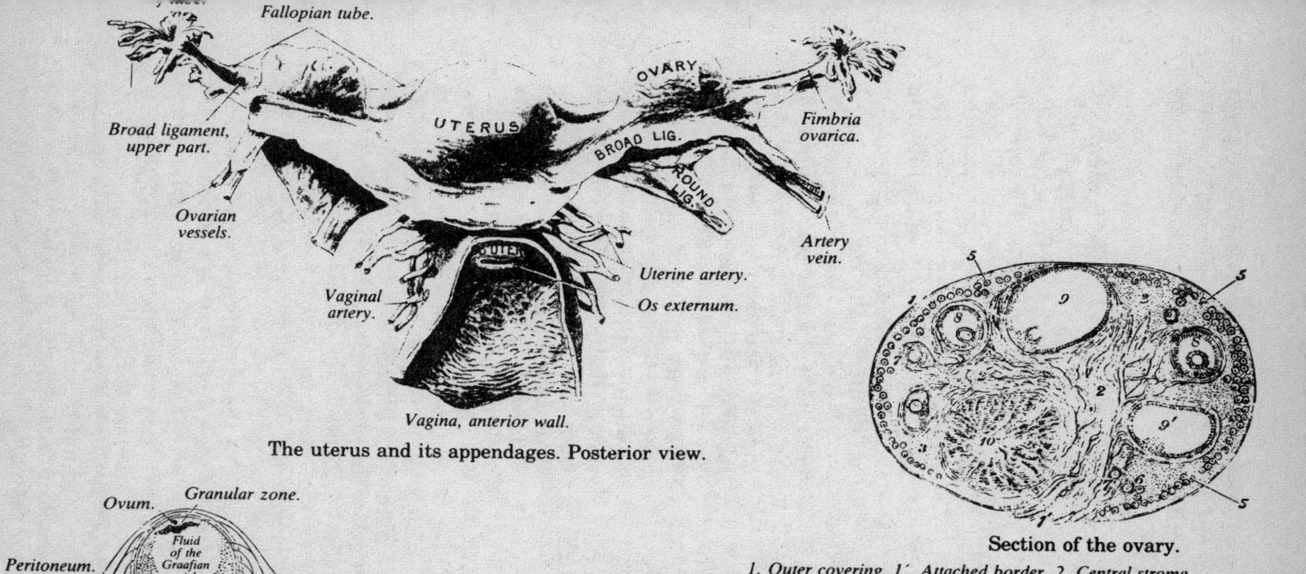

The uterus and its appendages. Posterior view.

Section of the Graafian vesicle.

Section of the ovary.

1. Outer covering. 1′. Attached border. 2. Central stroma. 3. Peripheral stroma. 4. Bloodvessels. 5. Graafian follicles in their earliest stage. 6, 7, 8. More advanced follicles. 9. An almost mature follicle. 9′. Follicle from which the ovum has escaped. 10. Corpus luteum.

projection of the cervix into the vagina, but in its stead there will be found a cup-shaped depression, the os, as a small, round, cartilaginous opening, occupying the superior or highest part of the vault.

The cavity of the uterus is made up of 3 triangles, when seen diagrammatically, as seen in the accompanying figure. The *larger* triangle, A, B, representing the cavity of the body or fundus of the uterus, the internal os, or narrowing (*os u´teri inter´num*), being at B. *The cavity of the neck* is represented by the two triangles, B, D and D, C, placed base to base, C representing *the external os*, or "mouth" of the womb.

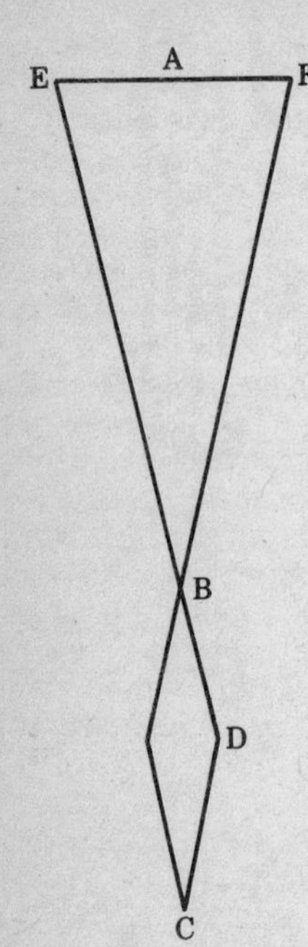

The normal size of the *os u´teri exter´num (C) and the os u´teri inter´num* is sufficient to allow passage of a sound-point having a diameter corresponding to No.11 of the French scale (No.7 English scale). The cavity at D will quite readily contain a No.18 sound-point (No. 12 English), while the broadest portion of the body cavity – extending from one Fallopian tube-entrance (E, F) to the other – measures from ½ to ¾ of an inch; this fundal cavity will contain about one drachm.

The combined length of the cavity of the fundus and the cavity of the neck is, usually, 2½ inches. In this case, the length of the fundal cavity is 1½ inches (A to B), and the length of the neck cavity will be 1 inch (from B to C). (It should be borne in mind that these measurements are for *normal* uteri; tumors, congestions, hypertrophies, misplacements, flexions, etcetera, will vary the ratios here given.

Usually the walls of the fundal cavity are in apposition, and at each superior angle, E, F, is a funnel-shaped cavity (the remains, primitive fœtal, of the cornua), and in these depressions will be found the minute openings of the Fallopian tubes, one on each side.

The cavity of the cervix is marked anteriorly and posteriorly by a longitudinal column, having several lateral branches, resembling somewhat the shape of the arbor vitæ leaf, hence the name *ar´bor vi´tæ uteri´na* has been applied to them. These folds (of mucous membrane mainly) are less distinct in the parous womb.

Ligaments: the uterine ligaments are eight in number, viz; 2 round; 2 anterior; 2 lateral and 2 posterior. The last *six* are formed by the different reflexions of the peritonæum.

The **2** *anterior*, called *ves'ico-u'terine*, are semi-lunar folds passing from the posterior surface of the bladder, and are attached to the lower anterior uterine surface.

The **2** *lateral*, or *broad ligaments*, pass outwardly from the uterine sides, forming a septum which divides the upper pelvic cavity into two portions (anterior and posterior), and finally are lost in the lateral walls of the pelvis. *In the anterior chamber*, formed by them, are found the vagina, bladder and urethra; *in the posterior chamber* is found the rectum. These ligaments are made of a double-fold of the peritoneal membrane, and between the folds there is cellular tissue, muscular fibres, besides the Fallopian tubes, ovarian ligaments, ovaries, parovaria, nutrient vessels and nerves, and the upper part of the round ligament. This is the usual seat of so-called cellulitis and peri-metritis. The upper borders of these ligaments are marked by **3** subsidiary folds, the anterior one admitting the passage of the round ligament; the middle one, the passage of the Fallopian tube; the posterior one contains the ovary. This ligament has been called the "bat's wing" (*ala vespertilionis*), from its fancied resemblance to this organ.

The muscle fibres found between the folds are, according to Rouget, continuous with those of the uterus, from which they are derived, and are divisible into two layers: *anterior*, from the anterior uterine surface, and these go to help form the round ligament; the *posterior*, from the posterior uterine surface, are continued outwards to the sacro-iliac synchondrosis. There thus seems to be a common muscular envelope surrounding the uterus, Fallopian tubes, round ligaments and ovaries, which aids in bringing the fimbriated extremity of the Fallopian tube into contact with the ovary to receive the extruded ovum; it also contracts the extended peritoneum following confinement, and aids in the general harmonious action of all the pelvic organs during the orgasm of sexual excitement and the congestion of menstruation.

The **2** *posterior*, or *rec'to-u'terine* ligaments, pass from the rectum to the sides of the uterus, and they form *the recto vaginal pouch*, or *cul-de-sac of Douglas* in this way: as the peritoneal covering leaves the anterior surface of the second part of the rectum it is reflected upon the posterior wall of the vagina, and from there it is continued on up the posterior uterine wall to near the fundus, thus leaving a pouch back of the upper portion of the vagina and the uterus, which has been named as above. This pouch, or Douglasian cul-de-sac, is frequently occupied by a small coil of the small intestine.

The **2** *round ligaments* are like rounded cords, their usual length being 4 or 5 inches. They commence, one on each side of the uterus, at the superior uterine angle, from the external muscular layer, just below the entrance of the Fallopian tube, between the two folds of the broad ligaments; they then pass downwards, forwards and outwards, through the internal abdominal ring, along the inguinal canal, *canal of Nuck*, (see page 220) into the labia majora, where they become lost. There seems to be three digitations in their labial terminations, the short or lower one going to the pubes, and the two longer continued down to the labium. They are essentially muscular in their structure, their upper portion being continuous with that of the upper part of the uterus, and is of the unstriped variety; farther down, receiving fibres from the transversalis muscle and the inguinal ring, this becomes of the striped variety, and covers over the unstriped fibres. In addition to this, they have elastic tissue fibres, connective tissue, bloodvessels (cremasteric) and nerves (genito-crural branches). Their action is to "steady" the uterus, much as the guy-ropes steady a tent, from a too backward action. They also contract during the sexual orgasm, thus bringing the fundus of the uterus forwards, and so aid in swinging the neck and mouth of the uterus into the "seminal lake," (a posterior pouch of the vagina), and so aid in impregnation. They partake of the general erethrism attending sexual excitement by their sympathetic contraction.

During pregnancy all these ligaments become greatly enlarged; but following the parturition, if the lying-in be normal, they gradually assume, through the contraction of their muscle-fibres, their condition seen in ordinary health. They all aid more or less, in keeping the uterus in its normal position in the pelvis, as it is supported in great part by the vaginal roof and intra-pelvic fascia.

Structure: the uterus has 3 coats, viz: the external or *serous*; the middle or *muscular*; the internal or *mucous*.

(1) THE SEROUS COAT is derived from the peritoneum, and covers the fundus, the whole posterior surface and the upper ¾ of the anterior surface.

(2) THE MUSCULAR COAT, which is the chief part of the uterine body, is of unstriped muscular fibre; the cells are long, accuminated, nucleated, and are arranged in more or less masses or layers. They are thickest about the Fallopian tubes. In the virgin, or unimpregnated state, it is of grayish-white color, and the uterus is dense and cuts like cartilage. In the impregnated state the uterine body becomes softer and of darker color, and the muscular elements show more plainly their three-layer formation.

The external one of these layers is thinner than the others; is closely adherent to the peritoneum, and covers over the fundus and neck, passing anteriorly and posteriorly, converging somewhat at the Fallopian tubes; it throws a mesh-work about them, and is also continued out upon the round ligaments and the ovarian ligaments, while some fibres go to the broad ligaments, and also from the cervix to the utero-sacral ligaments.

The middle fibre-layers have no regularity in direction, though they make up the bulk of the uterine tissue. They are composed of longitudinal, transverse and oblique layers or bundles, freely decussating with each other; in between these interlacements there courses the nutrient vessels, nerves and lymphatics.

The inner or deep layer is mainly of circular fibres, and is placed in a cone-like manner over and around the Fallopian tube entrances, being continuous with those upon the tubes; the fibres freely inosculate with the middle layer. These fibres are the remains of the early developed filaments of Müller. At the internal os there is a special reinforcement of these circular fibres, forming the so-called "sphineter" muscle at this point.

At the vaginal juncture of the cervix there is a considerable thickening of these muscular layers with a predominance of transverse arrangement; these enmesh circular nutrient vessels ("circular artery") and lymphatic spaces. Pregnancy increases the size of this "ridge."

The connective tissue of the cervix is in well-defined fibres and of the ordinary kind; while that in the body of the uterus is more of a wavy character and loosely-meshed. It surrounds the vessels and sends prolongations into the mucous coat.

The mucous membrane is smooth, of a reddish color, soft, with an average thickness of 1/25 of an inch, though thinner at the tubal portions and in the cervix.

It is covered with columnar, ciliated epithelium, with motion towards the Fallopian tubes. Arranged perpendicular to its surface are numerous tubular cells or *uterine glands*. The lining membrane is usually covered with a layer of *alkaline* mucus, while that secreted by the vagina has an acid reaction. In pregnancy these uterine glands become greatly enlarged, and the cilia are lost from the epithelium. These uterine glands are behind canals, dipping down through the mucous membrane, in a more or less tortuous course, and are lined by cylindrical epithelium. A capillary network surrounds these glands, and at the free surface of the mucous membrane assumes the form of venous radicles with delicate walls; these furnish the venous elements of the discharge at the menstrual epochs. The intermediate space is filled up with connective tissue, minutely and loosely arranged, and which Leopold claims to be lymph-spaces.

In the cervix the mucous membrane becomes of a lighter or faintly yellowish color, provided with mucous follicles and glands, and having numerous rugæ. The *glands of Naboth*, found in the mucous membrane, are sometimes found with stopped mouths, and this causes the little shot-like enlargements (*ovula of Naboth*) that are sometimes seen and felt covering the internal margins of the uterine lips. They are supposed to be simple inversions of the mucous membrane, and are lined with ciliated epithelium. The secretion is alkaline, from these cervical glands, and it plays an important function, as a lubricant. during labor, having become prominently developed during pregnancy. In the lower point of this canal the covering of the membrane is pavement epithelium.

Filiform papillæ are abundantly found in the mucous membrane of the lower cervix, and in that covering the vaginal portions. Their structure is similar to mucous membrane, and they each have a vascular loop. Kilian and Farre believe them to be the seat of sensibility in this part of the genital tract.

THE FALLOPIAN TUBES.

Named from before backwards, the appendages of the uterus will be found in this order: round ligaments; Fallopian tubes; ovaries and ligaments, besides nutrient vessels and nerves, and muscle fibres, all lying between the two folds of the broad ligaments.

The Fallopian tubes, the homologues of the vasa deferentia, in the male, are placed one on either side of the uterus; are about 4 inches in length, extending from the superior angles of the uterus, in a more or less curved direction, to the lateral walls of the pelvis, and are placed in the free margins of the broad ligaments. The canal is very minute especially at its uterine commencement (*os´tium inter´num*), where it barely permits the passage of a bristle; towards the abdominal orifice it gradually widens out till it reaches the trumpet-shaped expansion (*os´tium abdomina´le*) at the ovary. This expanded portion is fringed about with *fimbriæ*, giving to it the name of "*fimbriated extremity*; or, *mor´sus diab´oli*," referring to the way this portion of the tube contracts down and covers the ovary during sexual excitement, or at the bursting of the ovule from the ovary. One of the fimbriæ is always longer than the others, and this is indirectly united to the ovary by a peritoneal fold, thus forming a guide to the extremity when it is about to embrace the ovary. The tube, as a whole, has a cord-like feel to the touch.

The purpose of these tubes is to furnish a means of conveyance for the ripened ovules to the uterus, the action of the cilia of their lining cells aiding in this; and also to furnish a channel for the fertilizing elements of the semen to reach the ovaries.

STRUCTURE: the Fallopian tube has 3 coats, viz: serous, or external: muscular: mucous, or internal.

(1) *The serous coat* is derived from the peritoneum, and surrounds the tube to the extent of ¾ of its circumference; it also comes into conjunction with the mucous lining at the fimbriated extremity—the only instance of a serous and mucous membrane meeting.

(2) *The muscular coat* is in two layers of fibres, *longitudinal*, or external layer; *circular*, or internal layer. The former is in the lesser quantity, and is continuous with the external uterine layer; the circular layer is also continuous with the inner circular fibres of the uterus. Between this and the peritoneal coat is a layer of connective tissue, which bears a rich plexus of bloodvessels. Galvanization gives vermicular contraction to the tubes by its action upon the muscle-fibres of this layer.

(3) *The mucous coat*, continuous with that lining the uterine cavity, is covered with ciliated epithelium (the epithelium being continued upon the fimbriæ), the motion of the cilia being towards the uterus. It is also quite

vascular, and is thrown into numerous longitudinal folds, at the outer part of the tube, showing an adaptability for dilatation. The apposition of these folds form minute capillary tubes, which, from the motion of their cilia downwards, would seem to favor the journey of the ovule from the ovary to the uterus.

THE O´VARY.

The ovaries were called by Galen the *tes´tes mulie´-bres*, from the similarity, in function, to the male testes. They are two oval-shaped bodies occupying the sides of the pelvis, and are located between the folds of the broad ligament, to which they are connected at their anterior margin, one being on either side of the uterus, to which it is connected by the "ovarian ligament," at its inner extremity. Their outer extremity is connected with the fimbriated portion of the Fallopian tubes by a short digitation or cord.

Size, etc.: they are flattened from above downwards, have a somewhat uneven or puckered surface, and are of a whitish or reddish color outwardly, though darker on section. They are most convex on their posterior surface. They are about 1 to 1½ inches in *length*, ¾ of an inch in *width*, and ⅓ to ½ of an inch in *thickness*, though when diseased these measurements are variously altered. They also enlarge at the menstrual periods, and become atrophied in old age.

Their *weight* is usually 1 to 2 drachms.

Their position, in the pelvis, varies greatly, as they are subject to displacement from the enlargement of the uterus, due to pregnancy or tumors, and also to displacements of this organ. They also become, occasionally, prolapsed, and may sometimes be found in the Douglasian space.

In the fœtus, they are found in the lumbar regions near (to the front of) the kidneys, from where they gradually descend into the pelvis (see page 220).

Number: they are usually *two* in number, though occasionally but one is found, and sometimes there is no trace of any. On the other hand, *three* are sometimes met with, there being a small or supplementary one. In my opinion, however, this "supplementary ovary" is the hyper-development of the parovarium.

Ligament: the ovarian ligament is a muscular band about 1/5 of an inch in width, and 1½ inches in length, which attaches the ovary to the uterus at a point back of and between the uterine insertion of the Fallopian tube and the round ligament. It is made up from the prolongation of the superficial muscular fibres from the posterior uterine surface, covered with peritoneum. It is through this ligament that muscular fibres find entrance to the ovarian stroma.

Structure: the mass of the ovary is a collection of Graafian vesicles (the immature ova) imbedded in a stroma, with some muscular tissue, the whole covered in with a serous coat derived from the peritoneum.

THE SEROUS COAT, though derived from the peritoneum, differs from it in that it is covered with a layer of columnar epithelium (without cilia) instead of the squamous variety, as is the peritoneum. Waldeyer calls this the "germinal epithelium," because the ovules are formed from it in early fœtal life. This is what gives the grayish look to the ovary. Below this is the *tu'nica albugin'ea*, which does not exist as a distinct layer in the first three years of life, and it never is easily separable. It partakes more of the nature of a fibrous membrane, and is closely adherent to the stroma beneath.

THE STROMA or MEDULLARY SUBSTANCE, is of a reddish color, and when seen on cross-section, seems to be in two well-marked conditions, viz: an internal, or medullary vascular zone, and external and cortical or parenchymatous body.

The medullary substance is spongy-like, containing an abundance of spirally running bloodvessels with spindle-shaped cells, resembling muscle cells and connective tissue cells. There are also elastic fibres present. It is supposed these muscular elements play an important part, by their contraction, in the expulsion of the ovules and rupture of the Graafian follicles.

The cortical substance is of grayish color, and is the seat of development of the ovules and Graafian follicles; these are held together by meshes of connective tissue, the fibres radiating from the centre to the circumference.

Graafian vesicles: these are the numerous transparent vesicles seen in the ovarian stroma when it is cut across. In size they vary from the 1/100 of an inch in diameter (being smallest beneath the covering envelope) to (when near the menstrual periods) cysts of considerable size. It is estimated by Henle that there are 36,000 of these Graafian vesicles in each ovary; of course, the greater number are only visible under the microscope, some one or two being developed gradually at each menstrual period.

Each mature vesical consists of an *external fibro-vascular coat* connected by a network of bloodvessels the stroma and an *internal coat* (*ovicapsule*) lined with a coat of nucleated columnar cells called *membra´na gran´-ulo´sa*. These two coats enclose a transparent albuminous fluid, in which is suspended the ovum, a rounded vesicle but the 1/120 of an inch in diameter. The part of the Graafian vesicle nearest the surface of the ovary has the cells of the membrana granulosa collected into a mass—the *dis´cus prolig´erus*—in which the ovum is partially imbedded.

The ovule, or **ovum**, is formed from the germ-layer of epithelium on the ovary's surface, the depressions of the enlarging cells becoming deeper, as the ovule enlarges, and ultimately sinking into the stroma proper becomes entirely surrounded by it; the cell-wall forms the vitelline membrane, the nucleus forms the germinal area, and there soon appears the nucleolus or germinal spot.

Size, etc.: The human ovum varies from the 1/240 to the 1/120 of an inch in diameter. It has an enveloping membrane—the *zo´na pellu´cida*—which is transparent, colorless, and somewhat thick; it is also called the "vitelline membrane," and corresponds to the chorion, when impregnated.

The yolk, or *vitel´lus*, consists of granular protoplasm in a viscid fluid; the larger granules, which are near the vitelline membrane, look like fat globules, while the smaller ones, near the center, look like pigment; are not in large numbers.

The germinal vesicle is very small, only the 1/720 of an inch in diameter; its fine, transparent, structureless membrane encloses a watery fluid. It lies near the center of the yolk, before impregnation, but gradually approaches the vitelline membrane afterwards.

The germinal spot (*mac´ula germinati´va*) occupies the part of the germinal vesicle membrane that is nearest the enclosing vitelline membrane. It is from the 1/3600 to the 1/2400 of an inch in diameter; of yellow color, opaque, and made up of fine granular matter.

The ovule's discharge: after gradually re-approaching the surface of the ovary again, the enlarged and matured Graafian vesicle bursts its covering membrane, and the ovule, and the fluid about it, is thrown upon the ovarian surface to find its way into the fimbriated extremity of the Fallopian tube, to begin its journey to the uterus, and to its fertilization provided the conditions for this have been normally carried out. Sexual desire is usually most marked at this period of the discharge of the ovum, and the periods of menstruation are greatly controlled by it.

In impregnation, the spermatozoa penetrates the ovum, coming into contact with the germinal vesicle in the yolk, this taking place, usually, in the Fallopian tube. The effect is at once to set up cleavage, and finally a great multiplication of the yolk globules—by this cleavage process—the first physiological process of animal life.

Cor´pus lu´teum: after the rupture of the Graafian vesicle and the discharge of its contents, the cavity is filled with a blood-tinged fluid, which becomes harder and of a yellowish color as its age advances; this is the corpus luteum; and when pregnancy has taken place, the corpus luteum corresponding with it is more marked than those formed where the ovule has been unfertilized. The ovaries are found more or less filled with these various-aged corpora lutea when the ovaries of a child-bearing woman are examined.

The *true corpora lutea*, or those developed with the impregnation of the ovule discharged therefrom, are larger, project more from the surface, are more or less puckered, and contain a cavity in their earlier periods; are more vascular than

The false corpora lutea, which are of small size, contain no cavity, do not project from the ovarian surface, and their contents are soft, more resembling coagulated blood, and they present little or no cicatrix.

The true reach their maximum size in *two months*, remain stationary to the 6th month, and may measure ½ inch in diameter at the end of pregnancy; whereas,

The false reach their maximum size in *three weeks*, and at the end of 2 months cicatrization has occurred.

The parovarium, *Epooph´oron*, or *organ of Rosenmuller*, is a collection of 8 to 12 convoluted closed tubes spread out between the superior surface of the ovary and its Fallopian tube, within the folds of the broad ligament. They are arranged in a pyramidal form, the base corresponding with the ovary, and are lined with epithelium. One of them (the outside one) is usually bulbous in its termination. They do not have any special function so far as known; they are the remains of the Wolffian body of fœtal life.

ARTERIAL SUPPLY OF THE FEMALE GENITALIA.

UTERUS: This organ receives its supply, on each side, from three sources: I. The *Ovarian* (or spermatic) arteries, branches from the abdominal aorta. II. A *Branch*

of the Epigastric (which passes along the round ligament) which is from the external iliac. III. The *Uterine*, which is a branch of the internal iliac.

VAGINA: This organ receives its blood supply from 5 arteries, all *Branches from the Internal Iliac*, anterior trunk—except the circumflex uterine, which is a branch of the uterine—viz: 1st. Vaginal artery, a branch from the anterior trunk of the internal iliac, supplying lower third anterior wall; 2d. Vaginal branches from the uterine artery, supplying middle and upper third: 3d. Vaginal branches from circumflex uterine, supplying upper third. 4th. Vaginal branches from inferior vesical, supplying lower third. 5th. Vaginal branches of hemorrhoidal, supplying lower third.

OVA´RIA: The ovaries are supplied by the *Ovarian* arteries, branches of the abdominal aorta. These, having reached the inferior border of the glands, suddenly give off ten or twelve branches, which ascend, in a fan-shaped plexus, dividing and intertwining, to the hilum, or inferior portion; then they penetrate the substance proper of the ovary to still further subdivide and anastomose, and finally enmesh the walls of the Graafian follicles.

PERINÆ´UM and **PUDEN´DA**: These portions of the female anatomy are supplied with the following branches from the *Internal Pudic*: 1st. Inferior hemorrhoidal (2 branches); 2d.. Transverse perineal; 3d. Vulvar, or superficial perineal; 4th. Bulbar branches; 5th. Deep branch to crus; 6th. Dorsal branch to clitoris. Also the *Superficial External Pudic*, a branch of the femoral, arising about ½ inch below Poupart's ligament; after piercing the fascia lata at saphenous opening it runs inwards to supply the labia of that side, anastomosing with branches of the internal pudic. *Deep External Pudic*, another branch of the femoral, given off near the former, passes inwards, on pectineus muscle, to pierce fascia lata and be distributed to the labia of that side, anastomosing with the vulvar.

ABDOM´INAL AOR´TA: Extends from the diaphragmatic opening to the body of the 4th lumbar vertebra. See page 33. The branches being in order, 1st, Phrenic; 2d. Cœlic axis; 3d. Supra-renales; 4th, Superior mesenteric; 5th, Renales; 6th, *Ovarian* or *Spermaticæ*; 7th, Inferior mesenterica; 8th, Lumbales; 9th, Sacra media and the two terminal branches, Common Iliacs. **Ovaria´næ** or *Spermaticæ* arise from the front part of the aorta abdominalis, a little below the renal; *are long slender vessels* (though shorter than in the male) one on each side, and pass down and outwards across the ureter of their respective sides, beneath the peritoneum, lying on the psoas muscle; arriving at the pelvis each passes inwards between the broad ligament, laminæ to be distributed to the ovary. Two small branches are given to each Fallopian tube, and another to the uterus, which anastomoses with the uterine arteries. Other small branches are continued down the round ligament, through the inguinal canals, to integument of the groins and labia.

ILI´ACÆ COMMU´NES: See page 34.

Ili´acæ Exter´na: See page 35; from the junction of sacrum with the last dorsal vertebra around pelvic brim to femoral arch; terminal branches being epigastric and circumflex iliac. *Arte´ria Epigas´trica*: from a few lines above Poupart's ligament, it descends to this ligament, then ascends obliquely upwards and inwards, between peritoneum and transversalis fascia, to the umbilicus. It lies behind the inguinal canal, to the inner side of internal abdominal ring (just above the margin of the femoral ring) and in front of the round ligament. It here gives a *branch* to this ligament, that follows the round ligament back to and supplying the uterus, as well as ligament, with arterial blood.

Ili´acæ Inter´na: See also page 34. From sacro-dorsal vertebral junction to the great sacro-sciatic foramen, then dividing into its two terminal branches, I. the *Anterior* and II. *Posterior* trunks. Is a short, thick vessel.

POSTERIOR TRUNK. This is a short, rather thick artery dividing into 1st, *Superior*; 2d. *Ilio-lumbar*; 3d. *Lateral Sacral* branches, and which arteries supply the muscular parts in these regions.

ANTERIOR TRUNK; This is the most important branch. For the 3 *vesical branches* of this trunk, and *middle hemorrhoidal*, see page 34. ARTE´RIA UTERI´NA, a large branch on each side descending between the two laminæ of the broad ligaments, near their sacro-iliac attachment, in company with the ureter, to a point somewhat below the level of the ostium externum uteri, then it turns upwards, between the uterine attachments of the broad ligaments, and, in a tortuous course, close to the uterine body, it reaches the Fallopian tube, there inosculating with the terminal branches of the ovarian (spermatic) artery. In its course it gives numerous small branches to the substance of the uterus and adjacent organs. Its main branch is the circular artery. *Obtura´tor*: this is frequently a branch of the uterine artery, passing out of the foramen of same name, below the obturator nerve. *Vesica'lis Infe'rior*, with its "inferior vaginal branch," passes down to supply the lower portion of the bladder and vagina. *Vaginal Branches*, that supply the middle part of the vagina. *A. U'tero-cervica'lis*, the largest branch, which inosculates with its fellow and forms the so-called "circular artery" about the neck of the uterus. This is the artery that is liable to be wounded in performing trachelorrhaphy. *Fallopian branches*: these go to furnish the tube with a part of the arterial supply. A. VAGINA'LIS: descends to lower third of vagina to supply structures there adjacent. HEMORRHOIDA'LIS INFE'RIOR: passes down to lower portion of rectum, furnishing there a small *vesical branch* and numerous small *vaginal branches*. PUDI'CA INTER'NA: this is the smaller of the terminal branches of the anterior trunk, though the one that mainly supplies the pudenda. It passes down and outwards to greater sacro-sciatic foramen, and ascends, on the rami of the ischium and pubes, in a canal in the obturator fascia, to terminate in the dorsalis clitoridis. The branches in its course are, the *supe'rior vesica'lis*, sometimes with its remains of the impervious fœtal hypogastric. Two *inferior* or *external hemorrhoidal* branches that supply, in part, the lower portion of rectum. *A. perinæ'i transversa'lis*, runs transversely inwards, just beneath the transverse perinæal muscle, supplying the adjacent structures. *Vul'var branch* or *perinæ'i superficia'lis*, which is larger than corresponding branch in the male, ascends midway through the anterior perineal space, beneath the superficial fascia, supplying lower portion of vulva and superior perinæum. *Bulbar branch*, supplying the bulb of the vagina, lying beneath the labium majus. *Profun'da vul'var branch*, running deeply upwards to supply the crus clitoridis. *Arte'ria dorsa'lis clitori'dis* which runs downwards and forwards on the dorsum to the glans clitoridis, there inosculating with its fellow. A. SCIAT'ICA: this is the larger of the terminal branches of the anterior trunk, and is distributed to the muscles at the back of the pelvis; see pages 35 and 42.

NERVOUS SUPPLY OF THE FEMALE GENITALIA.

U´TERUS: This organ is poorly supplied with cerebro-spinal nerve fibres, the main nervous supply being from the sympathetic system; however, branches from the *second, third* and *fourth sacral* nerves, and a few filaments of the *sacral plexus*, go to help form the *Inferior Hypogastric Plexus* which is its main source of supply, through the *uterine nerves* that follow closely the course of the uterine arteries. The FUNDUS is supplied with branches from the Ovarian (spermatic) Plexuses. The MIDDLE portion, with branches from a prolongation of the Hypogastric Plexuses. The LOWER portion, with the anterior branches from the Inferior Hypogastric Plexus. All these plexuses inosculate freely with themselves, between the folds of the broad ligament and literally enmesh the uterus.

OVA´RIA: These organs are supplied, mainly, with branches from the *Ovarian Plexuses*; a branch, however of the *Hypogastric plexus*, by the way of the uterus and Fallopian tube (uterine nerve) also reaches each of them.

VAGI´NA: The main sentient nerve is the *Pudic*, and its branches; which see further on. The vagina is, moreover, completely enmeshed with the numerous inosculating branches of the *Inferior Hypogastric Plexus* of the sympathetic system, and which plexus has frequent communication with the cerebro-spinal system through the branches of the Internal Pudic.

PERINÆ´UM and **VUL´VA**: These parts are supplied by the various terminal branches of I. *Pudic* (from sacral plexus); II. The pudendal branch of the *Small Sciatic* (from lower part of sacral plexus); III. Terminal branches of the *Ilio-Inguinal* (branch of 1st lumbar); IV. Genital branch of the *Genito-Crural* nerve (branch 2d lumbar); V. The Sympathetic System through numerous branches from the *Inferior Hypogastric Plexus*.

Ner´vus Pudi´ca: This nerve arises from lower portion of the Sacral Plexus [this plexus is made up of five nerves, viz.; the *lumbo-sacral*, and the anterior branches of the *three upper*, and part of the *fourth, sacral nerves*] soon joins company with the internal pudic artery, and leaves the pelvis through the greater sacro-sciatic foramen, to cross the ischic spine and re-enter the pelvis through the lesser sacro-sciatic foramen. It then ascends, lying to the inside of its artery, the ischic and pubic rami, in a sheath of the obturator fascia, to end in the terminal branch, the dorsal nerve of the clitoris. *Hæmorrhoida´lis inferior*: from near origin of pudic [sometimes arises from the fourth sacral] traverses the ischio-rectal fossa, to be distributed to the external sphineter, and anal integument, inosculating with inf. pudendal and superficial perineal. *Muscula´res posterio´res*: several branches given off to supply the triform muscle, and upper border of sphineter ani. *Poste´rior Superficia´lis*: given off just below tuberosity of ischim, passes upwards and inwards to mesial line, supplying perineal muscles and integument, and lower portion of labia. *Ante´rior Superficia´lis*: given off at point about one-half inch above tuberosity, passes upwards and inwards to supply upper ⅔ of the labia. *Anastomot´ica*: small branch given off a little below level of meatus urinarius, to *inosc.* with the pudendal branch of the small sciatic. *Dorsa´lis Clitori´dis*: the pudic is continued up the pubic rami to the root of the clitoris, piercing the suspensory ligament, and becomes the dorsal nerve of the clitoris. It (one on each side) accompanies the dorsal artery to the glans, where it, through its numerous branches and inosculations with the sympathetic system, hoods over this organ. Savage asserts that the clitoris, in comparison with its size, has four or five times the nervous supply that the penis has.

Ner´vus Pudenda´lis: This nerve is a small branch of the *small sciatic* nerve (formed usually by the union of two branches from lower part of *sacral-plexus*, see page 50), and is given off from that nerve just at the back of the tuberosity ischii. It passes diagonally upwards and inwards, over the course of the ischic and pubic rami and gives branches to supply the labia, and overlying integument, gracilis muscle; its terminal branches reach the pubes, supplying the crus, clitoris and integument, anastomosing freely with the terminal branches of the ilio-inguinal, and pudic nerves.

Ilio-Inguina´lis: The ilio-inguinal nerve, a branch from the *1st lumbar*, runs around the abdominal walls, above the brim of the pelvis, and escapes at the external abdominal ring; its *terminal branch* runs down to the pubes supplying the mons and the superior portions of the labia and clitoris with its filaments, which here *inosc.* with the terminal branches of the pudic and pudendal branch of the small sciatic nerves.

NER´VUS SYMPATHET´ICUS. This is the nerve of all nerves that supplies the female (as well as male) genitalia. Through its perverted influence is to be attributed the thousand-and-one reflex ills that female flesh is heir to. **Plex´us Ovaria´nus** or *Spermat´icus*: this is derived from the *Renal Plexus* (the Renal is formed from branches from the solar plexus, semi-lunar ganglion, aortic plexus and greater and lesser splanchnic nerves); it also receives branches from the *aortic plexus*. It descends on the sides of the vertebræ to the ovaries, there supplying them; it then follows, down the tubes to the uterus (supplying both these organs) there inosculating with branches from the hypo-gastric plexus. **Plex´us Hypogas´tricus**: is situated in front of the promontory of the sacrum, between the common iliac arteries. It is formed by a union of filaments from the *aortic plexus*, all the *lumbar* and the *first two sacral ganglia*. It bifurcates below, forming the two inferior hypo-gastric (pelvic) plexuses. It supplies upper portion of rectum and a few branches to the uterus **Plex´us Hypo-gas´tricus Infe´-rior**: This is the pelvic plexus (one on each side) and is the source of chief nervous supply to the rectum, vagina, bladder and uterus. It is formed by the continuation downwards of the bifurcated *hypogastric plexus*, branches from the *second, third* and *fourth sacral nerves*, and a few filaments from the *sacral ganglia*. This plexus spreads out and enmeshes the viscera in the pelvic cavity with its inosculating filaments; these have received, from the placement of several aggregations of the filaments the names of *inferior hæmorrhoidal, vesical* and *vaginal plexuses*.

All the above mentioned plexuses are in free communication with all the minor plexuses (phrenic, cœliac axis, gastric, hepatic, splenic, supra-renal, superior and inferior mesenteric) and the large solar plexus, and the cerebro-spinal system, through their many inosculating branches, from one plexus to another; hence this furnishes the physiological and anatomical reason for the many reflex symptoms seen in distant organs when the uterus and ovaries are diseased.

POINTS WORTH REMEMBERING.

Largest artery—Abdominal aorta.
Largest nutrient artery—Tibial.
Largest synovial membrane—At the knee-joint.
Largest muscle—Glutæus maximus.
Largest nerve—Sciaticus magnus.
Largest vein—Vena cava.

Longest muscle—Sartorius.
Longest tendon—Plantaris.
Branchless artery—Common carotid (except the terminal branches.) There are also no branches from the *cervical portion* of the internal carotid.
Veins carrying arterial blood—Pulmonary. (In the fœtus, the veins carrying arterial blood are the umbilical, hepatic and inferior vena cava.)
Artery carrying venous blood—Pulmonary. (In fœtus, umbilical, also.)
Nerve perforated by an artery—Sciatic by the comes nervi ischiadici; the arteria centralis retinæ also pierces the optic nerve.
Nerve perforated by a vein—Those just named.
Muscle perforated by a muscle—Stylo-hyoid by the digastric.
Muscle perforated by a large nerve—Coraco-brachialis by the musculo-cutaneous.
Vein perforated by a nerve—Occasionally the axillary vein by the internal anterior thoracic nerve.
Ligament perforated by a nerve—Sacro-sciatic by the anterior branch of the coccygeal nerve.
Ligament pierced by an artery—The greater sacro-sciatic, by the coccygeal branch of the sciatic artery. The azygos articularis artery also pierces the posterior ligament of the knee-joint.
Membrane pierced by an artery—The thyro-hyoid by the superior laryngeal artery.
Tendons perforated by tendons—Those of the flexor sublimis digitorum, *of the hands*, for the passage of the tendons of the flexor profundus digitorum. *In the feet*, the tendons of the flexor brevis digitorum are split for the passage of the tendons of the flexor longus digitorum.
Largest branch of the internal carotid artery—Is the middle cerebral; this is the artery that is liable to become plugged by an embolus.
Bones with no muscular attachments—10; ethmoid, nasal, inferior turbinated, vomer, scaphoid, semi-lunar cuneiform, astragalus, middle cuneiform, incus.
Pillars of the palate—*Anterior*, formed by projection of palato-glossus muscle; *posterior*, by projection of palato-pharyngeus muscle.
False vocal cords—Formed by *superior* thyro-arytænoid ligaments.
True vocal cords—Formed by the *inferior* thyro-arytænoid ligaments.

Hamstrings—Outer formed by tendon of biceps; *inner*, by the tendons of the gracilis, sartorius, semi-membranosus, and semi-tendinosus.
The palatal veins and muscles, called azygos—Are *double*, although the term azygos signifies *not paired.*
Veins with only epithelial walls—Those of the diploe.

Amnios was a term given us by Empedocles (B.C. 450).
Aorta was named by Aristotle (B.C. 384), though he supposed it contained air.
Cataract—The first removal of the lens for this disease was made by Herophilus. Celsus cured the trouble by depressing the lens (couching?).
Dissection—First human dissection after Herophilus' time (Herophilus is said to have dissected 700 subjects) was by Mondini de Luzzi, Prof. of Anatomy at Bologna. Old Alexandria, in times before our era, was famous as being the possessor of *two* human skeletons; all Greece and Rome flocked there to see them. Montagana (A.D. 1460) boasted that he had examined *fourteen* human subjects.
Duodenum was named by Herophilus; he also showed the *heart* to be the beginning of *arterial circulation.* In fact, he is the father of anatomy. Fallopius (16th century) said of him, "That to contradict him, was like contradicting the gospels;" that he was "the evangelist of anatomists."
Gynæcologists—The most prominent ones of early date, so far as surgical procedures are concerned, were Paulus Ægineta (early part of the 7th century) though Aetius (close of the 5th century), Galen (A.D. 131), Soranus (A.D. 98-138), Celsus (about A.D. 60), and even Hippocrates (460 B.C.) treated quite lengthily of the subject. Indeed *five* Hippocratic treatises on female troubles, were, in early days, in the hands of the medical profession.
Leeches were first employed by Themison (B.C. 30)
Lexicographer, Medical. The first one was Rufus Ephesius, about A.D. 98 or 117.
Lithotomy was extensively practised in old Alexandria, and the famous oath of Hippocrates (460 B.C.) recognized it as undignified for the physician and surgeon.
Nerves—Their functions were discovered by Herophilus; he overthrew the doctrine that they sprang from the brain-membranes, and proved them to come from the brain itself; their *crossing*, near their cranial organ, was first proposed by Aretæus, and he, in this way,

accounted for a left-sided head injury resulting in a right-sided paralysis.

Physician—This term was first applied to doctors by the people of Charlemagne, A.D. 805.

Pharmacopæia—The first one was issued by an Arabian, Sabor-Ebr-Sahil (9th century) and was called Krabadin.

Rhinoplasty was devised by Vincent Vianso, an Italian, who lived in the 15th century; also performed by Brauca and Bojani.

Vein valves were first discovered by Frabricius, during the latter portion of the 10th century.

Tricuspid valves of the vena cava were discovered by Erasistratus, a contemporary of Herophilus. He called them *triglochine*.

Torcula Herophili first described (with *Calamus scriptorius*) and named by Herophilus (about 250 B.C.)

Tinctures were first introduced by Arnold, about the year 1315. He was then a professor at Barcelona.

CIRCULATION.

Cardiac and pulmonic: The venæ cavæ receive the systemic venous blood, and convey it into the right auricle; then it passes into the right ventricle through the tricuspid, or auriculo-ventricular valves, to be thrown into the pulmonic artery (going through the semi-lunar, or pulmonary valves); is then conveyed to the lungs and oxygenized in the capillary plexus about the intercellular structure and the air-cells, and returned, by the pulmonary veins (4 in number) to the *left* side of the heart, into the left auricle; it then passes into the left ventricle (through the semi-lunar valves) and from thence to support the system at large.

Fœtal: from the placenta through the umbilical vein to the liver; from thence, by the hepatic veins and *ductus venosus Arantii*, to the inferior vena cava, to the right auricle; the most of the current, guided by the Eustachian valve, passes through the foramen ovale into the *left* auricle, and from thence into the left ventricle, and from thence into the aorta and system at large. A part of the current, however, enters the *right ventricle*, is then forced into the pulmonary artery, and from the imperviousness of the fœtal lungs is most all conveyed to the aorta by the *ductus arteriosus Botalli*. The blood is at last conducted by the umbilical arteries (branches of the internal iliac) to the placenta for re-oxygenation.

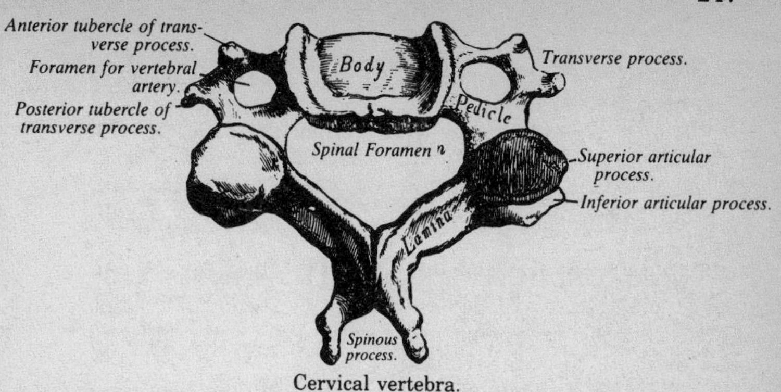

Cervical vertebra.

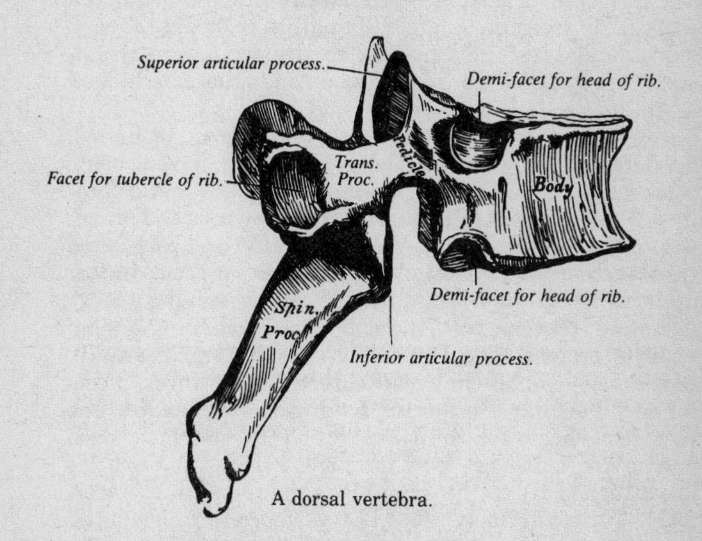

A dorsal vertebra.

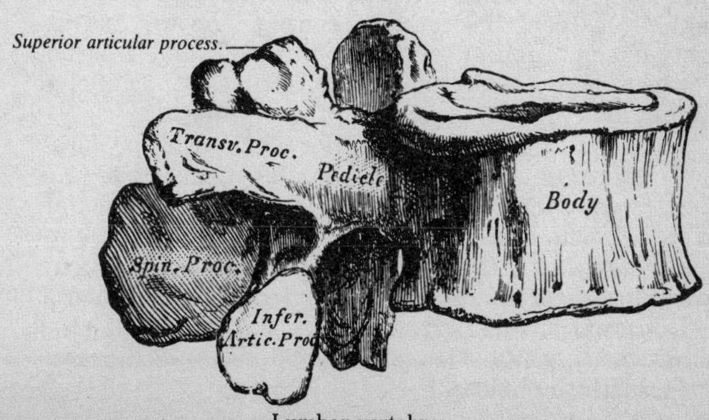

Lumbar vertebra.

OSTEOLOGY.

NOTE.—Muscles in *italics*, are muscles of insertion. Figures in [ ] show the primary number of ossific centres, and date of appearance of ossification.

COMPOSITION: Gelatine and blood-vessels, 33.30; calcic phosphate, 51.04; calcic carbonate, 11.30; calcic fluoride, 2.00; magnesic phosphate, 1.16; sodic chloride and oxide, 1.20; total, **100.00**.

NUMBER: vertebral column (including sacrum and coccyx) 26; cranium, 8; ossiculi auditus, 6; face, 14; hyoid, sternum and ribs, 26; upper extremity, 64; lower extremity, 60; total, **204**. To this may be added the patellæ and teeth, making a grand total of **238**.

SPINE has 33 vertebræ, viz: 7 cervical, 12 dorsal, 5 lumbar, 5 sacral, 4 coccygeal. They each have a body, 7 processes, 2 pedicles, 2 laminæ, 4 notches and a foramen. The **Cervical** are noted for the smallness and broadness of body, bifid spinous processes, bifid and perforated transverse processes, etc. The peculiar are the 1st, or *atlas*, which is like a "ring"; the 2d, or *axis*, having a large (odontoid) process; the 7th, or *prominens*, having a long, spinous process. The **Dorsal** have body largest antero-posteriorly, spinous processes directed downwards, facets for ribs. Peculiar are the 1st, having one whole facet; the rest demi-facets for the ribs; 10th, 11th and 12th, each one having a distinct facet for a rib. *Muscles*: to the atlas are attached 10; to the axis, 11; to the remaining (anteriorly) 10, (posteriorly) 22. [The vertebræ are developed from 3 centres by ossification, the first appearing at 6th week; at sixteen 4 secondary centres appear, and at twenty-one a circular plate for superior and inferior surfaces of body. A few exceptions, as atlas (2) primitive centers), axis (6), 7th cervical and the lumbar (5).]

Sa´crum: triangular, anterior and posterior foramina, lateral masses, lax, tubercular transverse processes, promontory, sacral canal and groove, auricular surface. *Articulations*, (4); 2 innominate, 5th lumbar, coccyx. *Muscles*, (5); pyriform, *coccygeus*, glutæus maximus, erector spinæ, latissimus dorsi. [35, 8th week.]

Coc´cyx: cornua. *Articulation*, (1); sacrum. *Muscles*, (4); *coccygeus*, glutæus maximus, sphineter and levator ani. [4, birth to puberty.]

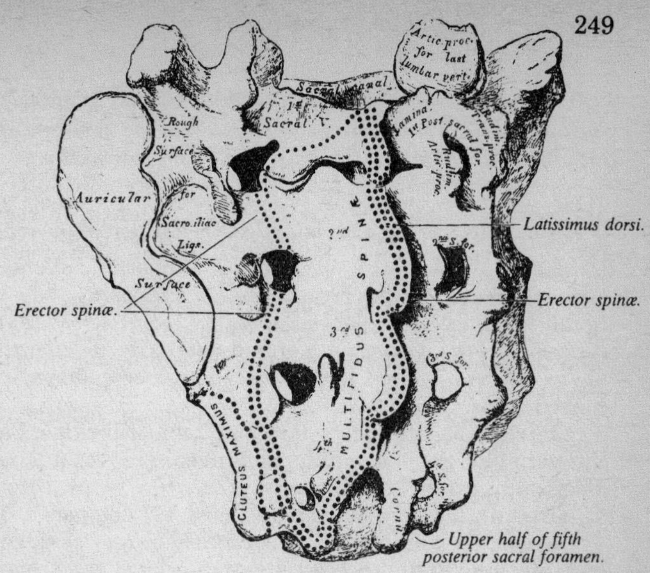

Sacrum, posterior surface.

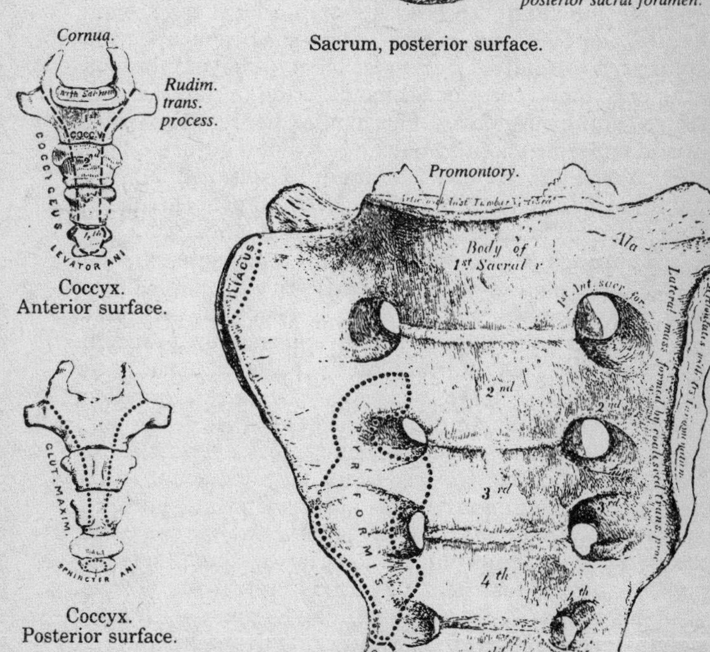

Coccyx.
Anterior surface.

Coccyx.
Posterior surface.

Sacrum, anterior surface.

Occipita´le: superior and inferior curved lines, crest, protuberance, foramen magnum, condyles, basilar and jugular processes, pharyngeal spine, anterior and posterior condyloid foramina; fossa cerebri et cerebelli, torcula protuberance, grooves for occipital, lateral, inferior, petrosal, superior longitudinal sinus and medulla, jugular fossa. *Artic*, (6); 2 parietal, 2 temporal, sphenoid, atlas. *Musc*. (12); occipito-frontalis, trapezius, *sterno-cleido-mastoid*, *complexus*, *splenius capitis*, *obliquus superior*, *rectus posticus major* and *minor*, *rectus lateralis*, *rectus anticus major* and *minor*, superior pharyngeus constrictor. [4, 10th w.]

Parieta´le: eminence, foramen, temporal ridge; Pacchionian depressions, middle meningeal groove, superior longitudinal and lateral sinus. *Artic*. (5); fellow, occipital, frontal, temporal, sphenoid. *Musc*. (1); temporal. [1].

Fronta´le: eminence, superciliary ridges, external and internal angular processes, supra-orbital notches and arches, temporal ridges and fossæ, nasal eminence and spine; orbital plates, lachrymal fossa, pulley depression, ethmoid notch, anterior ethmoid foramina, foramen cæcum, meningeal grooves, Pacchionian depressions, frontal and superior longitudinal sinus (frontal suture). *Artic*. (12); 2 parietal, sphenoid, ethmoid, 2 nasal, 2 superior maxillæ, 2 lachrymal, 2 malar. *Musc*. (3 pr.); corrugator supercilii, orbicularis palpebrarum, temporal. [2].

Tempora´le: zygoma, articular eminence, glenoid fossa, Glasserian fissure, vaginal, styloid, mastoid and auditory processes, mastoid foramen, superior and inferior petrosal and lateral sinus, aquæductus vestibuli, meatus auditorius internus, hiatus Fallopii, opening for smaller petrosal nerve, depression Casserian ganglion, carotid canal, openings for Jacobson's and Arnold's nerves, aquæductus, cochleæ, jugular fossa, stylo-mastoid foramen, auricular fissure, canal for Eustachian tube, and tensor tympani. *Artic*. (5); occipital, parietal, sphenoid, inferior maxilla, malar. *Musc*. (14); temporal, masseter, occipito-frontalis, *sterno-mastoid*, *splenius capitis*, *trachelo-mastoid*, digastric, retrahens aurem, stylo-pharyngeus, stylo-hyoid, stylo-glossus, levator palati, tensor tympani, stapedius. [4, 8th week.]

Sphenoi´des: ethmoid spine, optic groove, olivary process, sella turcica, anterior middle and posterior clinoid processes, cavernous groove; foramina opticum, lacerum anterius, rotundum, Vesalii, ovale, spinosum;

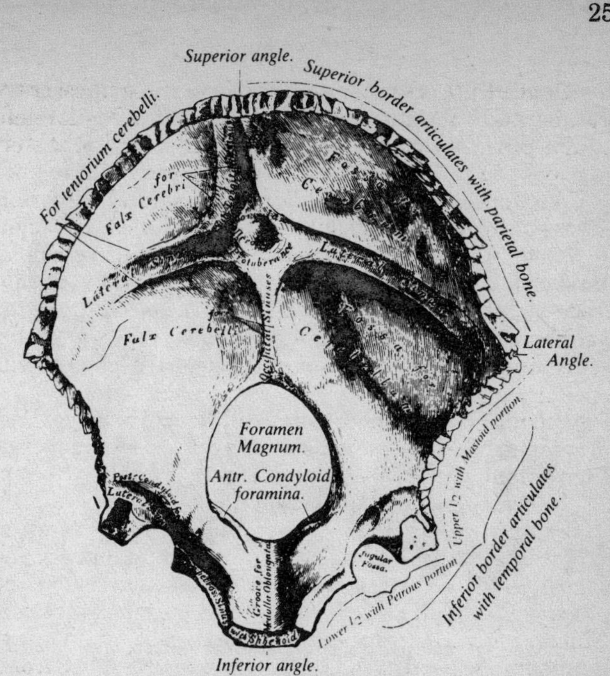

Occipital bone. Inner surface.

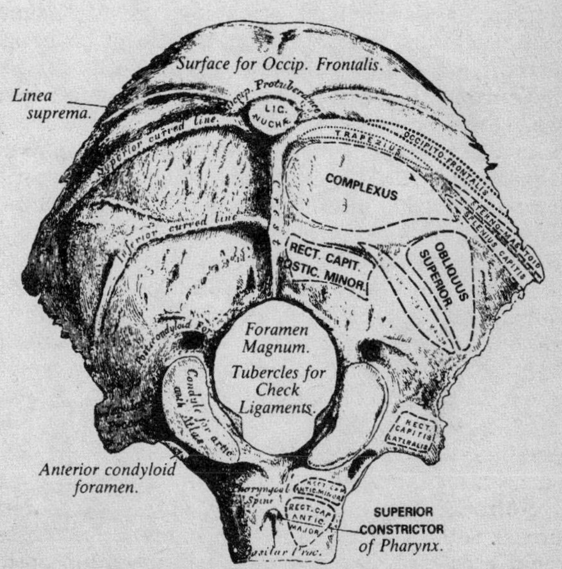

Occipital bone. Outer surface.

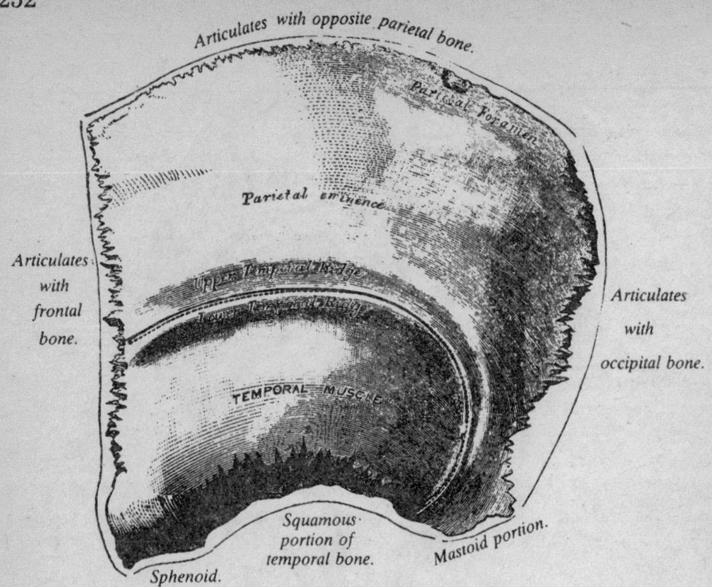

Left parietal bone. External surface.

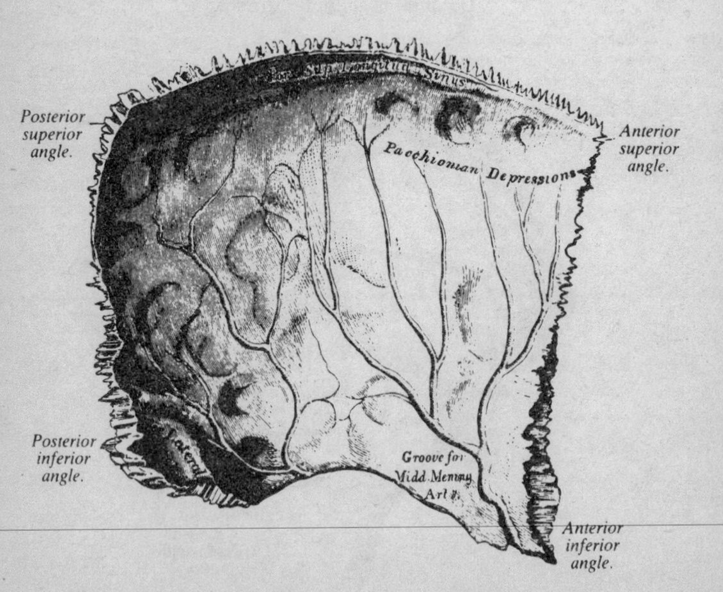

Left parietal bone. Internal surface.

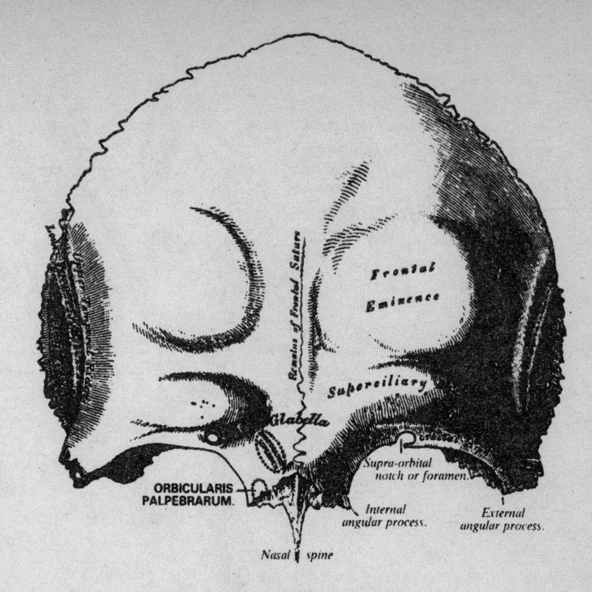

Frontal bone. Outer surface.

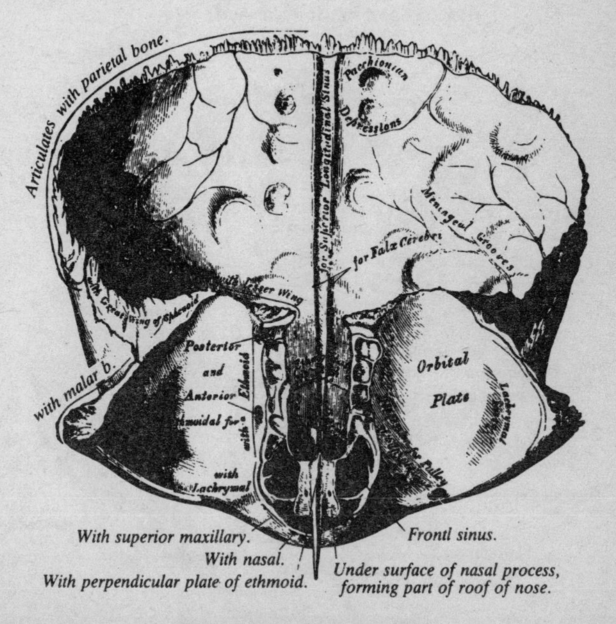

Frontal bone. Inner surface.

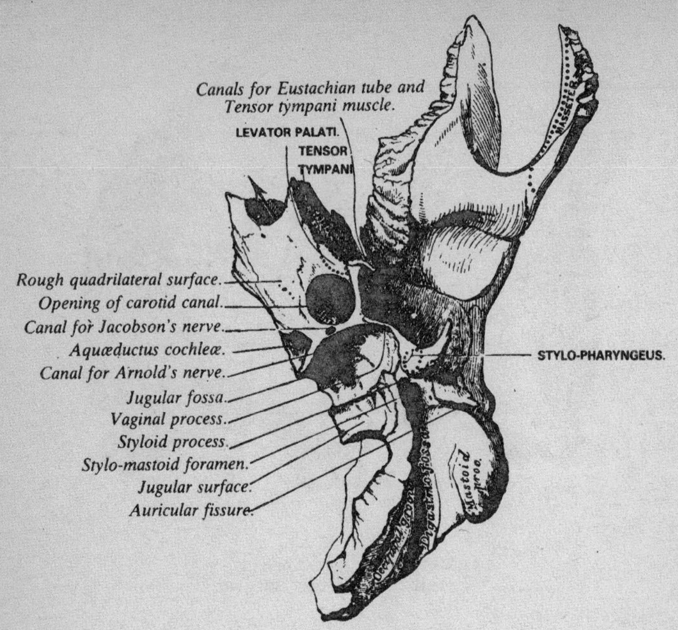

Petrous portion. Inferior surface.

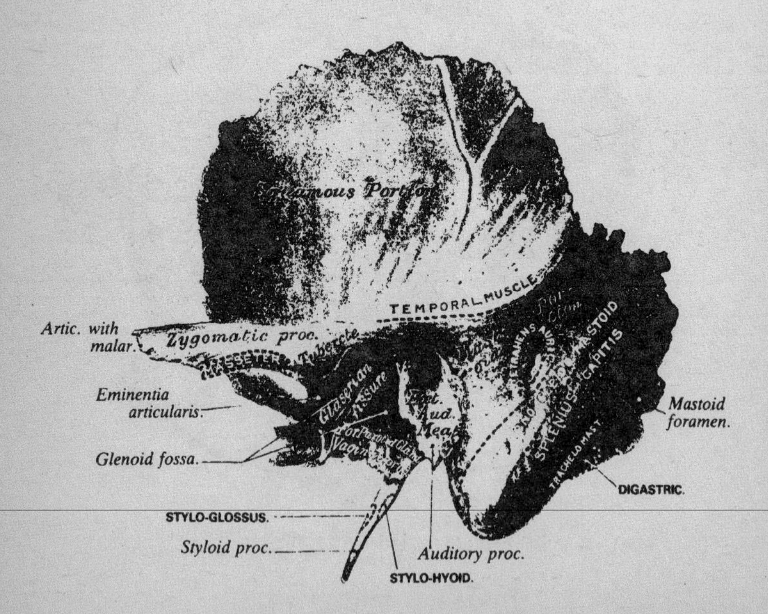

Left temporal bone. Outer surface.

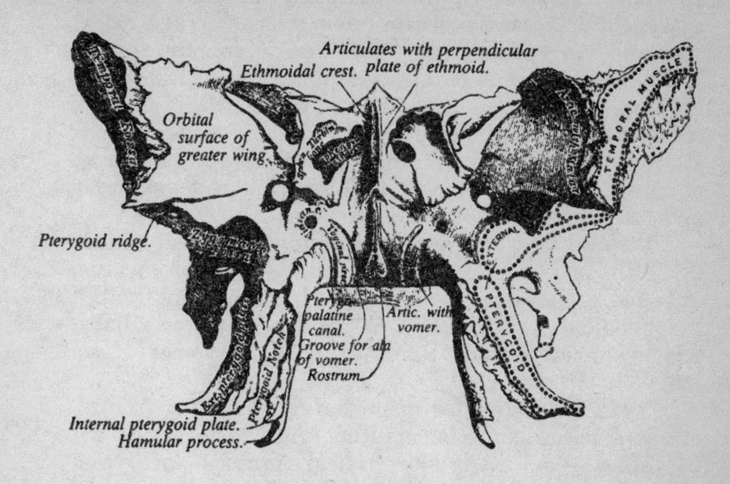

Sphenoid bone. Anterior surface.

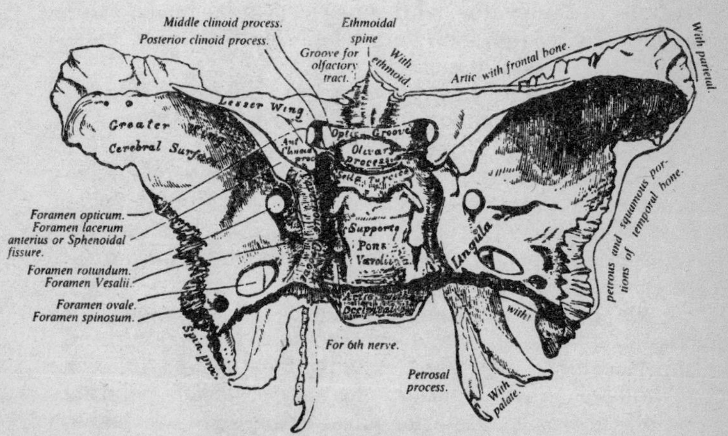

Sphenoid bone. Superior surface.

spinous, hamular, vaginal and external and internal pterygoid processes; rostrum, pterygoid notch and ridge, scaphoid, pterygoid, temporal and zygomatic fossæ, Vidian and pterygo-palatine canals. *Artic.* (12); all of cranium and 2 malar, 2 palate and vomer. *Musc.* (12 pr.); temporal, external and internal pterygoid, superior constrictor, tensor palati, laxator tympani, levator palpebræ, obliquus superior, internal and external recti, superior and inferior recti. [10, 8th w.]

Ethmoi´des: crista galli, infundibulum, os planum, unciform process, olfactory foramina, superior meatus, anterior and posterior cells. *Artic.* (15), sphenoid, frontal, 2 sphenoidal turbinated, 2 nasal, 2 superior maxillary, 2 lachrymal, 2 palate, 2 inferior turbinated, vomer. *Musc.* none. [3, 4th m.]

Nasa´le: groove for nasal nerve. *Artic.* (4); frontal, ethmoid, fellow, superior maxilla. *Musc.* none. [1, 8th w.]

Maxilla´re supe´rior: nasal process lachrymal tubercle, orbicular surface, infra-orbital groove and foramen, canine and incisive fossa, canine eminence, alveolar process, posterior dental canals, maxillary tuberosity, middle and inferior meatus, palate process, anterior and posterior palatine canals: antrum. *Artic.* (9); frontal, ethmoid, nasal, malar, lachrymal, inferior turbinated, palate, vomer, fellow. *Musc.* (9); orbicularis palpebrarum, inferior obliquus oculi, levator labii superioris alæque nasi, levator labii superioris proprius, levator anguli oris, compressor naris, depressor alæ nasi, masseter, buccinator. [4, early.]

Lachryma´le: lachrymal groove. *Artic.* (4); frontal, ethmoid, superior maxilla, inferior turbinated. *Musc.* (1); tensor tarsi.. [1, 8th week.]

Mala´re: frontal, zygomatic, orbital and maxillary processes, temporo-malar canal. *Artic.* (4); frontal, sphenoid, temporal, superior maxilla. *Musc.* (5); levator labii superioris proprius, zygomaticus major and minor, masseter, temporal. [1, 8th week.]

Os pala´ti: orbital, maxillary, and sphenoid processes, spheno-palatine foramen, superior meatus and superior turbinated crest, middle meatus and inferior turbinated crest, inferior meatus; posterior palatine canal, tuberosity, posterior nasal spine. *Artic.* (7); sphenoid, ethmoid, superior maxilla, inferior and superior turbinated, vomer, fellow. *Musc.* (4); tensor palati, azygos uvulæ, internal and external pterygoid. [1, —.]

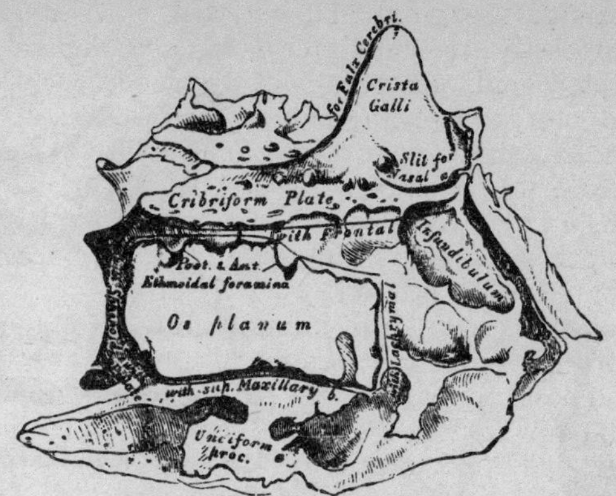

Ethmoid bone. Outer surface of right lateral mass (enlarged).

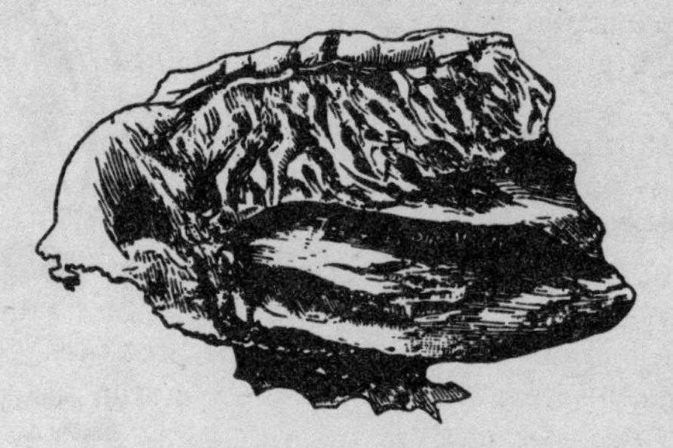

Ethmoid bone. Inner surface of right lateral mass (enlarged).

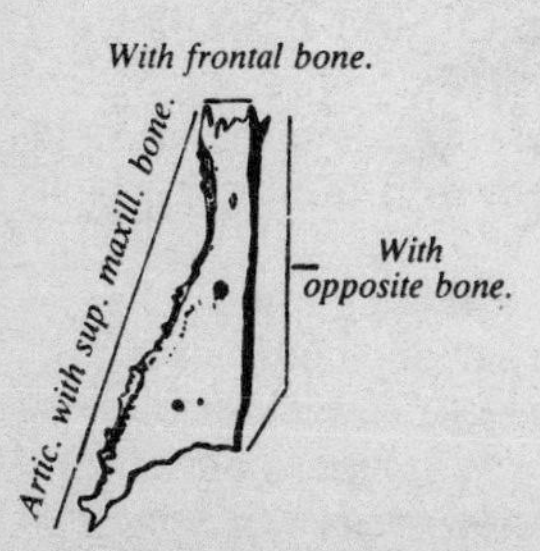

Right nasal bone. Outer surface.

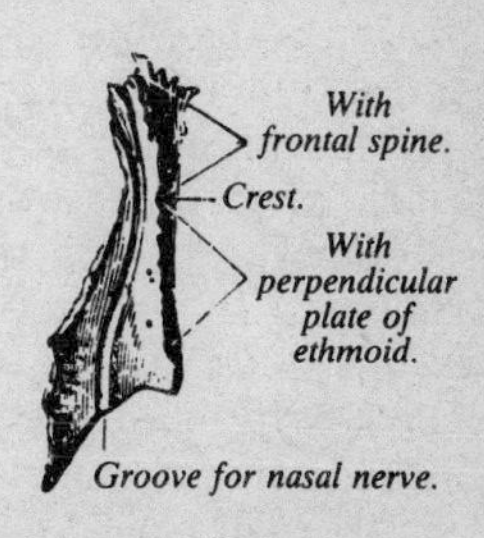

Left nasal bone. Inner surface.

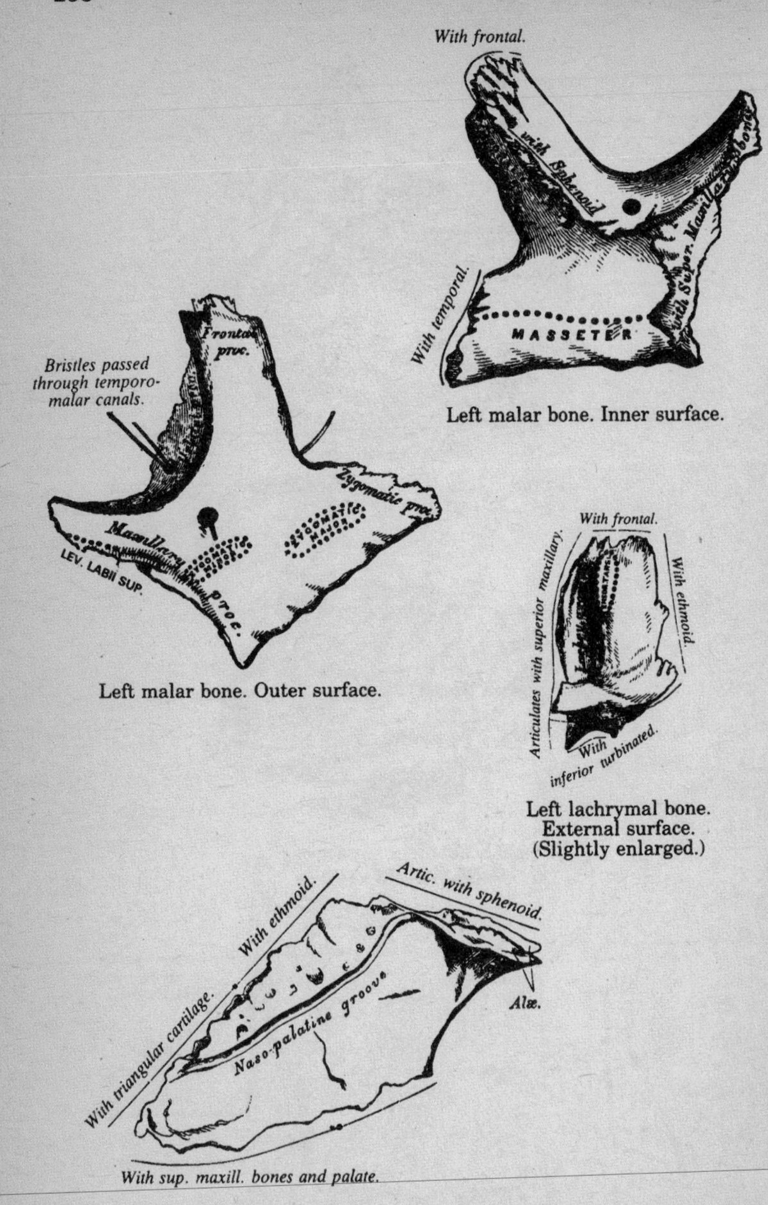

Left malar bone. Inner surface.

Left malar bone. Outer surface.

Left lachrymal bone. External surface. (Slightly enlarged.)

The vomer.

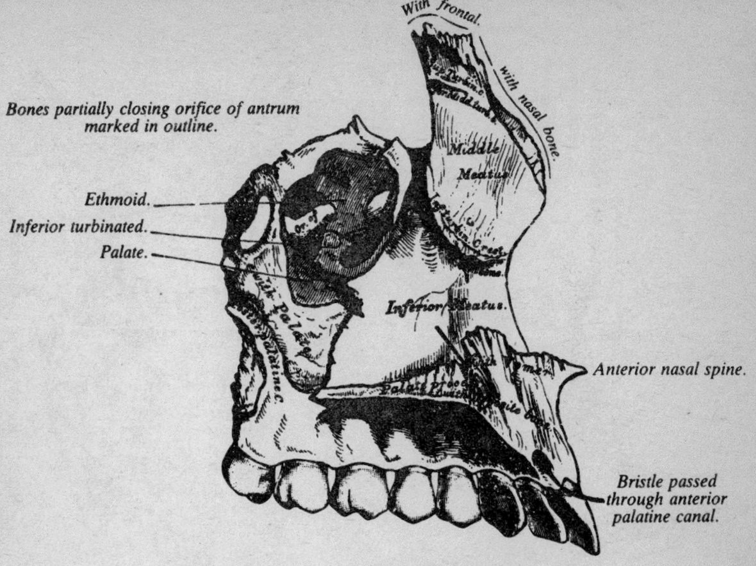

Left superior maxillary bone. Internal surface.

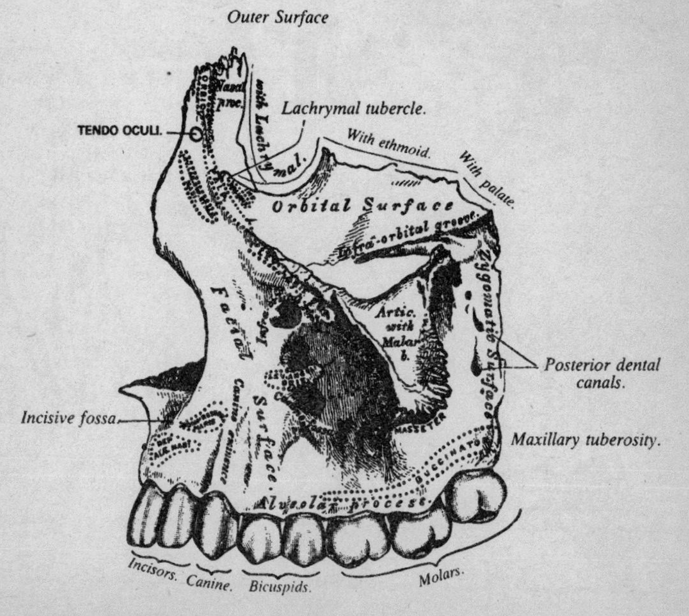

Left superior maxillary bone. Outer surface.

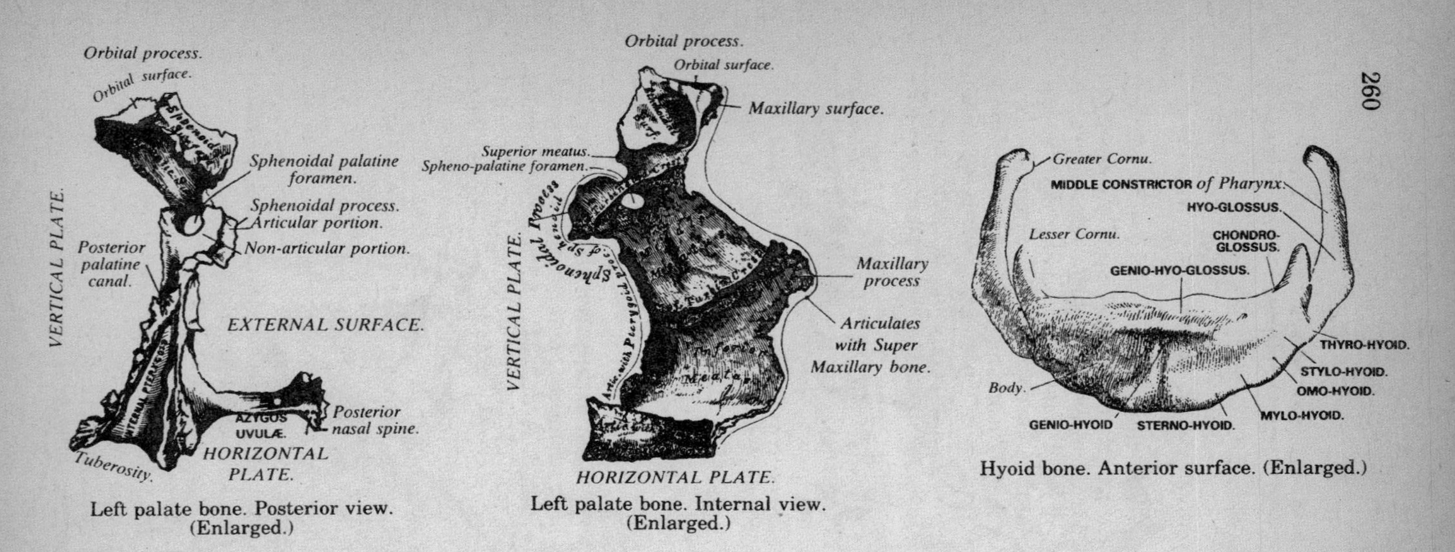

Left palate bone. Posterior view. (Enlarged.)

Left palate bone. Internal view. (Enlarged.)

Hyoid bone. Anterior surface. (Enlarged.)

With lachrymal bone.
With ethmoid.
With sup. max. bone.
Lac.proc.
Ethmoid. proc.
With palate.

Right inferior turbinated bone. Internal surface.

Right inferior turbinated bone. External surface.

Turbina´tum infe´rior: lachrymal, ethmoid and maxillary processes. *Artic.* (4); ethmoid, superior maxilla, lachrymal, palate. *Musc.* none [1, 4th month.]

Vo´mer: naso-palatine groove. *Artic.* (6); sphenoid, ethmoid, 2 superior maxilla, palate. *Musc..* none. [2, 8th week.]

Maxilla´re infe´rior: coronoid process, condyle, ramus, sigmoid notch, mental foramen and process, symphysis, groove for facial artery, inferior dental foramen mylo-hyoid groove and ridge, sublingual and submaxillary fossæ, genial tubercles. *Artic.* (2); 2 temporal. *Musc.* (14 pr.); levator menti, depressor labii inferioris, depressor anguli oris, *platysma*, buccinator, *masseter*; genio-hyo-glossus, genio-hyoid, mylo-hyoid, *digastric*, superior constrictor, *temporal, internal* and *external pterygoid. [2, early.]*

Hyoi´des: greater and lesser cornua, body. *Artic.* none. *Musc.* (11); *sterno-, thyro-, omo-, stylo-, mylo-* and *genio-hyoid*, genio-hyo-glossus, hyo-glossus, middle constrictor, lingualis, *pulley* of digastric. [5, 8th month.]

Ster´num: *manu´brium, gladi´olus, en´siform appendix*, facets for 7 superior ribs. *Artic.* (16); 7 pairs ribs, 2 clavicles. *Musc.* (10); pectoralis major, sterno-mastoid, sterno-hyoid and sterno-thyroid, triangularis sterni, obliquus externus and *internus*, transversalis, *rectus, diaphragm.* [6, 5th month.]

Cos´tæ (ribs); head, neck, tuberosity, articular and non-articular protuberances, angle, facets for superior and inferior vertebræ. *Artic.* (24); vertebræ and costal cartilages. *Musc.* (19); [3 each, save the last two, these but 2; early.] *Peculiar ribs*; 1st, shortest, most curved, horizontally placed, having grooves for subclavian artery and vein; 2d, some larger than 1st, is not twisted, etc.; 10th, single articular facet; 11th and 12th, single articular facet, no neck or tuberosity.

Costal cartilages: *artic.* with sternum and ribs. *Musc.* (10); subclavius, sterno-thyroid, pectoralis major, *internus obliquus*, transversalis, *rectus, diaphragm*, internal and external intercostal, triangularis sterni. (The last 3 are muscles of origin and insertion).

Clavic´ula: shape of letter *f*; sternal and acromial extremity; oblique line, tuberosity, rhomboid impression. *Artic.* (3); sternum, scapula, 1st costo-cartilage. *Musc.* (6); sterno-mastoid and sterno-hyoid, *trapezius*, pectoralis major, deltoid, *subclavius*. [2, first of all.].

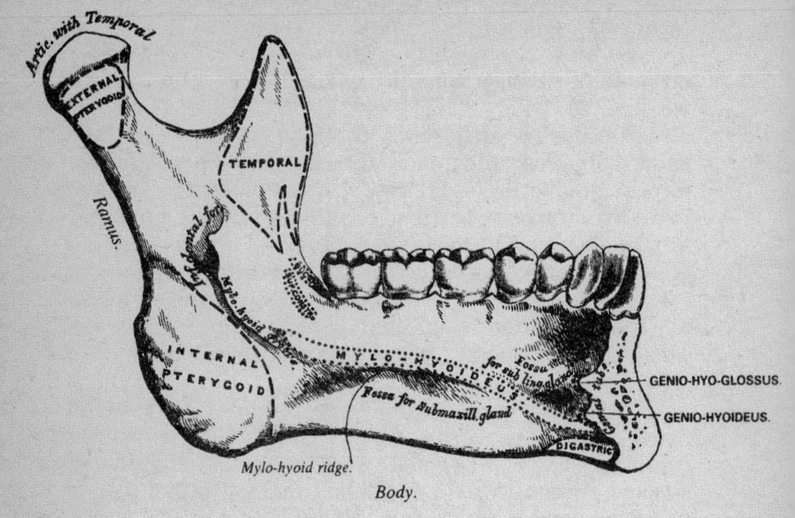

Inferior maxillary bone. Inner surface. Side view.

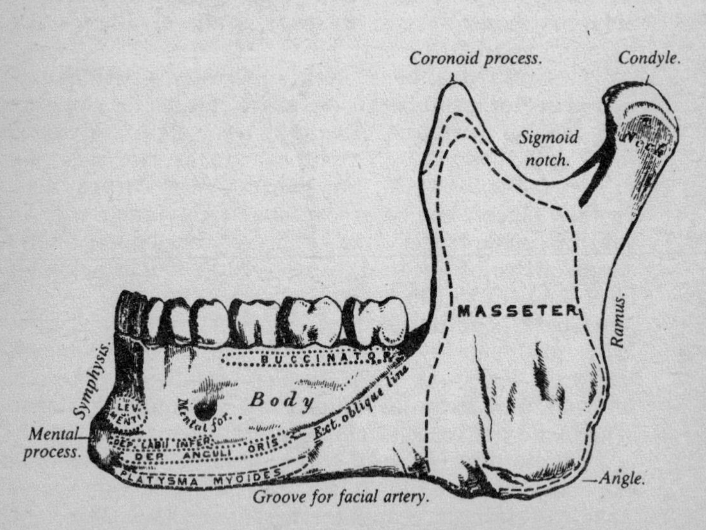

Inferior maxillary bone. Outer surface. Side view.

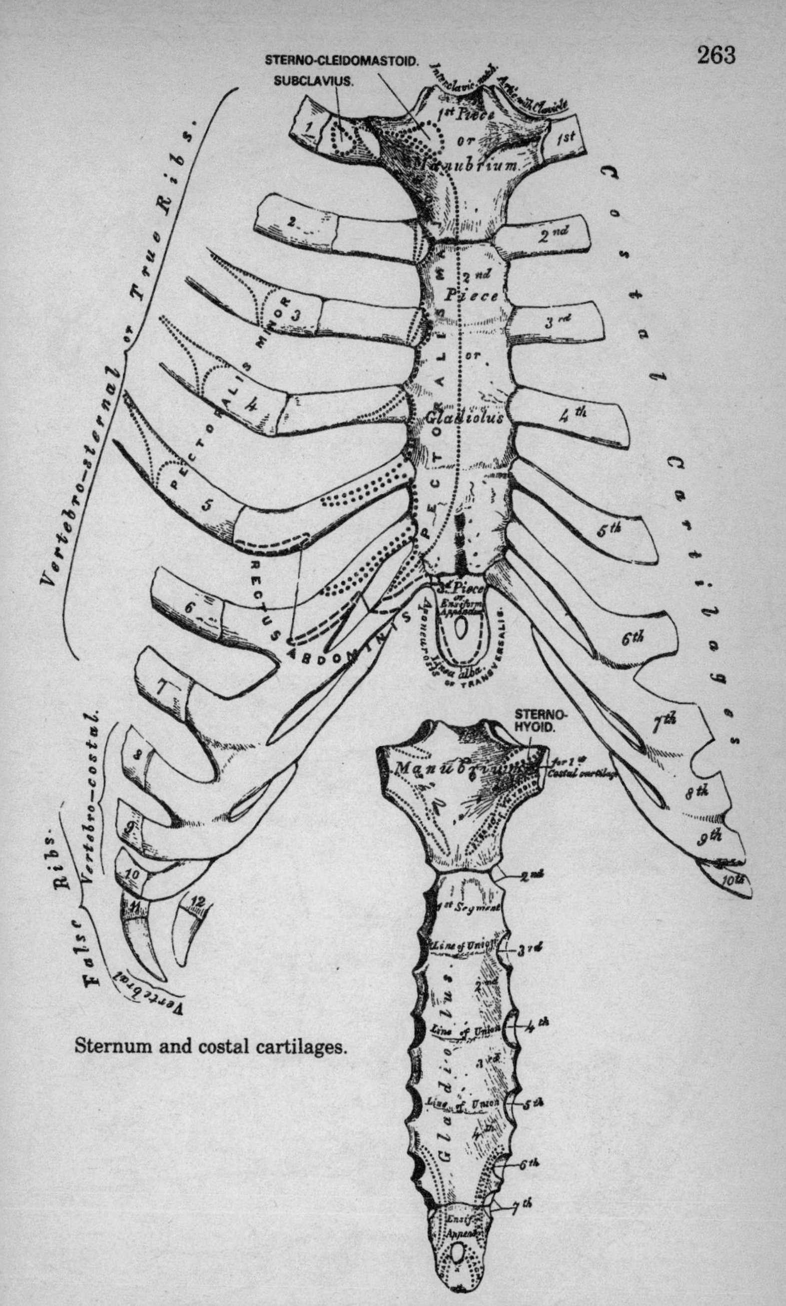

Sternum and costal cartilages.

Posterior surface of sternum.

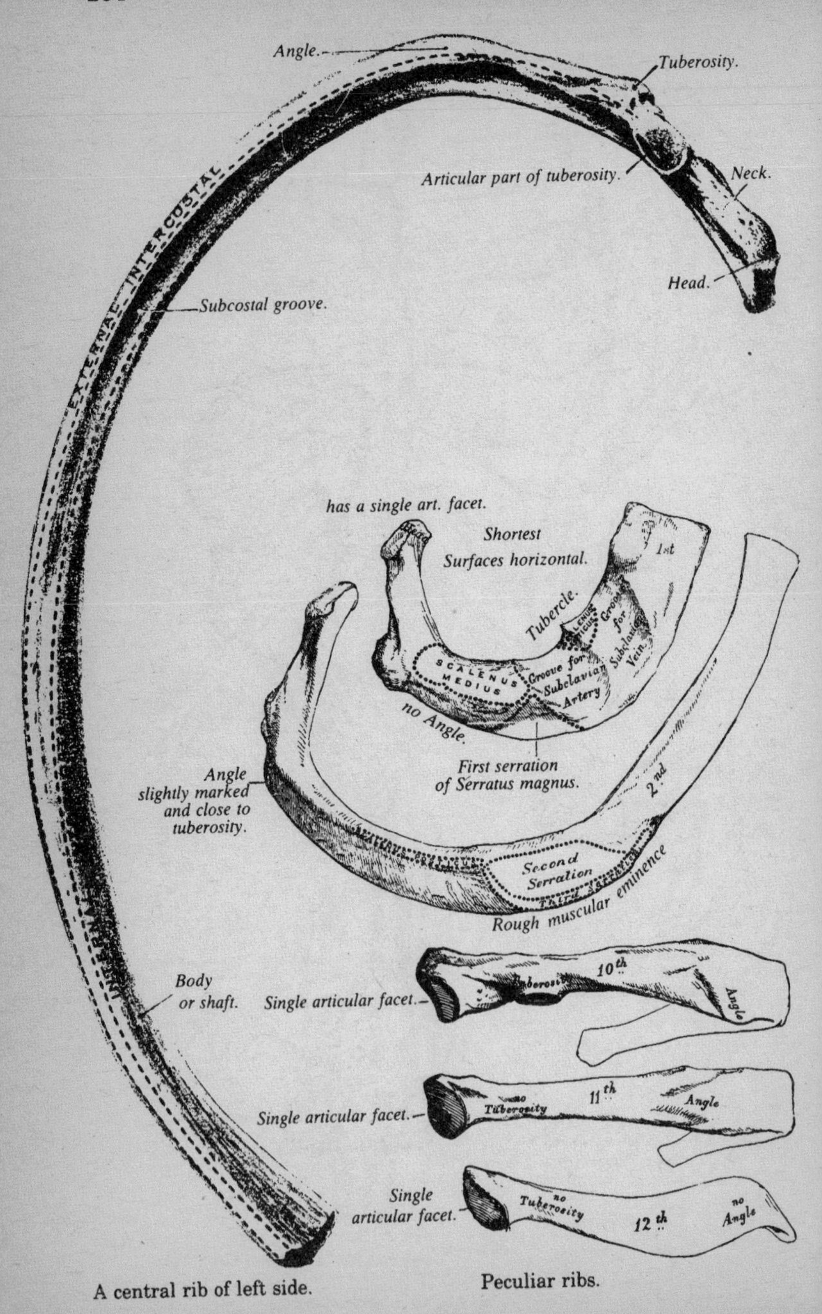

A central rib of left side.

Peculiar ribs.

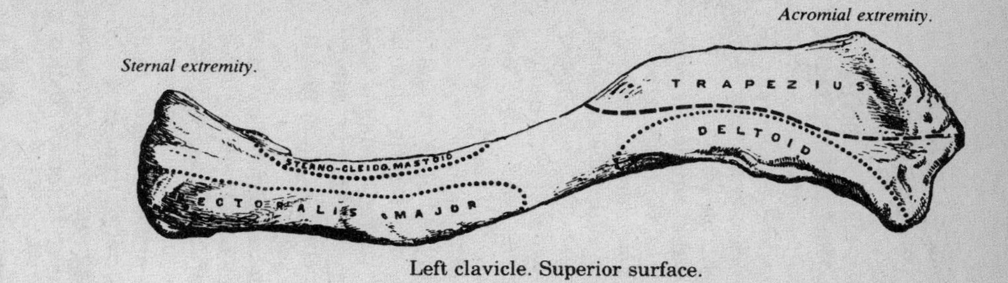

Left clavicle. Superior surface.

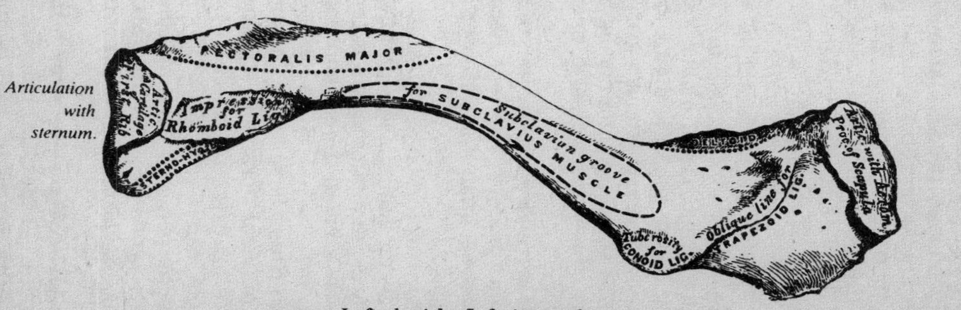

Left clavicle. Inferior surface.

Scap´ula: acromian and coracoid processes, glenoid cavity, neck, subscapular fossa, ridges; supra-scapular notch, supra- and infra-spinous fossæ, spine, groove for dorsalis scapulæ vessels. *Artic.* (2); clavicle, humerus. *Musc.* (17): subscapularis, supra- and infra-spinatus, *trapezius*, deltoid, omo-hyoid, *serratus magnus, levator anguli scapulæ, rhomboidens major* and *minor*, triceps, teres major and minor, biceps, coraco-brachialis, *pectoralis minor*, latissimus dorsi. [7, 8th week.]

Hu´merus: head, anatomical and surgical necks, greater and lesser tuberosities, bicipital ridge and groove, posterior bicipital ridge, rough deltoid surface, internal and external condyles, coronoid and radial depressions, radial head, trochlear surface; musculo-spiral groove, olecranon depression. *Artic.* (3); scapula, ulna, radius. *Musc.* (24); *supra* and *infra-spinatus, teres major* and *minor, subscapularis, pectoralis major, latissimus dorsi, deltoid, coraco-brachialic*, brachialis anticus, triceps; pronator radii teres, flexor carpi radialis, palmaris longus, flexor sublimis digitorum, flexor carpi ulnaris; supinator longus, extensor carpi radialis longior and brevior, extensor communis digitorum, extensor minimi digiti, extensor carpi ulnaris, anconeus, supinator brevis. [7, early.]

Ul´na: olecranon, greater and lesser sigmoid cavities, coronoid process, nutrient foramen, styloid process; oblique line, groove for extensor carpi ulnaris. *Artic.* (2); humerus, radius. *Musc.* (13); *triceps, anconeus*, flexor and extensor carpi ulnaris, *brachialis anticus*, pronator radii teres, flexor sublimis and profundus digitorum, pronator quadratus, supinator brevis; extensor ossis metacarpi and extensor secundi internodii pollicis, extensor indicis. [3, 5th w.]

Ra´dius: head, neck, bicipital tuberosity, oblique line, nutrient foramen, styloid process, 2 grooves; 4 grooves for extensor muscles. *Artic.* (4); humerus, ulna, scaphoid, semi-lunar. *Musc.* (9); *biceps, supinator longus* and *brevis*, flexor sublimis digitorum, flexor longus pollicis, *pronator quadratus*, extensor ossis metacarpi pollicis, extensor primi internodii pollicis, *pronator radii teres.* [3.]

CAR´PUS: (8); [1 after birth]: **Scaphoi´des**: *artic.* (5); radius, trapezium, trapezoid, magnum, semi-lunar. **Semi-luna´re**: *artic.* (5); radius, magnum, unciform, scaphoid, cuneiform. **Cuneifor´me**: *artic.* (3); semi-lunar,

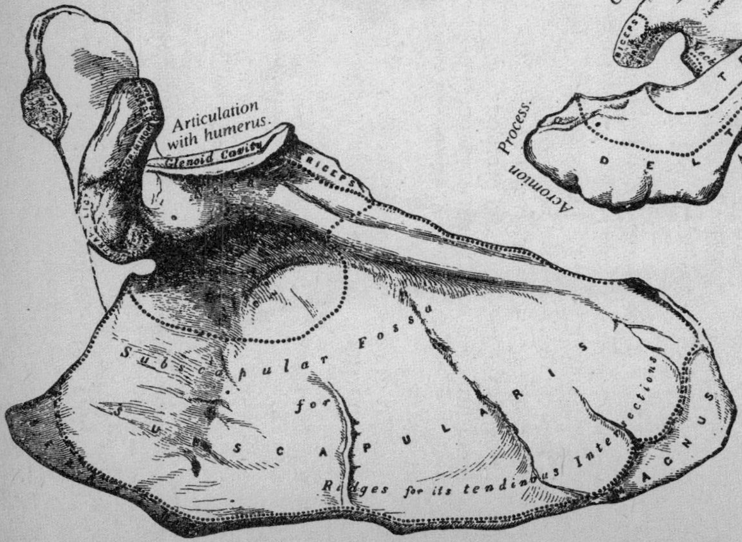

Left scapula, anterior surface, or venter.

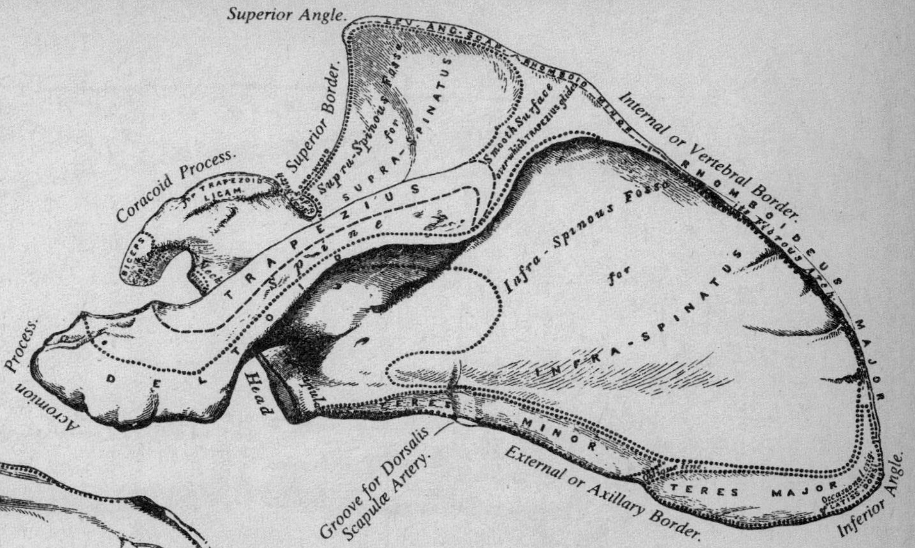

Left scapula. Posterior surface, or dorsum.

Left humerus. Posterior surface.

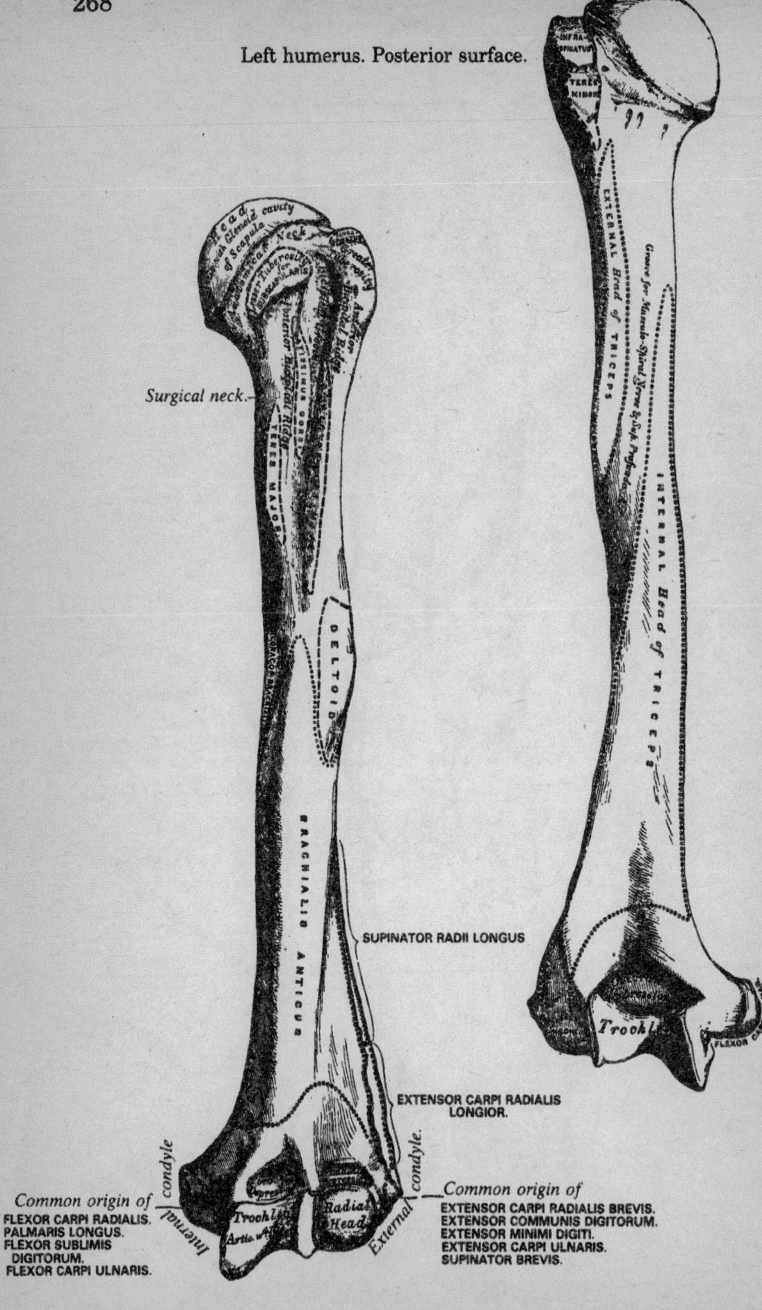

Left humerus. Anterior view.

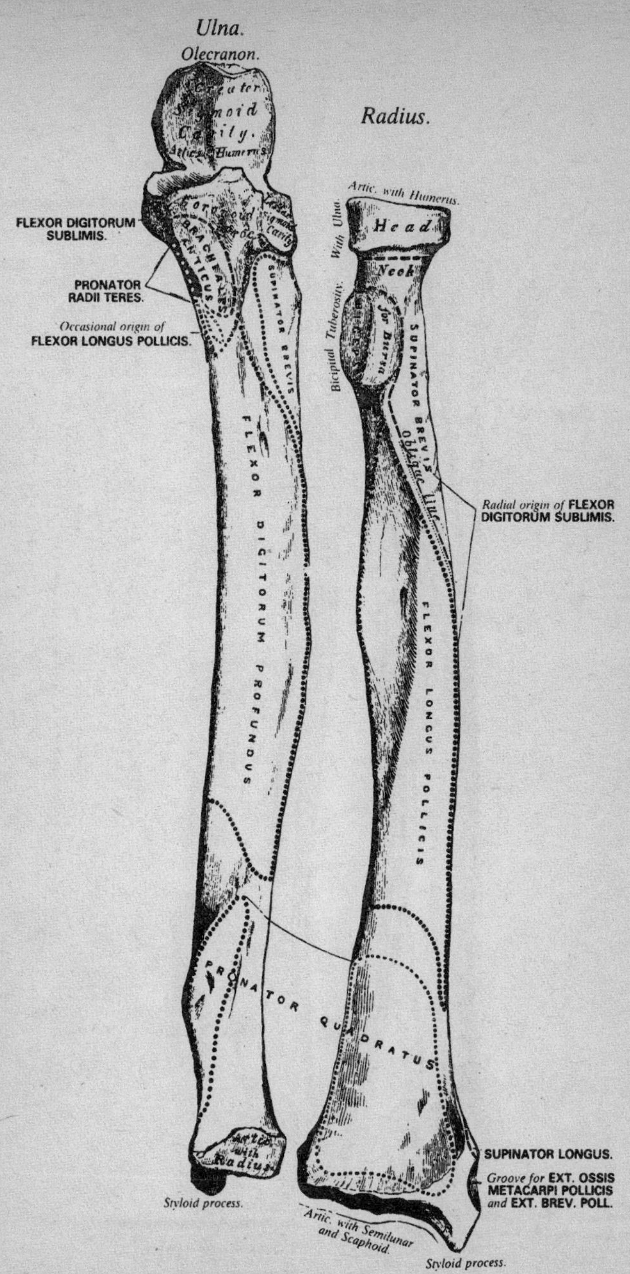

Bones of the left forearm. Anterior surface.

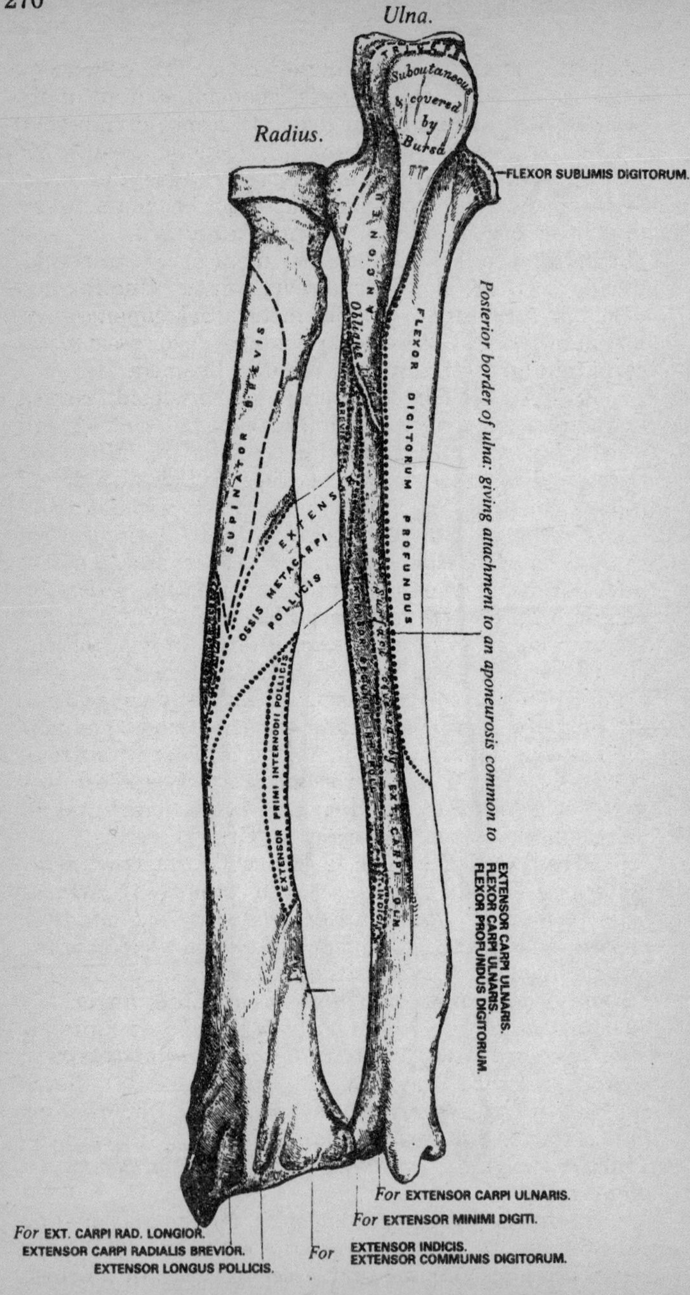

Bones of the left forearm. Posterior surface.

pisiform, unciform. **Pisifor´me**: *artic.* (1); cuneiform. *Musc.* (2); *flexor carpi ulnaris*, abductor minimi digiti. (LOWER ROW.) **Trape´zium**: *artic.* (4); scaphoid, trapezoid, 1st and 2d metacarpal. *Musc.* (3); abductor, flexor ossis metacarpi and flexor brevis pollicis. **Trapezoi´des**: *artic.* (4); scaphoid, 2d metacarpal, trapezium, magnum. *Musc.* (1); flexor brevis pollicis. **Os mag´num**: *artic.* (7); scaphoid, semi-lunar, 2d, 3d, 4th metacarpal, trapezoid, unciform. *Musc.* (1); flexor brevis pollicis. **Uncifor´me**: *artic.* (5); semi-lunar, 4th, 5th metacarpal, cuneiform, os magnum. *Musc.* (2); flexor brevis and flexor ossis metacarpi minimi digiti; anterior annular ligament.

METACAR´PI: (5); bones are prismoid, curved longitudinally, convex behind, concave in front. [2, 6th week.] **1st:** *artic.* (2); trapezium; 1st phalanx. *Musc.* (3); *flexor* and *extensor ossis metacarpi pollicis*, 1st dorsal interosseous. **2d:** *artic.* (5); trapezium, trapezoides, magnum, 3d metacarpus, 2d phalanx. *Musc.* (5); *flexor carpi radialis, extensor carpi radialis longior*, 1st and 2d dorsal interosseous, 1st palmar interosseous. **3d:** *artic.* (4); magnum, 2d and 4th metacarpal, 3d phalanx. *Musc.* (5); *extensor carpi radialis brevior*, flexor brevis pollicis, abductor pollicis, 2d and 3d dorsal interosseous. **4th:** *artic.* (5) magnum, unciform, 3d and 5th metacarpal, 4th phalanx. *Musc.* (3); 3d and 4th dorsal and 2d palmar interosseous. **5th:** *artic.* (3); unciform, 4th metacarpal, 5th phalanx. *Musc.* (5); *flexor* and *extensor carpi ulnaris, flexor ossis metacarpi minimi digiti*, 4th dorsal and 3d palmar interosseous. (An error in "Gray" here.)

PHALAN´GES: (14); [2, 6th w.] **First row:** *artic.* metacarpal and 2d row. *Musc.* 1st or thumb, (4); *extensor primi internodii, flexor brevis, abductor* and *adductor pollicis*. Index, (2); 1st *dorsal* and *palmar interosseous*. Middle finger, (2); 2d and 3d *dorsal interosseous*. Ring 2, 4th *dorsal* and 2d *palmar interosseous*. Little finger, (3); 3d *palmar interosseous, flexor brevis* and *abductor minimi digiti*. **Second row**: thumb (2); *flexor longus* and *extensor secundi internodii pollicis*. To the others, (4); *flexor sublimis* and *extensor communis digitorum*, with *extensor indicis* to index and *extensor minimi digiti* to little finger. **Third row**: *flexor profundus*, and *extensor communis digitorum*.

Innomina´tum: crest, superior, middle and inferior curved lines, anterior and posterior superior and inferior spinal processes, greater and lesser sacro-sciatic notches,

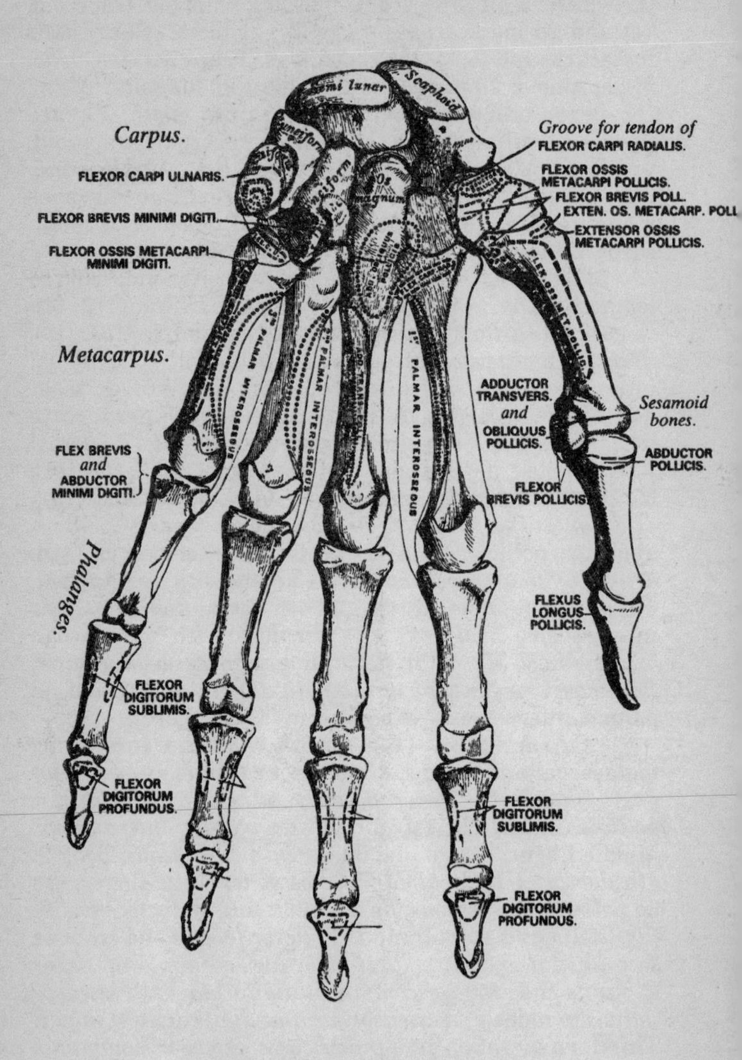

Bones of the left hand. Palmar surface.

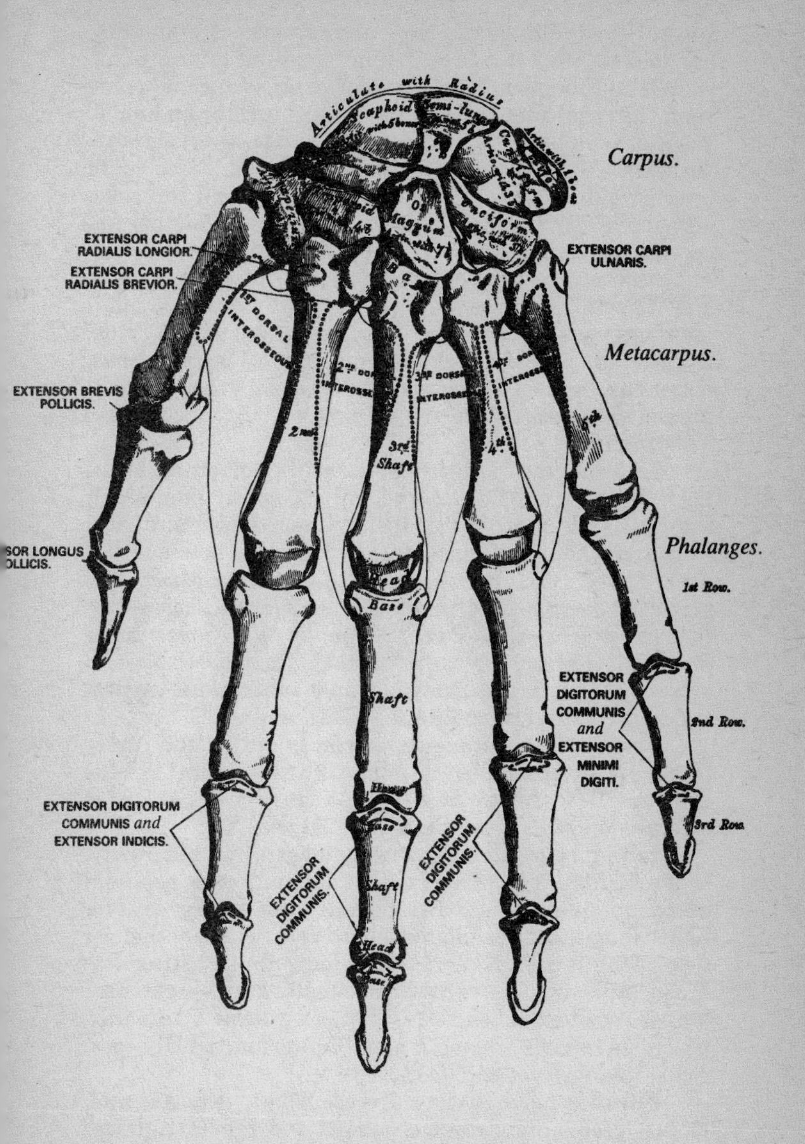

Bones of the left hand. Dorsal surface.

ilio-pectineal eminence line and groove, acetabulum, cotyloid notch; body, crest, spine, angle of pubes, ischic spine and tuberosity, obturator foramen, ischic and pubic rami; internal iliac fossa, groove for obturator and pubic vessels, symphysis pubis, auricular and sacro-iliac rough surfaces. *Artic.* (3); fellow, sacrum, femur. *Musc.* (33); tensor vaginæ femoris, *obliquus externus* and internus, latissimus dorsi, *transversalis*, quadratus lumborum, erector spinæ; 3 glutæi, rectus, pyriformis, iliacus, sartorius; (ischium) obturator externus and internus, levator ani, 2 gemelli, coccygeus, biceps, semi-tendinosus, semimembranosus, quadratus femoris, adductor magnus, transversus perinæi, erector penis; (pubes) *psoas parvus*, pectineus, adductor longus and brevis, gracilis, compressor urethræ (accelerator urinæ). [8; 3 primary, 5 secondary.]

Fe´mur: head, depression for ligamentum teres, neck, greater and lesser trochanters, spiral line, shaft, internal and external tuberosities and condyles; digital fossa, trochanteric line, inter-condyloid notch, linea aspera. *Artic.* (3); innominatum, tibia, patella. *Musc.* (23); *glutæus medius* and *minimus, pyriformis, obturator internus* and *externus*, 2 *gemelli, quadratus femoris; psoas magnus, iliacus;* 2 vasti, *flutæus maximus*, biceps, 3 *adductors, pectinæus*, crureus and subcrureus, gastrocnemius, plantaris, popliteus. [5, 5th w.]

Patel´la: subcutaneous surface; outer and inner facets. *Artic.* condyles of femur, (ligamentum patellæ attaches it to tibia.) *Musc.* (4); *rectus, crureus, vastus externus* and *internus*. [Sesamoid, 3d year.]

Tib´ia: head, spine, internal and external tuberosity, tubercle, fibular facet, crest, internal malleolus; popliteal notch, oblique line, nutrient foramen, common groove for flexor longus digitorum and tibialis posticus, another for flexor longus pollicis. *Artic.* (3); femur, fibula, astragalus. *Musc.* (10); *semi-membranosus*; tibialis anticus, extensor longus digitorum; *sartorius, gracilis, semi-tendinosus*; popliteus, soleus, flexor longus digitorum, tibialis posticus, *ligamentum patellæ*. [3, 5th w.]

Fib´ula: head, styloid process, shaft, external malleolus; groove for peroneus longus and brevis, nutrient foramen. *Artic.* (2); tibia, astragalus. *Musc.* (9); *biceps*, soleus, 3 peronei; extensor longus digitorum and pollicis, tibialis posticus, flexor longus pollicis. [3, 6th w.]

TAR´SUS: (7): **Cal´cis**: greater and lesser processes,

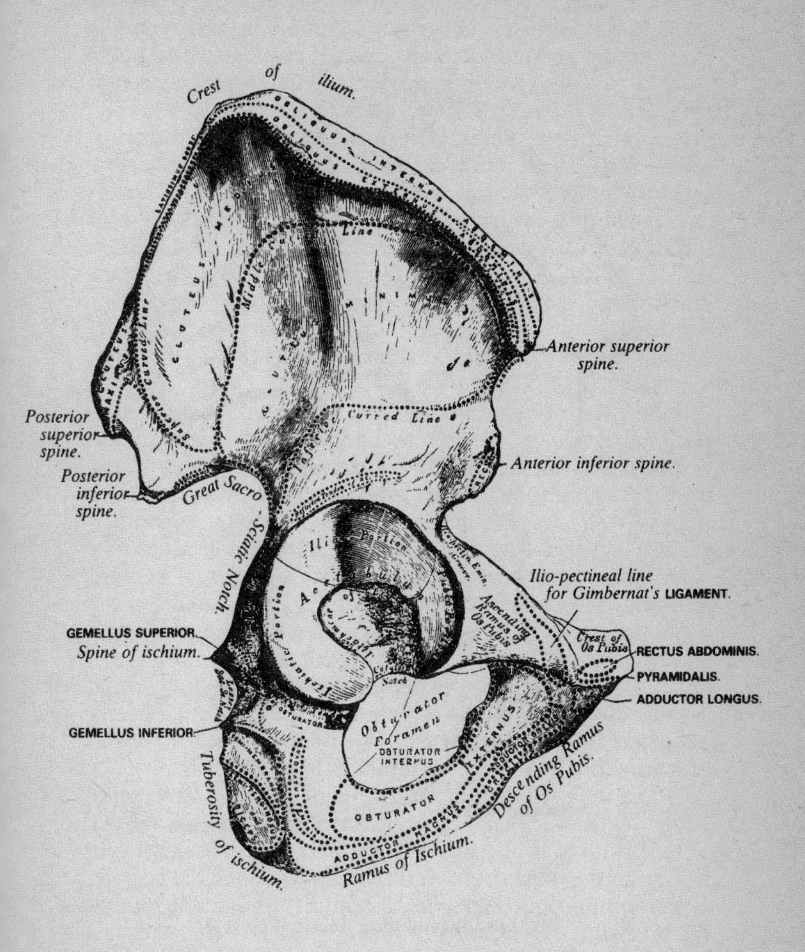

Right os innominatum. External surface.

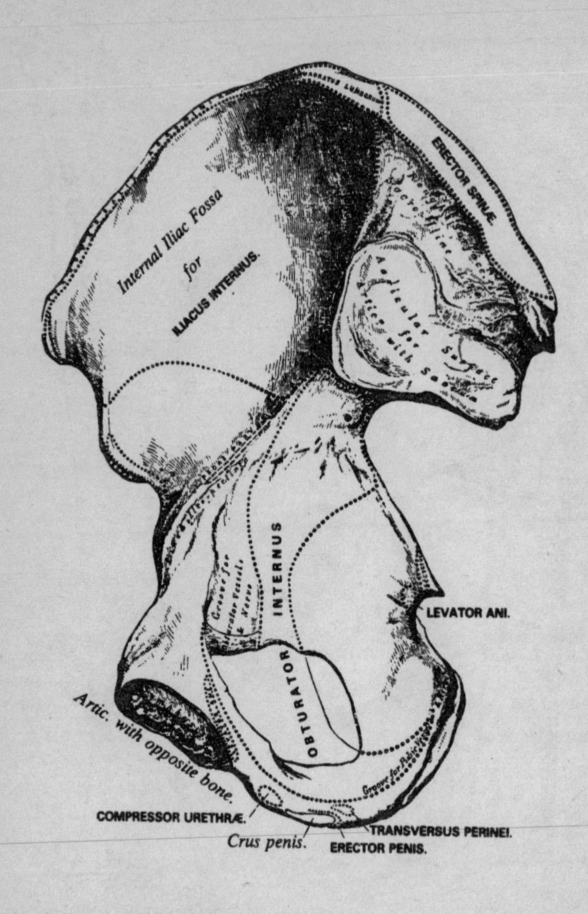

Right os innominatum. Internal surface.

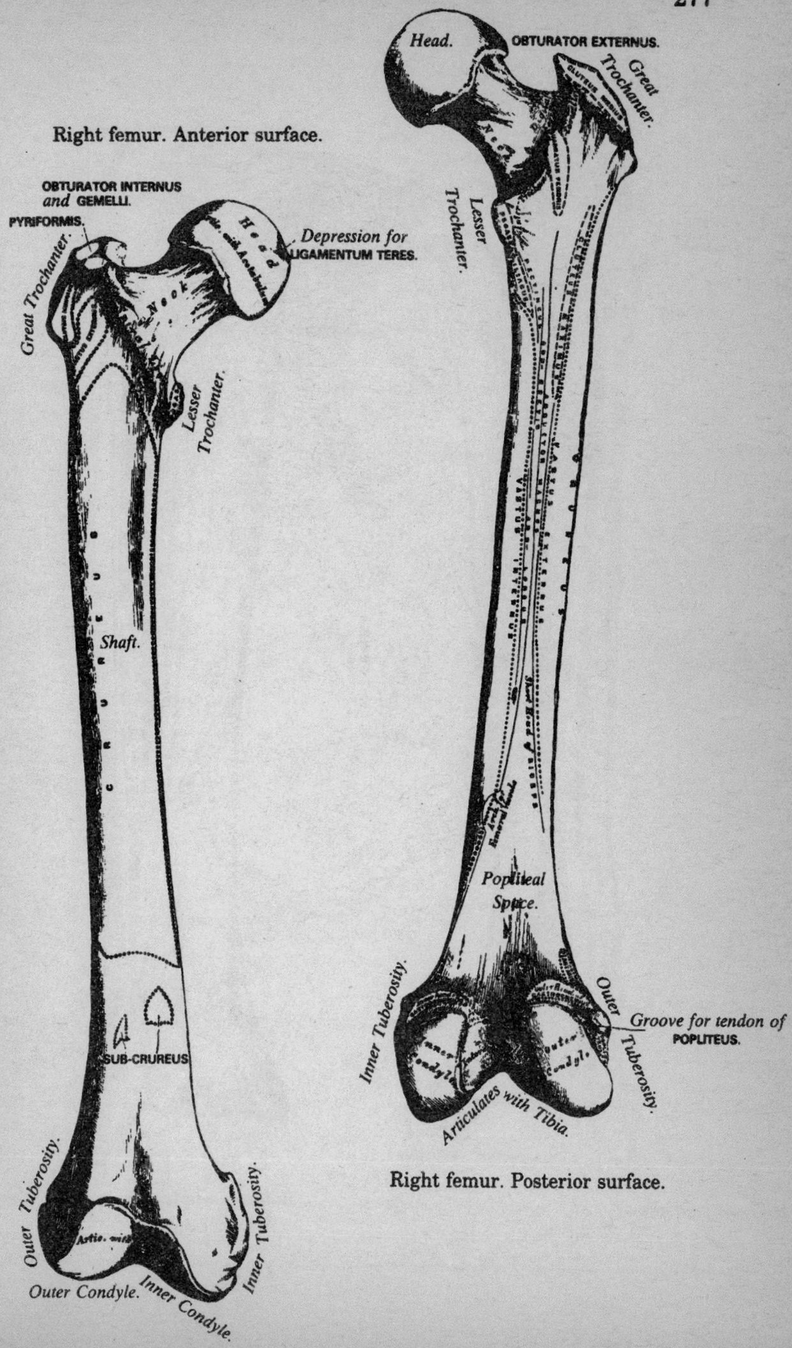

Right femur. Anterior surface.

Right femur. Posterior surface.

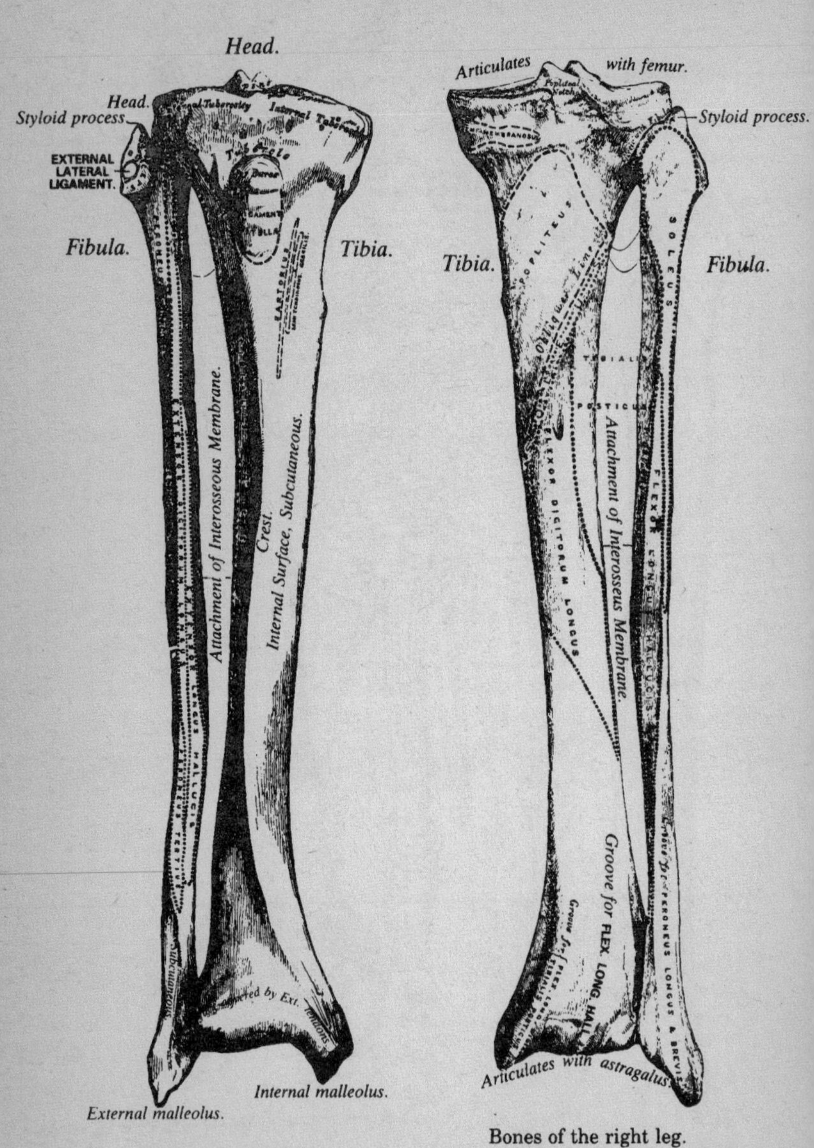

Bones of the right leg.
Anterior surface.

Bones of the right leg.
Posterior surface.

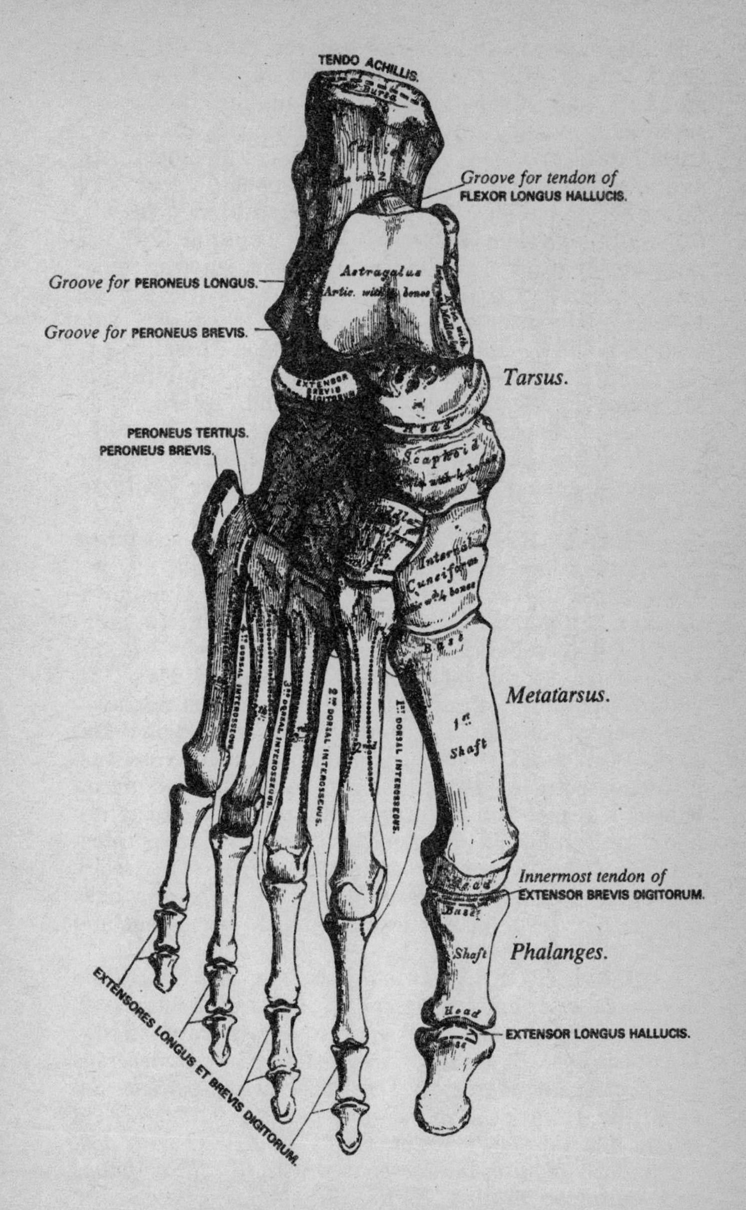

Bones of the right foot. Dorsal surface.

tubercle, superior and inferior grooves. *Artic.* (2); astragalus, cuboid. *Musc.* (8); *tibialis posticus, tendo Achillis, plantaris*, abductor pollicis and minimi digiti, flexor and extensor brevis digitorum flexor accessorius. [1, 6th m.] **Cuboi´des:** *artic.* (4); calcis, external cuneiform, 4th and 5th metatarsi (occasionally scaphoid.) *Musc.* (1); flexor brevis pollicis. [1, 9th m.] **Astrag´alus:** *artic.* (4); tibia, fibula, calcis, scaphoid. *Musc.* none. [1, 7th m.] **Scaphoi´des:** *artic.* (4); astragalus, 3 cuneiform (sometimes cuboid.) *Musc.* (1); *tibialis posticus.* [1, 4th y.] **Cuneifor´me inter´nus:** largest of the three; *artic.* (4); scaphoid, middle cuneiform, 1st and 2d metatarsal. *Musc.* (2); *tibialis anticus* and *posticus.* [1, 3d y.] **Cuneifor´me me´dius:** smallest; *artic.* (6); scaphoid, internal and external cuneiform, 2d metatarsal. *Musc.* none. [1, 4th y.] **Cuneifor´me exter´nus:** *artic.* (6); scaphoid, middle cuneiform, cuboid, 2d, 3d, 4th metatarsi. *Musc.* (2); *tibialis posticus*, flexor brevis pollicis. [1, 1st y.]

METATAR´SI: (5): shaft straight; posterior extremity wedge-shaped, anterior rounded. [2, 8th w.] **1st:** greater size, shortest. *Artic.* (3); internal cuneiform, phalanx, 2d metatarsus. *Musc.* (3); *tibialis anticus, peroneus longus*, 1st dorsal interosseus. **2d:** longest; *Artic.* (6); 3 cuneiform, 1st and 3d metatarsi, 2d phalanx. *Musc.* (3); adductor pollicis, 1st and 2d dorsal interosseous. **3d:** *artic.* (4); external cuneiform, 2d and 3d metatarsi, 3d phalanx. *Musc.* (4); 2d and 3d dorsal and 1st plantar interosseous, adductor pollicis. **4th:** *artic.* (5); external cuneiform, cuboid, 3d and 5th metatarsi, 4th phalanx. *Musc.* (4); adductor pollicis, 3d and 4th dorsal and 2d plantar interosseous. **5th:** tubercular eminence. *Artic.* (3); cuboid, 4th metatarsus, 5th phalanx. *Musc.* (5); *peroneus brevis* and *tertius* flexor brevis minimi digiti, 4th dorsal and 3d plantar interosseous.

PHALAN´GES: (14): shaft convex above, concave below; posterior extremity concave, anterior is convex. [2, after metatarsus.] **1st row:** *artic.* metatarsal and 2d row. *Musc*: big toe, (5); *extensor brevis digitorum, transversus pedis, abductor, adductor* and *flexor brevis pollicis.* Second, (2); 1st and 2d *dorsal interosseous.* Third, (2); 3d *dorsal* and 1st *plantar interosseous.* Fourth, (2); 4th *dorsal* and 2d *plantar interosseous.* Fifth, (3); *flexor brevis* and *adductor minimi digiti*, 3d *plantar interosseous.* **2nd row:** *artic.* 1st and 3d phalanges. *Musc.* big toe, (2); *extensor* and *flexor longus pollicis.* Remaining toes,

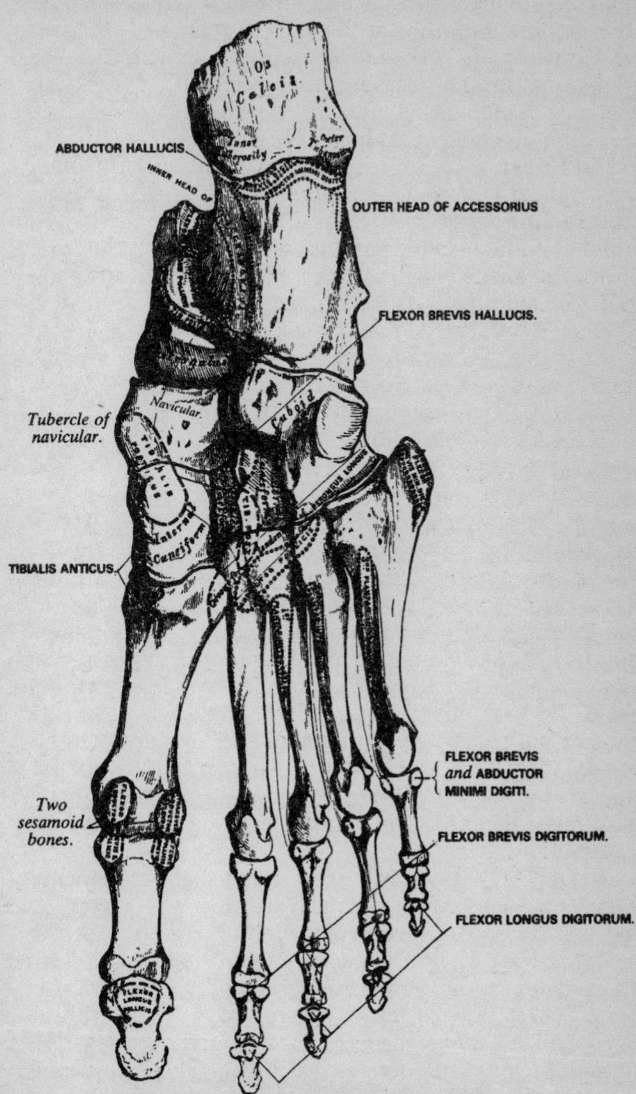

Bones of the right foot. Plantar surface.

(4 each); *flexor brevis digitorum*, and *ext. longus* and *brevis digitorum, lumbricales.* **3d row:** *artic.* 2d phalanges. *Musc.* (3 each); *extensor longus* and *brevis*, and *flexor longus digitorum.*

OSSIC´ULA AUDI´TUS, (3): Mal´leus: head, neck manubrium (handle), processus brevis and gracilis. *Artic.* (1); incus. *Musc.* (3); *laxator major* and *minor tympani, tensor tympani.* **In´cus**: body, short and long processes, os orbiculare. *Artic.* (2); malleus, stapes. *Musc.* none. **Sta´pes**: head, neck, base, crura. *Artic.* (1); incus. *Musc.* (1); *stapedius.*

RÉSUMÉ OF OSTEOLOGY.

Name of Bone	Number of Articulations.	Number of Muscles attached.	Primary Developmental Centres.
Occipital	6	12	4
Parietal	5	1	1
Frontal	12	3	2
Temporal	5	14	4
Sphenoid	12	12	10
Ethmoid	15	none	3
Nasal	4	none	1
Maxillary Sup	9	9	4
Lachrymal	4	1	1
Malar	4	5	1
Palate	7	4	1
Turbinated Inf	4	none	1
Vomer	6	none	2
Maxillary Inf	2	14	2
Hyoid	none	11	5
Sternum	16	10	6
Ribs (12)	24	19	34
Clavicle	3	6	2
Scapula	2	17	7
Humerus	3	24	7
Ulna	2	13	3
Radius	4	9	3
Scaphoid	5	none	1
Semi-lunar	5	none	1
Cuneiform	3	none	1
Pisiform	1	2	1
Trapezium	4	3	1
Trapezoid	4	1	1

Name of Bone.	Number of Articulations.	Number of Muscles attached.	Primary Developmental Centres.
Os Magnum	7	1	1
Unciform	5	2	1
Metacarpal (5)	19	18	10
Phalanges (14)	23	20	28
Vertebræ	72	39	85
Sacrum	4	5	11
Coccyx	1	4	4
Innominatum	3	33	3 and 5
Femur	3	23	5
Patella	1	4	sesamoid
Tibia	3	10	3
Fibula	2	9	3
Calcis	2	8	1
Cuboid	4	1	1
Astragalus	4	none	1
Scaphoid	4	1	1
Int. Cuneiform	4	2	1
Mid. Cuneiform	4	none	1
Ext. Cuneiform	6	2	1
Metatarsal (5)	21	13	10
Phalanges (14)	23	23	28
Malleus	1	3	?
Incus	2	none	?
Stapes	1	1	?

ACTION OF MUSCLES

Head is moved *forwards* by platysma myoideus, sterno-mastoid, rectus capitis anticus major, rectus capitis anticus minor (assisted by, when jaw is fixed), mylo-hyoid, genio-hyoid, genio-hyoglossus, digastricus. *Backwards* by trapezius, splenius capitis, complexus, trachelo-mastoid, rect. capt. post. maj., rect. cap. post. min., obliquus cap. superior. *Sideways* by platysma myoideus, sterno-cleido-mastoid, trapezius, splenius capitis, splen. colli, trachelo-mastoid, complexus.

Neck: *forwards* platysma myoideus, sterno-cleido-mastoid, digastricus, mylo-hyoid, genio-hyoid, genio-hyoglossus, omo-hyoid, sterno-hyoid, thyro-hyoid, rect. cap. ant. major and minor, longus colli. *Backwards* by trapezius, rhomboideus minor, serratus posticus superior splenius capitis, splenius colli, complexus, trachelo-mastoid, transversalis colli, inter-spinales colli, rect. cap. post maj. and minor, obliquus capitis superior and inferior, scalenus posticus, levator anguli scapulæ. *Sideways* by the above in conjoined action, and the scaleni, inter-transversales, recti-laterales.

Trunk: *forwards* by rectus abdominis, pyramidalis, obliquus externus and internus abdominis, psoas magnus and parvus; assisted by (when arms are carried forwards) pectoralis major and minor, serratus magnus. *Backwards*, trapezius, rhomboideus major, latissimus dorsi, serratus posticus superior and inferior, sacro-lumbalis, longissimus dorsi, spinales dorsi, semi-spinalis dorsi, multifidus spinæ, inter-transversalis dorsi et lumborum. *Laterally,* obliquus externus and internus, quadratus lumborum, longissimus dorsi, sacro-lumbalis, serratus posticus, latissimus dorsi.

Scapula: *forwards* by pectoralis minor, serratus magnus. *Backwards*, trapezius, rhomboidei, latissimus dorsi. *Upwards*, trapezius, levator scapulæ, rhomboidei. *Downwards*, trapezius, latissimus dorsi, pectoralis minor.

Humerus: *forwards*, deltoid, pectoralis major; assisted, sometimes, by biceps, coraco-brachialis. *Backwards*, deltoid, teres major and minor, triceps (long head), latissimus dorsi. *Inwards*, pectoralis major, latissimus dorsi. *Rotated inwards* subscapularis, assisted by pectoralis major, lat. dorsi, teres major. *R. outwards*, supra-spinatus, infra-spinatus, teres minor.

Forearm: *forwards*, biceps, brachialis anticus, pronator radii teres; assisted by flex. carpi rad., flex. sublimis digitorum, flex. carpi ulnaris, supinator longus. *Backwards*, triceps, anconeus. *Rotated inwards*, pronator radii teres, flex. carpi radialis, palmaris longus, flexor sublimis dig., pronator quadratus. *R. outwards*, biceps, supinator brevis, extensor secundi internodii pollicis.

Carpus: *forwards*, flex. carpi radialis, palmaris longus, flex. sublimis and profundus dig., flex. carpi ulnaris, flex. longus pollicis. *Backwards*, ext. carpi rad. long. and brev., ext. secundi internodii pollicis, ext. indicis, ext. com. dig., ext. prop. pollicis. *Outwards*, flex. carpi rad., ext. carpi rad. long. and brevior, ext. ossis metacarpi pol., ext. primi internodii pol. *Inwards*, flex. sublim. and profund. digitorum, flex. and ext. carpi ulnaris, ext. com. dig., ext. min. digiti.

Thumb: *inwards* and *forwards*, opponens, flex. brevis and flex. long. pollicis. *Outwards* and *backwards*, ext. ossis metacarpi, ext. primi and secundi internodii pollicis. *Upwards* and *away from fingers*, abductor, flex. brev. pollicis. *Backwards* and *towards fingers*, adductor, ext. primi and secundi pollicis.

Fingers: *flexed*, flex. sublimis and profundus dig., lumbricales, flex. and abductor minimi digiti. *Backwards, ext. communis, ext. minimi digiti and indicis. Outwards*, interossei, abductor indicis and minimi digiti. *Inwards*, interossei, abductor minimi digiti.

Thigh: *forwards*, psoas mag., iliacus, tensor vaginæ fem., pectineus, adductor longus and brevis. *Backwards*, glut. max. and med., pyriformis, obdurator intern., add. mag., biceps, semi-tend., semi-membranosus. *Inwards*, psoas mag., iliacus, pectineus, gracilis, the 3 adductors, obturator extern., quad. femoris. *Outwards*, tens. vag. fem., the 3 glutæi, pyriformis. *Rotated inwards*, tens. vag. fem., glut. med., and, if leg extended, sartorius, semi-tendinosus. *R. outwards*, glut. max. and med., pyriformis, gemelli, obturatores, quad. fem., psoas mag., iliacus, the 3 adductors, biceps femoris.

Leg: *flexed*, semi-tendinosus, biceps, semi-membranosus, gracilis, sartorius, popliteus. *Extended*, rectus fem., crureus, 2 vasti.

Foot: *inwards*, ext. prop. pollicis, flex. long. dig., flex. long. pol., tibialis posticus. *Outwards*, the 3 peronei, ext. long. dig. *Flexed*, tibialis anticus, ext. prop. pol., ext. long. dig., peroneus tertius. *Extended*, gastrocnemius plantaris, soleus, flex. long. dig., flex. long. pol., tib. posticus, peroneus longus and brevis.

Toes: *flexed*, adductor, abductor, flex. longus and brevis pollicis, abductor and flex. brev. minimi digiti. flex. brev. and longus digitorum, flex. accessorius, lumbricales, interossei. *Extended*, ext. long. and brevis digitorum, ext. prop. pollicis. *Inwards*, abductor pollicis, interossei. *Outwards*, add. pollicis and min. digiti, interossei.

TRIANGLES AND SPACES.

ANTERIOR TRIANGLE OF NECK. The anterior triangle of the neck is the space in front of the anterior border of the sterno-mastoideus, and is limited by the following boundaries:—*in front*, median line of the neck from chin to top of sternum; *behind*, the anterior border of sterno-mastoideus; *above*, body of lower jaw, and a line continued from its angle to mastoid process of temporal bone, forming the *base* of the triangle, the apex being at top of sternum; *the floor*, is formed by the following muscles: sterno-thyroideus, sterno-hyoideus, thyro-hyoideus, inferior and middle constrictors of pharynx, anterior belly of digastricus, stylo-hyoideus, mylo-hyoideus and hyo-glossus. The floor is crossed by the anterior belly of the omo-hyoideus and posterior belly of the digastricus, which subdivide the anterior triangle into three smaller ones, viz.—(1) INFERIOR CAROTID TRIANGLE; (2) SUPERIOR CAROTID TRIANGLE; SUBMAXILLARY TRIANGLE; *Roof*, this triangle is covered in by integument, superficial fascia, platysma myoides, and deep fascia. Between the layers forming the roof are the cutaneous branches of the facial and superficial cervical nerves. The contents will be enumerated in the description of the subdivisions. **(1) Inferior Carot´id Triangle.** This is the lowermost subdivision of the anterior triangle of the neck, and has the following boundaries; *in front*, median line of neck; *behind*, anterior border of the sterno-mastoideus; *above*, anterior belly of the omo-hyoideus; the *muscles* met with on the floor of space are sterno-hyoideus and sterno-thyroideus; it is covered in by integument, superficial fascia, platysma myoides muscle, and deep fascia. CONTENTS: This space contains the following structures: thyroid gland, lower part of larynx and trachea; internal jugular and inferior thyroid *veins*; common carotid and inferior thyroid *arteries*; pneumogastric, recurrent laryngeal, descendens noni, communicans noni, and sympathetic *nerves*. **(2) Superior Carot´id Triangle.** This is the middle of the three subdivisions of the anterior triangle of the neck, its boundaries being; *behind*, anterior border of sterno-mastoideus; *above*, posterior belly of digastricus; *below*, anterior belly of omo-hyoideus; the muscles forming the *floor* are the thyro-hyoideus, hyo-glossus, and the inferior and middle constrictors of the pharynx. The *roof* is formed by the same structures as cover in the inferior carotid triangle. CONTENTS: Upper part of larynx and lower part of pharynx; Internal jugular, and those which open into it, viz.—lingual, facial, superior thyroid, pharyngeal and sometimes the occipital *veins*; termination of common carotid, external carotid, internal carotid, superior thyroid, lingual, facial, ascending pharyngeal, and occipital *arteries*; pneumogastric, superior laryngeal, external laryngeal, hypo-glossal, descendens noni, spinal accessory, and sympathetic *nerves*; **(3) Sub-maxillary Triangle.** This is the most superior of the three subdivisions of the anterior triangle, and has the following boundaries; *behind*, posterior belly of digastricus; *above*, lower border of the jaw, and line continued from angle of jaw to the mastoid process; *in front*, median line of neck from the chin to the hyoid bone. (Some anatomists limit this space in front by the anterior belly of the digastricus.) The muscles forming the *floor* are the anterior belly of the digastricus, the mylo-hyoideus, and the hyoglossus, and its *roof* is formed by the same structures as cover in the superior and inferior carotid triangles. CONTENTS: Portion of parotid and submaxillary (salivary), and submaxillary lymphatic *glands* and vessels; internal jugular, commencement of external jugular and venous radicles of anterior jugular, the facial, submental, submaxillary, inferior palatine and ranine *veins*; external carotid, internal carotid, facial, sub-mental, mylohyoidean (and several smaller branches) *arteries*; within this space are the mylo-hyoid (branch of inferior dental), the infra-maxillary branches of facial, and the ascending branches of the superficial cervical *nerves*. (The two latter, strictly speaking, are not contents of the triangle, as they ramify in the structures which form its roof.) Deeply situated at the back part of the space are the pneumogastric and glosso-pharyngeal *nerves*. (That portion of the hypoglossal nerve, which lies on the hypoglossus muscle, should be included as one of the contents.) The stylo-hyoideus, the stylo-glossus, origin of the stylopharyngeus *muscles*, and stylo-maxillary ligament, may also be given as contents of the space. The stylo-hyoideus is sometimes given as a part of the posterior boundary.

POSTERIOR TRIANGLE OF THE NECK. The posterior triangle of the neck is the space behind the posterior border of the sterno-mastoideus, and has the following boundaries; *in front*, posterior border of sterno-mastoideus; *behind*, anterior border of trapezius; *below* (*base*), upper border of the middle third of clavicle; *apex*, meeting of anterior and posterior boundaries at the occiput; *floor* (*from above downwards*), splenius capitis, levator anguli scapulæ, scalenus medius, scalenus posticus and upper digitation of serratus magnus. The space *is covered in* by the platysma myoides. The contents will be named in the two following subdivisions of this space which are made by the crossing of the space by the posterior belly the omo-hyoid, about 1 inch above the clavicles. **(1) Occip´ital Triangle.** This is the larger of the two divisions. Is bounded *in front* by sterno-mastoid; *behind*, by trapezius; *below*, by omo-hyoid. Its *floor* is formed by (from above downwards) splenius, levator anguli scapulæ, by middle and posterior scaleni. Is *covered* by integument, platysma (below), superficial and deep fascia. CONTENTS: Spinal accessory nerve, transversalis colli artery and vein, and chin lymphatic glands. **(2) Subcla´vian Triangle.** So called because best situation for tying subclavian artery in the third part of its course. Is bounded, *above*, by posterior belly of omo-hyoid; *below*, by clavicle; base (directed forwards) by posterior border of sterno-mastoid. Varies greatly in size in different subjects, and different positions of same subject. Is *covered in* by same structures as the Occipital. CONTENTS: Descending branches of superficial cervical plexus; brachial plexus nerves; subclavian artery (third part of its course); transversalis colli artery and vein; transversalis humeri (supra-scapular) artery and vein; external jugular vein, and communicating branch with cephalic vein; lymphatic vessels and glands.

SUB-OCCIP´ITAL TRIANGLE. This is situated immediately below the occipital bone, and beneath the upper part of the complexus muscle. Its boundaries are as follows: *above*, obliquus superior; *below*, obliquus inferior; *behind*, rectus capitis posticus major; the *roof* is formed by the complexus muscle, and the *floor* by the posterior occipito-atloid ligament and posterior arch of the atlas. CONTENTS: Vertebral artery and sub-occipital nerve (post. br. of first cervical).

TRIANGLE IN FRONT OF ELBOW-JOINT. This is bounded, *externally*, by the supinator longus; *internally*, pronator radii teres; *above* (*base*) a line—imaginary—drawn across the arm two inches above the condyles; *apex*, meeting of the supinator longus and pronator radii teres. This space is *covered in* by skin, superficial fascia and bicipital fascia; the *floor* is formed by the lower part of the brachialis anticus and the oblique fibres of the supinator brevis muscles. CONTENTS: (from within outwards): Median nerve; brachial artery and venæ comites (about the centre of the space the artery divides into radial and ulnar); tendon of biceps; musculo-spiral nerve. (The supinator longus and brachialis anticus must be slightly separated in order to expose this nerve.)

SCARPA'S TRIANGLE. This is situated at the upper part of the anterior surface of the thigh, with apex downwards, immediately below the fold of the groin, and has the following boundaries; *Externally*, sartorius; *internally*, adductor longus; *above* (*base*), Poupart's ligament; *apex*, meeting of the sartorius and adductor longus muscles. The space is *covered in* by skin, superficial fascia, fascia lata, and cribriform fascia, and the *floor* is formed (from without inwards) by the iliacus, psoas, pectineus, and small portion of adductor brevis muscles. CONTENTS: Femoral sheath (derived from the iliac fascia and fascia transversalis); femoral artery (giving off cutaneous branches and a large deep branch—the profunda femoris); femoral vein (here joined by the saphena and profunda veins); anterior crural nerve and its branches; deep lymphatic glands and vessels and fatty tissue. This is the best point for ligation of femoral artery, the artery lying between the vein (inside) and nerve (outside.)

HES'SELBACH'S TRIANGLE. This space is situated at the lower part of the abdominal wall, on either side, and is of surgical importance as being the spot where direct inguinal hernia makes its escape from the abdomen. Its boundaries are: *Externally*, epigastric artery; *internally*, outer margin of rectus abdominis muscle; *below* (*base*), Poupart's ligament. The structures entering into the formation of the abdominal wall at this spot are (from without inwards); skin; superficial fascia; inter-columnar fascia; conjoined tendon of internal oblique and transversalis muscles; fascia transversalis; subserous cellular tissue; peritoneum. *These seven structures form the coverings of direct inguinal hernia.*

AXILLA'RY SPACE. This is of conical form, and is

situated between the upper part of the side of the chest, and the inner side of the arm, and has the following boundaries; *In front*, pectoralis major and minor muscles; *behind*, subscapularis, teres major, and latissimus dorsi; *inner side*, upper four ribs and intercostal muscles, and upper part of serratus magnus; upper part of the humerus, the coraco-brachialis and biceps; the *apex* of the cone is directed upwards, and is formed by an interval between the first rib, the clavicle and the upper border of the scapula; its *base* is formed by the skin and axillary fascia stretched across from the lower border of the pectoralis major to the lower border of the latissimus dorsi. CONTENTS: Axillary artery and vein and their branches; brachial plexus of nerves, and branches of distribution below the clavicle; a few branches of the intercostal nerves; about ten or twelve lymphatic glands, and a quantity of loose fat and areolar tissue.

POPLITE´AL SPACE. This space is situated at the back of the knee-joint; and forms what is called the ham. It is lozenge-shaped, and has the following boundaries; *Externally, above the joint*, biceps; *below the joint*, outer head of gastrocnemius and plantaris; *internally, above the joint*, semi-tendinosus, semi-membranosus, gracilis and sartorius; *below the joint*, inner head of gastrocnemius. The *floor* is formed by the lower part of the back of the femur, the posterior ligament of the knee-joint (ligamentum posticum Winslowii) and the popliteus muscle covered by its fascia. The space is *covered in* by skin, superficial fascia and fascia lata. CONTENTS: Popliteal vessels and their branches; termination of external saphenous vein; internal and external popliteal nerves and branches; branch of small sciatic nerve; articular branch of obturator nerve; four or five small lymphatic glands, and a quantity of fat and loose areolar tissue.

THE MEDIASTI´NUM. This is the space in the middle line of the thorax, formed by the approximation of the pleura on either side, and extends from the sternum in front to the bodies of the vertebræ behind. In no place do the reflected pleuræ come in contact with each other, so that the space between them forms a complete septum, dividing the two pulmonary cavities. The mediastinum is divided into three portions. (1) *anterior*, (2) *middle* and (3) *posterior* which contain all the viscera of the chest, with the exception of the lungs. The boundaries and contents of the three divisions are as follows. **(1) Anterior**

Mediasti´num. BOUNDARIES: *In front,* the sternum; *behind,* pericardium; *laterally,* pleuræ; contains origin of sterno-hyoideus muscles; origin of sterno-thyroideus muscles; triangularis sterni muscle; left internal mammary artery and venæ comites; (the right internal mammary vessels being covered by pleura, are not included among the contents of the space); remains of thymus gland; lymphatic vessels from convex surface of liver, and loose areolar tissue. **(2) Middle Mediasti´num.** BOUNDARIES: *In front,* anterior mediastinum; *behind,* posterior mediastinum; *laterally,* pleuræ. Contains the heart enclosed in pericardium; ascending portion of aorta; superior vena cava; bifurcation of trachea; pulmonary artery and veins; phrenic nerves (from third, fourth and fifth cervical); arteriæ comites nervi phrenici (from internal mammary). **(3) Posterior Mediasti´num.** BOUNDARIES: *In front,* pericardium and root of lungs; *behind,* vertebral column; *laterally,* pleuræ. Contains descending aorta; vena azygos major; vena azygos minor; superior intercostal veins; pneumogastric nerves; greater splanchnic nerves; œsophagus; thoracic duet and lymphatic glands and vessels.

ROOT OF LUNG. This is formed by bronchus; pulmonary artery; pulmonary veins; bronchial vessels; bronchial glands; anterior and posterior plexuses of nerves; connective tissue. The following are the relations of the pulmonary veins, pulmonary artery and bronchus; RIGHT SIDE. *from before, backwards:* veins, artery, bronchus. *From above, downwards:* bronchus, artery, veins; LEFT SIDE. *from before, backwards,* same as right side. *From above, downwards:* artery, bronchus, veins.

THE IN´GUINAL CANAL and HER´NIÆ. The inguinal or spermatic canal commences at the *internal abdominal ring,* and terminates at the *external abdominal ring,* its length being about one and a half inches. It serves for passage of the *spermatic cord,* with its vessels, in the male, and the *round ligament* in the female. This canal is bounded in *front,* by the integument, superficial fascia, aponeurosis of external oblique and partly by the outer third of the internal oblique; *behind,* by the conjoined tendon, triangular ligament, fascia transversalis, areolar tissue, fat, peritoneum; *above,* by the arch of the internal oblique and transversalis; *below,* by union of fascia transversalis with Poupart's ligament. It is of great surgical importance on account of being the

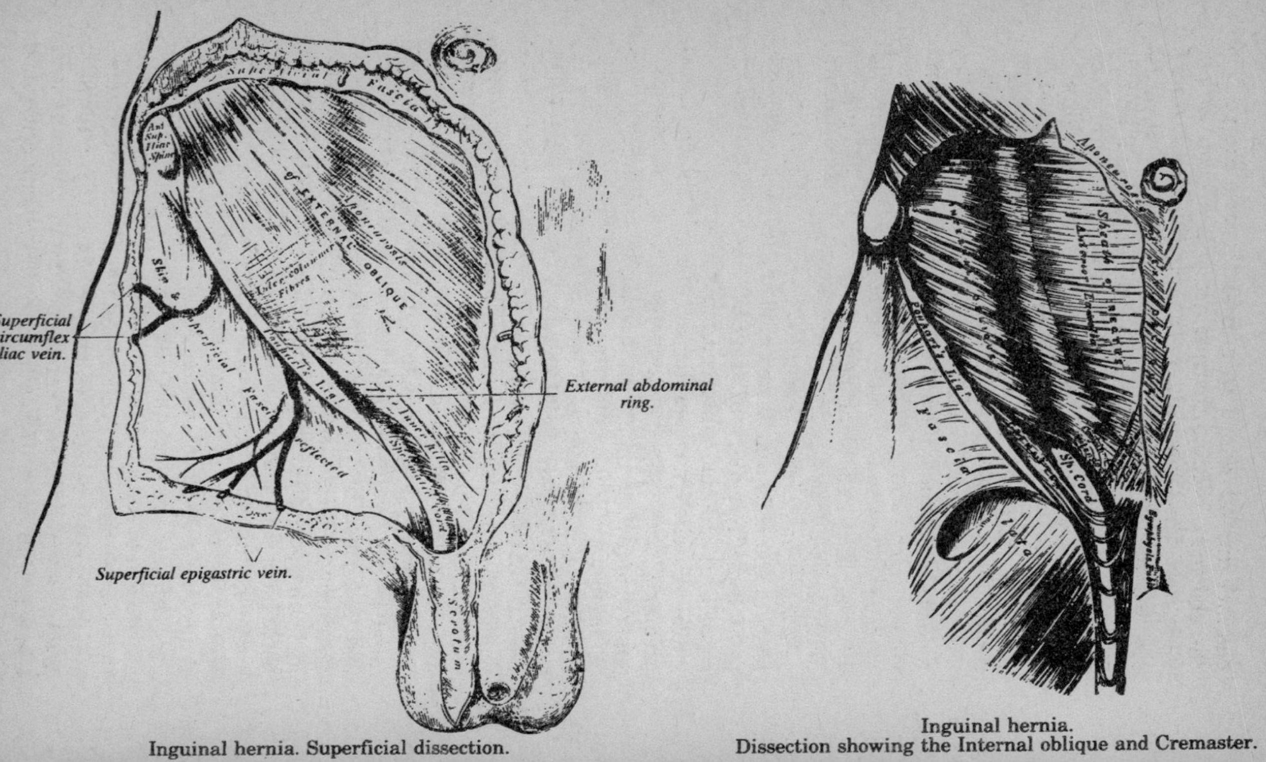

Inguinal hernia. Superficial dissection.

Inguinal hernia.
Dissection showing the Internal oblique and Cremaster.

channel through which **In'guinal Herniæ** escape from the abdomen. Inguinal herniæ are of two kinds, *oblique* and *direct.* The former enters the inguinal canal through the internal abdominal ring, passes obliquely along the canal and through the external ring to descend into the scrotum. *Direct* inguinal hernia escapes from the abdomen at Hesselbach's triangle, and then passes through the external ring. **External ring** is 1½ inches above Poupart's ligament; has for its *inner pillar* the fascia of the external oblique; for its *outer pillar*, Poupart's ligament and fibres of fascia. The *intercolumnar* fascia extends between the pillars at their lower portion. **Internal ring** is ½ inch above Poupart's, in the transversalis fascia, between pubes and anterior spine of ilium. Is an oval opening, long axis being perpendicular. On the internal margin, just above peritonæum, are the epigastric vessels; the *transversalis fascia* here gives the *infundibuliform* to the cord and testes, and transversalis covering to hernia. **Coverings of Hernia.** *Oblique*: Integument; superficial fascia; intercolumnar fascia; cremaster muscle; fascia transversalis, or infundibuliform fascia; areolar tissue and peritonæum. *Direct*: Integument; superficial fascia; intercolumnar fascia; conjoined tendon of internal oblique and transversalis muscles; fascia transversalis.; areolar cellular tissue; peritoneal sac.

CRU'RAL, or FEM'ORAL CANAL, and FEMORAL HERNIA. This canal is a funnel-shaped interval which exists within the femoral sheath between its inner wall and the femoral vein; it is of great surgical importance as being the space into which the sac of femoral hernia is protruded. It is ¼ to ½ inch long, extending from Gimbernat's ligament to saphenous opening. Its *anterior wall* is formed by transversalis fascia, falciform process and Poupart's ligament; its *posterior wall*, by iliac fascia and pubic portion of fascia lata: *outer wall*, by the septum, between it and femoral vein; *inner wall*, by junction of transversalis and iliac fascia; *upper orifice*, closed by septum crurali and a small gland; *lower*, or *saphenous* orifice, by cribriform fascia. It is limited above by the *crural* or *femoral ring*, and is lost below by the adhesion of the sheath to the coats of the vessels. In the normal state the canal is occupied by loose cellular tissue, and numerous lymphatic vessels which perforate the *cribriform fascia*, covering the saphenous opening in the fascia lata, and the walls of the sheath to reach a lymphatic gland

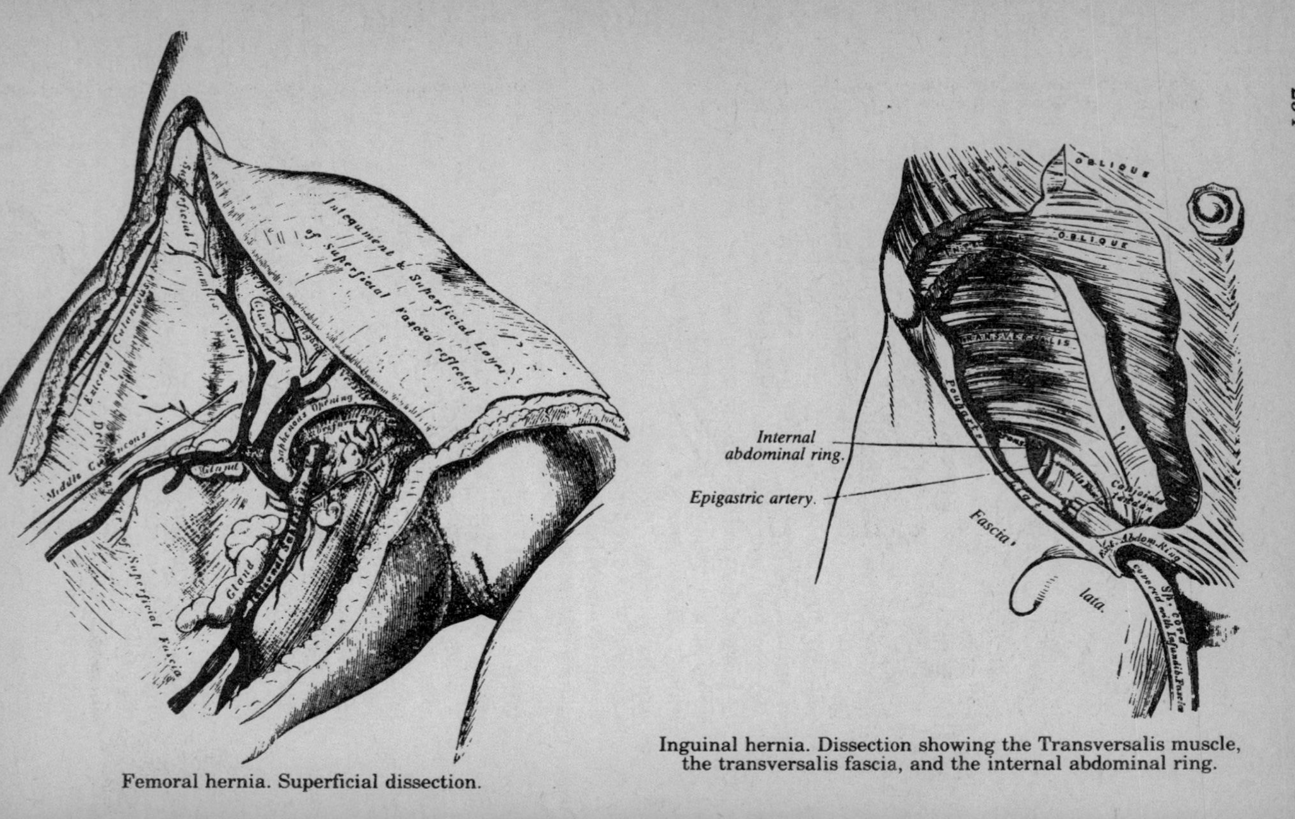

Femoral hernia. Superficial dissection.

Inguinal hernia. Dissection showing the Transversalis muscle, the transversalis fascia, and the internal abdominal ring.

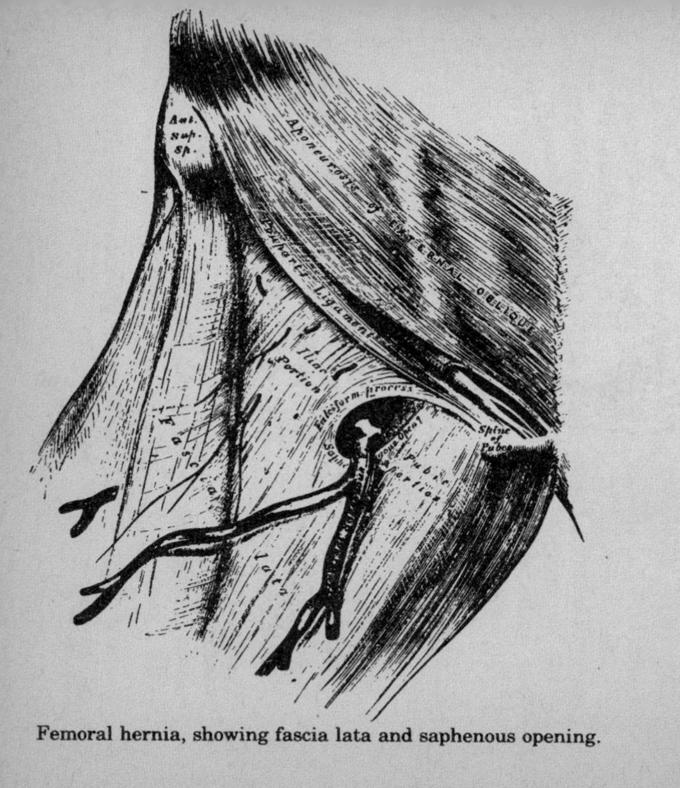

Femoral hernia, showing fascia lata and saphenous opening.

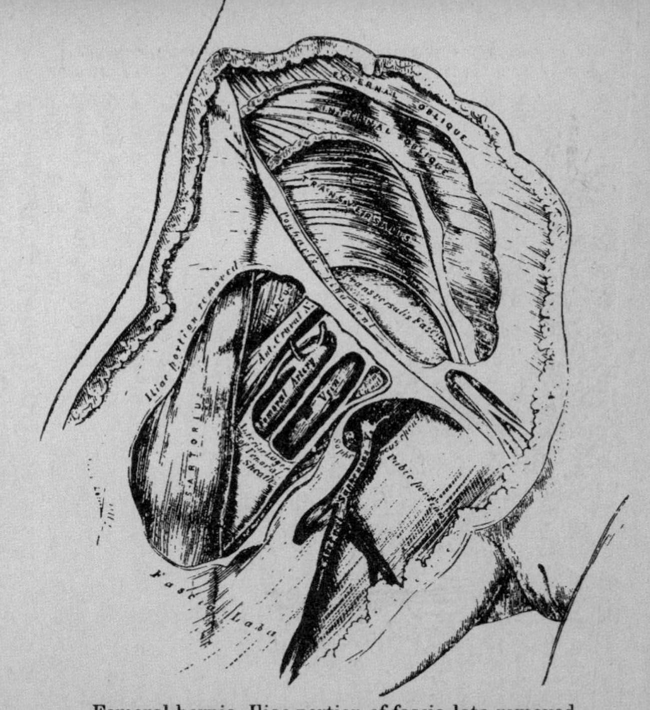

Femoral hernia. Iliac portion of fascia lata removed,
and sheath of femoral vessels and femoral canal exposed.

situated at the crural ring. This gland is retained in its position by a thin layer of sub-serous cellular tissue—*sep´tum crura´le*—which, together with the peritoneum, separates the canal from the abdominal cavity. The **Cru´ral ring** is the point where femoral herniæ leave the abdomen, and is the most frequent seat of strangulation; its boundaries are: *In front*, Poupart's ligament; *behind*, pubes, covered by pectineus and pubic portion of fascia lata; *externally*, septum separating femoral vein; *internally*, the sharp margin of Gimbernat's ligament, conjoined tendon, transversalis fascia and deep crural arch. **Fem´oral sheath** is formed, *anteriorly*, by transversalis fascia, and anterior portion of fascia lata; *posteriorly*, by iliac fascia and pubic portion of fascia lata. **Crib´riform fascia** is from deep layer of superficial fascia; is attached to falciform process. **Superficial fascia:** Superficial layer, over Poupart's ligament connecting abdominal fascia with fascia lata; deep layer up to Poupart's, above, from its connection with femoral sheath. **Fascia lata:** The *iliac portion* from spine ilium and Poupart's, throwing *falciform process* over pubic portion, almost transversely to the pubes; at top it is adherent to the femoral sheath; the *pubic portion*, from Poupart's ligament and pubes, lying below internal saphenous vein, passes beneath femoral sheath to become attached to pectineal line and capsule of hip-joint. **The Coverings of Fem´oral Hernia,** commencing at the surface, are; Integument; superficial fascia; cribriformn fascia; femoral sheath or fascia propria; septum crurale; areolar tissue and fat; peritoneal sac.

ERUPTION OF TEETH.

DECIDUOUS, 20 in number: *central incisors*, 7th mo.; *lateral incisors*, 7—10 mo.; *ant. molars*, 12—14th mo.; *canine*, 14—20 mo.; *post. molars*, 18—36th mo.

PERMANENT, 32 in number: *first molars*, 6½ years; *two mid. incisors*, 7th year; *two lat. incisors*, 8th year; *first bicuspids*, 9—10th year; *sec. bicuspids*, 10—11th year; *canine*, 11-12th year; *sec. molars*, 12—14th year; *wisdom*, 17—21st year.

Those of the lower jaw generally precede those of the upper by one or two months.

INDEX